T0212100

Engineering Thermodynamics

Engineering Thermodynamics

Michael Horsley

Faculty of Engineering, Portsmouth University, Portsmouth, UK

CHAPMAN & HALL

London · Glasgow · New York · Tokyo · Melbourne · Madras

Published by Chapman & Hall, 2-6 Boundary Row, London SE1 8HN

Chapman & Hall, 2-6 Boundary Row, London SE1 8HN, UK

Blackie Academic & Professional, Wester Cleddens Road, Bishopbriggs, Glasgow G64 2NZ, UK

Chapman & Hall GmbH, Pappelallee 3, 69469 Weinheim, Germany

Chapman & Hall USA., One Penn Plaza, 41st Floor, New York, NY10119, USA

Chapman & Hall Japan, ITP - Japan, Kyowa Building, 3F, 2-2-1 Hirakawacho, Chiyoda-ku, Tokyo 102, Japan

Chapman & Hall Australia, Thomas Nelson Australia, 102 Dodds Street, South Melbourne, Victoria 3205, Australia

Chapman & Hall India, R. Seshadri, 32 Second Main Road, CIT East, Madras 600 035, India

First edition 1993
Reprinted 1994

© 1993 Michael Horsley

Typeset in Palatino by Falcon Graphic Art Ltd, Wallington, Surrey
Printed in Great Britain by Page Bros. Norwich

ISBN 0 412 44520 4

A Catalogue record for this book is available from the British Library

Library of Congress Cataloging-in-Publication Data available
Horsley, Micheal
    Engineering thermodynamics / Micheal Horsley. --1st ed.
      p.   cm.
    Includes index.
    ISBN 0-412-44520-4
      1. Thermodynamics.   I. Title
TJ265. H68   1993
621.402'1--dc20   93-12578
                                             CIP

♾ Printed on permanent acid-free text paper, manufactured in
   accordance withthe proposed ANSI/NISO Z 39.48-199X and
   ANSI Z 39.48-1984

Contents

Introduction – worth reading

This is not yet another book on thermodynamics for the applied science or engineering student. It is a text designed specifically to help you to help yourself master some of the engineering thermodynamics found in the early parts of higher diploma or degree courses. The text does that by its progressive development of ideas and concepts and by setting you tasks, called progress questions (PQs) as you read through. In the main (but not always), these PQs are designed so that you should work through them before you can move on successfully.

The 'progress' part means that you are deciding whether you have understood so far because you are using the PQ as a lead-in to the next topic or to explain a concept to yourself in your own words or something similar. If you are happy with the outcome of the PQ, then you carry on. If not, then that is the right time to go back a pace. Answers to PQs are provided, of course, but they are at the end of the chapter, quite separate from the questions so that you do not just read on.

There are illustrative examples in each chapter, complete with fully worked solutions and relevant discussion of various points as they arise. This discussion is really an integral part of the text so do follow the examples through carefully! Some tutorial questions are also included at the ends of chapters for practice in the usual fashion.

The topic sequence more or less follows a typical syllabus so that you can go through a course. You will, however, be able to use the individual parts of the text if you wish, such as where a particular topic is giving difficulties. This does mean, though, that some things are repeated in various chapters so that the chapters can be used individually. If you are following the whole text, then use these repetitions as opportunities to revise.

The text is not intended to be exhaustive nor exclusive, nor is it a reference book in the usual style. Its prime aim is to lay down, clarify and develop the early concepts of thermodynamics in the engineering context. Wherever possible and solely in the interests of simplicity, the use of calculus is avoided or separated from the main text in a manner which does not detract from relevance or accuracy. If the subject is understood in this manner, then it will be a simple matter to transfer various relationships to the calculus form when necessary. Where the calculus has been avoided in the main text, those derivations are grouped together in Appendix B, at the end of the book. In some places however, the calculus route is the easiest to follow and where this is so, the calculus derivations stay in the main text.

Accepted conventions, for there are several in thermodynamics as with most subjects, are also introduced and used in a simple manner. When the reasoning behind the conventions is understood, then again their use and applicability can be tested in any advanced fashion. By the same token, various laws or relationships may be stated in a less formal manner than found in other texts. It is their sentiment and value which is of real importance and, if understood, their more formal definitions can be better appreciated. These are also included, in Appendix C, at the end of the book. The emphasis is on you grasping the main ideas and so on, then refining them as necessary.

The subject is a logical one, provided you build on knowledge gained and do not try to skip bits. SI units are used throughout and imperial conversions are listed because imperial units still occur in real-life engineering. The more common equations or relationships are also listed as a reference point. Both appear in Appendix C.

Some background knowledge has been assumed before you start. For example, it has been assumed that you are familiar with terms such as energy, kinetic energy, gravitational potential energy, temperature, specific heat capacity, heat and work; that units such as joules, pascals, newtons, megawatts and so on are already understood; that you have already met the laws of Boyle, Charles and Avogadro and that you are familiar with molar mass and the 'mol' – kmol, kilogram mole or similar. If there is any need to revise these at any stage, most school physics textbooks will provide this sort of background.

Good luck with your studies!

Back to basics 1

1.1 UNITS AND DIMENSIONS

Like many scientific and engineering terms, the words 'units' and 'dimensions' have both everyday and precise meanings, which sometimes get confused. Here we need to use their precise, engineering meanings. So 'units' means units of measurement – **metres, seconds, kilograms** – and **dimensions** means the fundamental parameters of **mass, length** and **time**, for example. So if I have some apples on a weighscale in a shop, I might record that I have five kilograms of apples. The unit involved is the kilogram and the dimension is mass. If I have a piece of rope two metres long, then the unit is the metre and the dimension is length. Kilograms, metres and seconds are, of course, the basic measurements used in the SI scheme and are thus the base units of SI.

I have been careful to say 'for example' because there are many straightforward units and dimensions derived from or which are combinations of these simple ones. For instance, the basic SI unit of velocity is metres per second (m/s) and so the dimensions of velocity are length ÷ time. Similarly, the basic unit of volume is the cubic metre, m^3, dimensions (length)^3; density is kilograms per cubic metre, kg/m^3 and dimensions mass ÷ (length)^3; the newton (N, unit of force) combines basic units and the pascal (Pa, unit of pressure) goes further, to combine the newton with area. You will meet others in your various studies.

- PQ 1.1 What, in words, are the units and dimensions of acceleration?

Just as there are standard abbreviations for the units kilogram kg, metres m and seconds s, there are standard abbreviations for their dimensions. These are mass M, length L and time T. There is additionally the use of the Greek letter theta, θ for temperature with some others in other scientific disciplines. Those four, M, L, T and θ will be adequate for our needs, however. Note that it is common for the dimensions or combinations of dimensions to be presented in **square brackets**, such as time T may be shown as [T]. The square brackets help to emphasize that dimensions are being discussed or used, should there be any chance of confusion. For instance, M on its own is the abbreviation for mega, meaning one million, as in megawatts, MW. However [M] represents the dimension mass. They are quite different!

Example 1.1 What, in symbols, are the dimensions of velocity?

Velocity, metres/second or m/s in SI units, has the dimensions of length/time. So the dimensions of velocity can be represented as [L/T]. Do remember the square brackets, whether individually as [L] and [T] or as the group [L/T].

● PQ 1.2 Using the symbols, express acceleration, density and area in SI units and in dimensions. What would be the dimensions if imperial units (pounds, feet and so on) were used?

Note carefully that dimensions refer simply to the fundamentals of length, mass, time and so on, not to the arbitrary measurements of kilograms or miles or minutes. You see then that dimensions ([M], [L], [T] or whatever) can be used to describe any property to which numbers could be attached, such as acceleration, viscosity or area, irrespective of the measurement system (SI, imperial, centimetre–gram–second or any other) in use.

The need for units is fairly apparent. Anyone who is designing or making something needs to know real sizes and real rates and real times. The problems that could arise by not knowing whether an area was in square feet or square metres, a velocity in metres per second or miles per hour can be imagined! It is vital too that the **units used are stated clearly** with each value or measurement so that other people – suppliers, engineers, accountants – can use the information properly.

So, for our purposes, why use dimensions? After all, they are simply fundamental statements and cannot be used to tell other people how big, how long or how heavy. Their prime value to us here is exactly that – they are quite independent of the actual measuring system and one elementary but important use is to see if an equation might make sense.

Equations are unavoidable in everyday life just as they are in engineering and other scientific work. A shopping list is a good example. It may say something like: 1 kg apples; 2 bags of sugar; 2 bottles of lemonade – total spent £2.80p.

The list makes use of a type of shorthand because the shopper knows exactly what products are wanted and their units of supply. The shopper doesn't have to write down '2 × 1 kg bags of sugar at 60p/kg' because the standard bag is 1 kg and the price is well established. If we now write this list as an equation we have

1 kg apples + 2 bags sugar + 2 bottles lemonade = £2.80p.

and it makes sense because the shopper knows what it means. Being critical though, it could make nonsense to someone else because (Figure 1.1) it says

1 kg + 2 bags + 2 bottles = £2.80p.

The units are quite inconsistent. Kilograms and bags and bottles alone like this just do not mix. If, however, the list was written out fully with the units of each item written down completely, then the shopping list would be as accurate as any scientific equation and the overall units would be consistent.

- **PQ** 1.3 Rewrite the shopping list equation so that it becomes scientifically sensible or consistent, with money values throughout. Assume that apples are 80p per kg, lemonade 40p per bottle and sugar 60p per kg.

So, an equation can be sensible if the units are ultimately consistent. Notice carefully though what this says – it **can** be sensible if it is consistent. It does **not** say that it **will** be sensible. If the shopping list had been written with wrong values – apples at 10p per kg, sugar at 15p per bag, lemonade at £5 per bottle for instance – then the outcome would have been wrong however consistent the units of the items in the shopping list equation.

A complete equation has to have **consistent units throughout** to be correct. Without consistency, the equation will be incorrect. However, unit consistency doesn't make up for any other errors. Checking unit consistency helps to avoid mistakes and is essential but it cannot guarantee accuracy in other respects.

While everyday equations – shopping lists, cake recipes, football league tables and so on – obviously don't have to be written out fully, technical equations certainly do have to be seen to be consistent because they are likely to be used by a wider audience. The units of any measurement must be quoted quite clearly for all technical, scientific

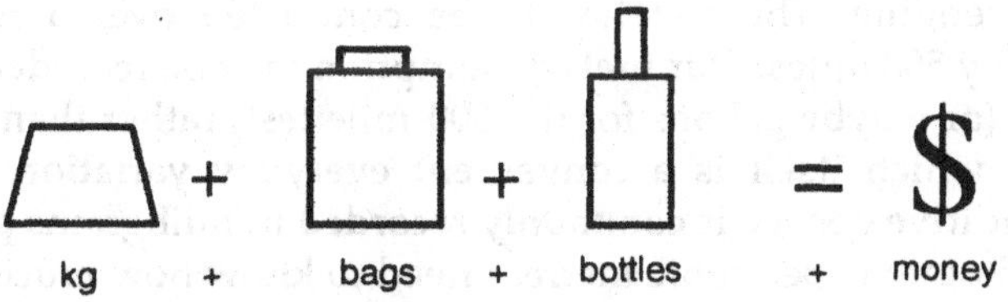

Figure 1.1 The shopping list equation.

and engineering records and communications.

In simple equations, it is not too difficult to make sure that the units align. If I say 'I drive along the motorway at a steady 100 km/h. How far will I travel in 2 hours?' the equation is simply

$$100 \text{ km/h} \times 2 \text{ h} = 200 \text{ km,}$$

and the units are consistent on either side of the 'equals' sign as

$$\text{km/h} \times \text{h} = \text{km.}$$

If the units had not been consistent in their complete form, then the result would have been incorrect. Remember though that the correctness of the units does not guarantee correctness of the associated numbers.

- PQ 1.4 What are the dimensions of this travel equation?

So, the equation is **dimensionally consistent** too. This is an important point – that all valid equations have to be dimensionally consistent. The dimensions or sets of dimensions on either side of the 'equals' sign have to come down to the same individual value or group of values. The information which this gives is just the same as the units information – if the dimensions are not consistent, then the equation is wrong. Also, like the units information, it does not say the equation is inevitably correct just because the dimensions align. For an equation to be correct:

- the dimensions must balance;
- the units must balance;
- but this does not guarantee that the answer is correct!
- if they don't balance, it will guarantee that the answer is wrong!

This still leaves the question of 'Why are both units and dimensions needed in engineering or scientific work?' After all, if the units are consistent, then the dimensions must be consistent because one is simply the fundamental representation of the other – they are inseparable. The first reason is mainly one of convenience in the real world of science and engineering. However widespread the acceptance of a standard system of measurement – SI for example – there are always little convenient deviations when making measurements.

Take a practical example of testing a fuel additive in the petrol of a motor car engine. The test has to be conducted over a reasonable distance, say 500 miles. Our fuel consumption may be recorded in miles per gallon (or maybe gallons for the 500-mile test) rather than litres per kilometre, which itself is a convenient everyday variation of the SI values. Additive dosage is commonly recorded in milligrams per litre of fuel, not kilograms per cubic metre. I need to know how much additive to put in the petrol tank, so I write down an equation relating miles to cover, fuel consumption and dosage, but first I must convert all the units to a **consistent** scheme. However, to check that my equation is sensible, I can compare dimensions very easily. The equation will say something like

amount for 500 miles = dosage per unit of fuel ×
fuel used in 500 miles.

Now we have dosage in mass per unit volume (mg/l), distance (miles) and fuel consumption in gallons for the test distance and an amount of additive used in mass (mg) – quite a mixture of convenient measures. First, though we can check the validity of our equation without reference to the conversion factors needed to bring all our measurements to one consistent set.

● PQ 1.5 Write down the dimensions used in this dosage equation.

That was a very simple illustration of the use of dimensions to check the validity of an equation, but even the most complex equations would follow a similar pattern. This use of dimensions rather than units is a quick way of avoiding simple mistakes, even before the actual units used in real-life measurements have all been brought into line for consistency. Table 1.1 gives some common property units and dimensions, and we meet a further use of dimensions later.

1.2 PROPERTIES AND SYSTEMS

As with units and dimensions, these two words 'properties' and 'systems' have both everyday and more precise meanings. Again, it is the more precise meaning that is of interest here and it is worth laying

Table 1.1 Some properties, units and dimensions. Note that the newton, N, resolves to kgm/s^2, [ML/T^2]

Property	Units		Dimensions
Velocity	m/s		[L/T]
Acceleration	m/s^2		[L/T^2]
Density	kg/m^3		[M/L^3]
Pressure	N/m^2	(=Pa)	[M/LT^2]
Volume	m^3		[L^3]
Temperature	K		[θ]
Work	Nm	(=J)	[ML^2/T^2]
Rate of work transfer	Nm/s	(=W)	[ML^2/T^3]
Heat	J	(=Nm)	[ML^2/T^2]
Rate of heat transfer	J/s	(=W)	[ML^2/T^3]
Specific heat capacity	J/kgK		[L^2/T^{2}Mθ]
Thermal conductivity	W/mK		[ML/T$^3\theta$]
Dynamic viscosity	kg/ms	(=Ns/m^2)	[M/LT]

down a reasonable and convenient definition of each.

A **property** can be regarded as anything which may be used to describe or identify something. Some properties are measurable, like the length of a plank or the temperature of a liquid, while others are recorded more loosely – smell or taste, for instance. In the study of thermodynamics, the measurable properties dominate by a large margin and so relevant calculations can be performed.

A **system** for our purposes can be taken as a coherent grouping of things which are interdependent. There are many common examples of both large and small systems including the human body, an aeroplane, a city, or a clock. Each of these is a group of interdependent parts (Figure 1.2).

• **PQ 1.6** Think of a small steel bar sitting on your desk. I can recognize various of its properties such as length, mass, density and temperature. I now cut a piece off the end of the bar and give it away. What has happened to the properties of the remaining piece of bar?

So, some properties are to do with the *material* – whether it's a gas or a liquid or a solid that we are considering makes little difference. The steel's density and specific heat didn't change, just as the viscosity and refractive index of water will not change if I pour it from a jug into two glasses. These properties which are inherent – to do with the material itself – are called **intensive properties**.

Some properties are not primarily to do with the material but with the **form** in which they are used. As I have cut some off the steel bar, the length and mass of the remaining bar have changed. This sort of property is called an **extensive property**. In general, the grouping of properties in this way is a matter of common sense but it is the ability to make use of the properties rather than a knowledge of their grouping that is really important.

Figure 1.2 A system is a group of interdependent parts.

● PQ 1.7 Is temperature an intensive or an extensive property? If I have some gas in a rigid container like an oxygen cylinder, is its pressure an intensive or extensive property?

As a little aside to this PQ, the fact that the water has been sloshed around in the pouring will make a tiny difference to its temperature. It is very small, almost immeasurably small, and is of no real consequence here.

This word 'systems' also has various extra words attached to it – open systems, closed systems, system boundaries are terms that are often met. They are descriptive and, for our purposes, their values are in deciding on what approaches or equations to use in solving problems.

Boundaries are often easily recognized – the casing of a machine, the geographical limits of a town – and it is tempting to regard these boundaries as all that is necessary when examining any system. However, there is frequently a lot to be gained by putting in a more convenient and almost arbitrary boundary. A **system boundary** can be regarded as a box – not necessarily a solid box, an imaginary one will do and it can have whatever shape we wish – into which the system to be studied is put. The walls of the box now form the edge or boundary of the system. For a clock, the box will be small; for a town it will be big but it is still a box, real or imaginary (Figure 1.3)

● PQ 1.8 Why put something in a box when it has its own casing?

Once we imagine the system in a box, then we can imagine also **open** and **closed** systems – effectively, does the box have doors through which something may pass to take part in the workings of the system?

● PQ 1.9 Is the idea of a closed system realistic?

In real life then, there is no such thing as a completely closed system. There is always some sort of influence passing across the system boundary. Even with the working clock in a box, the energy of the spring or battery which drives the clock must be dissipated somehow, somewhere. The 'somehow' is that the spring or battery energy finally ends up as heat energy (because components are driven and there is friction), not very much heat but still heat. Since there is no such thing as a perfect insulator, the heat will escape so the 'somewhere' is across the system boundary to the outside world.

However, there are plenty of systems where the main events or where selected events are not particularly influenced by nor contribute to the outside world. Pretend that I am in my kitchen making a milkshake. I have the milk in the kitchen and the flavouring; I've got a hand mixer and a jug. My mixing the milk and flavouring and producing a milkshake has all the components of a coherent grouping of interdepen-

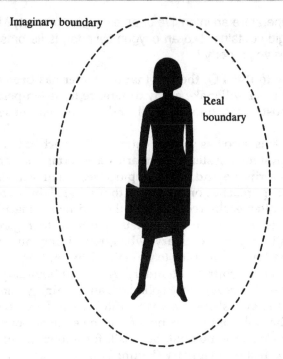

Imaginary boundary

Real
boundary

Figure 1.3 A boundary can be real or imaginary.

dent items – a system – but I am fairly independent, for that purpose, of the world outside the kitchen. I can regard this as a closed system.

Actually, I have eaten some food which has given me the energy to do the mixing; I have brought into the kitchen the milk and flavouring; someone outside can hear the mixing noise; someone outside may drink the milkshake but, for the purpose of studying the preparation of a milkshake from assembled constituents, the system is closed. I have chosen the system boundary to suit my needs and the main or selected events for the immediate purpose are isolated from the rest of the world. The closed system is then one of convenience and common sense – we know that it is not really closed but the open bits have no significant influence on our study. That is an important feature – are the things or influences which cross the system boundary significant? If not, then the simplicity of a closed system can be assumed (Figure 1.4).

So, just to be quite clear – there is no such thing as a closed system because there will always be something which can cross the system boundary. However, there are plenty of examples where the **main events** or contributors to the system are fairly independent of the outside world. If these dominate, then the system can be taken as closed. We will meet some systems examples in later chapters which will help to illustrate the idea.

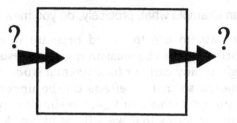

Figure 1.4 Think of the system as closed if anything crossing the boundary does not really matter.

The open system is perhaps easier to accept or recognize. As its name suggests, an open system does have a boundary but important things or influences are expected to cross that boundary. If my milkshake mixer had been a mains electric one, then electricity would flow across the system boundary. If I take a wider view of the whole operation, then water for washing the mixer blades and the jug would flow into (clean) and out of (dirty) the system. It is an open system. The words **inputs** and **outputs** are also found and they are descriptive and obvious uses of the words.

- PQ 1.10 Identify some inputs and outputs of my 'open system' kitchen.

It is useful to have this idea of open and closed systems, even if the closed one is a bit artificial because it can make you think about what is important in a thermodynamic (and plenty of other disciplines) problem. If you can highlight the important parts of a problem, then the closed system approach may be applicable and may simplify matters. Do not, however, treat something as closed when it is very clearly an open system – if in doubt, consider possible inputs and outputs. Are they important or not? Do they affect that which you are studying mainly or not? If still in doubt, treat whatever it is as an open system until you are sure. That's the safe way.

Rather like intensive and extensive properties, there is more value in using the systems terms and ideas to encourage you to think clearly and logically about topics and problems than in simply being able to offer definitions.

1.3 SOME DEFINITIONS

We have already met some words that have both an everyday and a more precise meaning. Even when some words are used scientifically, their meaning can sometimes be a bit vague – what they represent may be easier to recognize than to define.

- **PQ 1.11 As an example what, precisely, do you mean by pressure?**

So, it is quite easy to use the word pressure in an almost loose scientific sense because it can be measured and the result recorded. It is important, though, in any early or fundamental study to be quite clear what the term means so that its effects can be understood properly. Rather than go through a stream of PQs, the simple way at this stage is to define a few of the terms that we will be using. As with previous words or terms, the reason is to encourage clear thought, not to develop an ability to repeat definitions. The latter is just a use of memory, not an application of intelligence. I am keeping the list short, enough for the next chapter, because other terms will be defined as they are needed.

Pressure Force per unit area. For a solid resting on a horizontal surface, the combination of the mass of the solid, the earth's gravitational attraction for the solid and the amount of surface supporting the solid determines the pressure on the surface. The total pressure effect is, to all intents, vertical upon the surface.

For a liquid in an open container – a watering can for instance – there are pressure effects upon the retaining surfaces of the can, the sides as well as the base. Left to its own devices, the liquid would flow away from whatever shape the can held it and down to the lowest available surface – a leaking watering can will soon have a puddle around it. The pressure on the base of the can is directly dependent upon the depth of water in the can. If pressure measurements were taken up the sides, the values would decrease the higher up the sides they were taken – a reflection of the height of water above the measuring point. For a liquid in an open-topped can, then, there are pressure effects on the base and wetted sides of the can.

- **PQ 1.12 Any upward pressure in this case?**

With a gas in a container – an inflated party balloon is a good example – then all the walls of the container, bottom, sides and top, are holding the gas in check and all will therefore feel the pressure exerted. Since a gas does not usually have a well defined free surface like a solid or a liquid has, then a gas can be imagined as a lot of molecules moving freely, rapidly and randomly in all directions. This is one of the ideas of the kinetic theory of gases which we shall meet briefly later. So, any wall or surface that the molecules hit will have to exert a retaining force

$$\text{Pressure,} \quad Pa \; = \; \frac{\text{Force}}{\text{Area}}, \quad \frac{N}{m^2}$$

Figure 1.5 Pressure is force per unit area.

– that is, withstand the gas pressure.

● **PQ 1.13** What of gravitational effects here? Will the mass of the gas have any influence on pressure readings taken at various places on the balloon surface?

An example of a large container of gas in the last PQ is the earth's atmosphere. The weather forecast will give an atmospheric pressure reading, often at sea level. This pressure is mainly to do with the amount of air above the earth's surface, with weather effects making small variations. If you climb a mountain, there is progressively less air above you pressing down under gravitational effects. If you climb to about 1000 metres, the atmospheric pressure will be about 12% less than at sea level.

So whether we are thinking about solids, liquids or gases, whether gravitational force or molecular movement is the dominant effect, the definition of pressure as force per unit area still applies (Figure 1.5).

Pressure is measured fairly easily and there are many designs of pressure gauge. The more common ones, such as you see for reading motor vehicle tyre pressures, actually indicate the pressure above the current atmospheric pressure – simply a feature of their design and quite adequate for many purposes. After all, if the air in the tyre was at atmospheric pressure, the tyre on the vehicle would be flat, so the only reading of any value is that of pressure above atmospheric! This gives rise to the term **gauge pressure** – the pressure that the gauge reads above the pressure of the surroundings in which it is used. For a lot of work, that relative reading (pressure above the surroundings) is all that is needed.

However, something better is needed when calculations are to be done. Gauge pressure is a little arbitrary – atmospheric pressure varies from place to place and day to day – so for many calculations it is necessary to work in terms of the total pressure, which is the sum of the gauge pressure and the atmospheric or surroundings pressure. This total is the **absolute pressure** or the total pressure above zero. The safe way in all pressure calculations, unless you are quite sure, is to use absolute pressure.

By convention, the letter small p is used to represent pressure in mathematical form and, as an SI reminder, pressure is measured in Newtons per square metre, N/m^2. The unit of 1 N/m^2 is also called a pascal, Pa and 100 000 pascals make 1 bar. The reason is one of convenience, as 1 Pa is rather small compared to many real-life or industrial pressures. Note, however, that this is a unit of convenience and that the basic unit of N/m^2 or Pa is necessary in many calculations. If in doubt, always use the fundamental units and avoid simple errors.

Temperature A measure of molecular kinetic energy. The higher the temperature of a substance, the higher is the kinetic energy of its molecules. This is of major importance in many areas of science, engineering and technology because it helps to explain why other properties may be temperature-dependent.

It was noted earlier that some properties are easier to recognize than to define and, to some extent, temperature is one of those. In this case, you can imagine the difficulty of attempting to measure molecular kinetic energy on an everyday basis.

Temperature is an important feature of many activities and processes, so there has to be some way of assessing it and thus controlling it. One way, which gives rise to the common idea or perception of temperature, is by noting what effect a body's or substance's temperature has on an indicating device. That is a rather long way of saying 'use a thermome-ter' but the words are chosen carefully. If I immerse a mercury-in-glass thermometer in a hot liquid, the mercury column rises up the glass tube to indicate a temperature on a scale. The common scales, however scientific, are fairly arbitrary and the derivations of these scales are now a matter of history. It is these commonplace scales that provide the routine idea of temperature, without reference to its real identity.

• PQ 1.14 When I immerse the thermometer, what is happening to the mercury in the tube? Why does it stop rising up the tube? Think – has some sort of relationship developed between the thermometer and the hot liquid?

So, the thermometer and the hot liquid have achieved **thermal equilibrium**. At that point there will be no further heat transfer from one to the other – the term strictly is 'net heat transfer' but we shall meet that later. Any other substance at the same temperature would give the same reading of course. It too would be in thermal equilibrium with the thermometer and its molecular kinetic energy – its temperature – could thus be assessed.

This gives rise to a thermodynamic law – the **Zeroth Law**. It is a clumsy title but before this law was laid down as such, the First and Second Laws of Thermodynamics had already been stated. Since this law should really have preceded the next two, there was little alternative but to call it the Zeroth Law! It says that any two systems which are in temperature or thermal equilibrium with a third system must be in temperature or thermal equilibrium with each other (Figure 1.6).

It sounds obvious but it is the foundation of thermometry. At its simplest, if I use a thermometer (take this as the third system) to read the temperature of two different bodies (take these as the any two systems) and get the same reading, then the two bodies must be in thermal equilibrium. Its formal statement helps, among other things, to

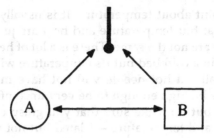

Figure 1.6 If A and B are each in thermal equilibrium with the thermometer, then they are at the same temperature.

get rid of ideas about perpetual motion or of heat transfer without temperature gradients.

Note that any temperature measurement, whatever thermometer type is used, is effectively a measure of molecular kinetic energy relative to some datum or base line. The longest standing datum is the freezing point of water but that is just a very handy reference and it is quite arbitrary. This gives rise to the idea of an **absolute zero** of temperature, where there would be no molecular kinetic energy to measure. Just as with the use of absolute pressure as the basis of calculation, so the absolute scale of temperature is the safe one to use in calculation also, not the one which refers to the freezing point of water!

This then gives us a usable overview of temperature with some sort of workable definition. The common identifier in mathematical relationships is the letter capital T – be careful not to confuse it with [T] – and temperature is measured above absolute zero in degrees Kelvin, K. It is **vital** in calculations or records to note very carefully whether the temperature refers to absolute zero (K) or to the more common but arbitrary (convenient) Celsius (°C).

For the vast majority of everyday calculations, the round-number conversion from Celsius to Kelvin of °C + 273 = K is quite adequate. This is because the accuracy of any conversion can only be as good as the basic data. If, for example, a temperature is recorded as 25°C (that is, rounded to the nearest degree) then it is very sensible to say 25°C + 273 = 298K.

However, the temperature was given more accurately as 25.1°C, then the yet more accurate conversion would be 25.1 + 273.2 = 298.3K. Further, if the temperature was recorded as 25.08°C, then the yet more accurate conversion would be 25.08 + 273.16 = 298.24K.

In **any** working, accuracies throughout have to be comparable. For most of the calculation examples in this text, the value 273 will suffice but do note the accuracy of the supplied data and then use the appropriate conversion value.

Just one final point about temperature. It is usually, quite correctly, associated with heat but temperature and heat are just that – they are associated but they are not the same. There is a lot of heat available from an electric blanket in a cold bed but its temperature will be low enough for comfort. A small red hot needle will not have much heat but its temperature will be be high enough to be very uncomfortable! This is a simple illustration but do make sure that you grasp clearly the difference between heat and temperature – related but not identical – as we will use the difference later.

Volume This is the last of our introductory property definitions and is the easy one. It is a fairly basic word and it means, of course, the amount of space occupied by a substance or a system or an activity. The main reason for introducing it here is to implant the idea that pressure, temperature and volume are often associated, frequently inseparably.

● PQ 1.15 If I have an air-inflated party balloon, how may this demonstrate a relationship between pressure, temperature and volume? Use words, not numbers.

The word volume sometimes has other words attached to it and one of them is *'specific'*. Specific volume is the volume of unit mass. Again by convention, volume in general is represented by the letter capital V and specific volume by small v, usually with some suffix as in v_s to make it quite clear.

● PQ 1.16 Are volume and specific volume intensive or extensive properties? In SI, what are the units and dimensions of specific volume?

There are other properties which we will meet and use but, for the present, these are sufficient.

PROGRESS QUESTION ANSWERS

PQ 1.1

Acceleration is usually defined as the rate of change of velocity and its SI units are metres per second per second or m/s^2. The dimensions involved are length (the m part) and time (the s part) and thus the dimensions of acceleration are (length)/(time)^2.

PQ 1.2

The SI units of acceleration, density and area are m/s^2, kg/m^3 and m^2, respectively. Their dimensions are $[L/T^2]$, $[M/L^3]$ and $[L^2]$ also respectively. The dimensions are independent of the system of units used, so they would still be $[L/T^2]$, $[M/L^3]$ and $[L^2]$ even if you were working in imperial units.

PQ 1.3

The rewriting entails every relevant unit being included in the equation. So, it may start out as something like

1 kg apples × 80p/kg of apples + 2 bags sugar × 60p/bag of sugar +
2 bottles lemonade × 40p/bottle lemonade,

and the amounts and commodities (kg apples, bags of sugar, bottles of lemonade) now cancel out just like any other coherent equation to give

$$1 \times 80p + 2 \times 60p + 2 \times 40p = £2.80p.$$

The units are now self-consistent with money on either side.

PQ 1.4

The two dimensions involved are distance and time. The speed (100 km/h) dimensions are $[L/T]$; the time (2 h) dimension is $[T]$ and the distance (200 km) dimension is $[L]$. On the left-hand side of the equals sign, the dimensions can cancel out just as if they were numbers, so that $[L/T] \times [T]$ can become $[L]$. The dimensions on either side of the equation are thus consistent.

PQ 1.5

In words, what is being said is 'The amount of additive used per 500 miles is equal to the milligrams of additive per litre of fuel times the amount of fuel used for the 500-mile test.' In the form of an equation, this becomes

$$\frac{\text{mass of additive}}{\text{500 miles}} = \frac{\text{mass of additive}}{\text{unit volume of fuel}} \times \frac{\text{volume of fuel}}{\text{500 miles}}$$

$$\frac{[M]}{[L]} = \frac{[M]}{[L^3]} \times \frac{[L^3]}{[L]} = \frac{[M]}{[L]}$$

So the equation is dimensionally correct and it can be used with confidence to include the real numbers and the necessary conversion factors to make the units consistent throughout.

PQ 1.6

Some of the properties do not change – it is still a piece of steel; it will still melt at the same temperature; it has the same colour, for example. Some of the properties do change – the bar weighs less than before, it is shorter than before, for instance. Since the properties can be used to describe the bar, they can still do that in a very objective way. Some properties have changed, some have not but they are still properties.

PQ 1.7

Again using a jug of water as an example and assuming that the jug and any glasses are at the same temperature, then pouring half the water from a jug into a glass will not alter the temperature of the remaining water. Temperature is an intensive property.

If, though, I let half the gas out of the gas cylinder, then the pressure inside the cylinder will certainly fall and the pressure of the released gas will also be quite different from its initial value. Thus pressure is an extensive property.

PQ 1.8

The reason for doing this is that these imaginary boundaries can be moved around to include or exclude various things as necessary. For instance, we could put the clock in a box and effectively isolate the clock from the rest of the world. If it was working, it would still be a coherent system and still telling the time but little else. Now, if we make the box bigger and put something else in the box, such as a cooker switch that is activated by the clock, we can look at another form of system – the clock and the cooker switch interacting. Then we could add the cooker and so on. What we have achieved by putting in our own boundaries is flexibility – we are not constrained by the clock's case.

PQ 1.9

No – it is impossible to isolate anything absolutely. There will always be something that can get through the supposedly closed system boundary. Heat will always get through the best insulation, for instance. The insulation will delay it but that's all. If a machine is the system in the box, sooner or later it will require maintenance, so something has to cross the system boundary. The closed system does not exist.

PQ 1.10

We have already met some inputs – electricity for the mixer, water for the washing up – and others can be added such as light through the window as well as my personal contribution or that of the dairy.

Outputs may include heat and noise, water from the sink or prepared food.

PQ 1.11

Pressure has the precise definition of force per unit area. It may be used freely in a general fashion – the atmospheric pressure is 1050 mbar for instance in the weather forecast – but even then the measure is one of forces and areas. The word also has the everyday meaning of someone being under stress. Here, though, force per unit area is the meaning that matters.

PQ 1.12

The liquid – say water by way of example – as such will not be pushing upwards to exert a pressure. There will, however, be some vapour pressure effect as the liquid water strikes a concentration balance with any water substance in the surrounding atmosphere. It is a very small pressure compared to that experienced by the sides and the base of the can and, for the present, is ignored. Vapour pressure does have its own importance as a separate issue.

PQ 1.13

Strictly, yes – the gas in the balloon will experience the effects of gravity just as if it were a solid or a liquid. Compared to the gas pressure holding the balloon inflated, though, the gas gravitational effect is very small. If we were dealing with a very large container of gas, then a careful assessment of its importance may have to be made.

PQ 1.14

The mercury rises up the tube because it expands as it gets warmer. It stops rising because it has stopped expanding – the influence of the hot liquid temperature is at its limit. In fact, the mercury is now in thermal equilibrium with the liquid whose temperature is being measured. The measure of temperature in this case is one of noting the equilibrium condition – here highlighted as some marks on a very handy scale scribed on the thermometer. As an aside, the tube will also change because it, too, gets warmer and this is why thermometers have to be calibrated, to allow for the combined effects.

PQ 1.15

There must be several ways of doing this and you may well have devised a different one from mine. It is the principle that matters though, not the detail.

If I immerse the sealed balloon in hot water, it will expand – that is, the air inside will and so cause the balloon to swell. There is a relationship between temperature and volume.

The pressure inside the balloon keeps the balloon skin taut. If I let some air out, the balloon skin is less taut, maybe to the point of being floppy. The change of air volume – I've let some out – has had an effect on the pressure.

If I blow the balloon up almost to its limit and now immerse it in hot water (not a good idea, don't do it!) the balloon will rupture as the internal pressure rises beyond that which the skin can withstand. The relationship between temperature and pressure is the cause.

PQ 1.16

Volume, meaning the overall volume, is extensive. If I have a litre of paint and I use half a litre, the volume of the remainder has changed. Total volume is certainly extensive, therefore. Specific volume however is, by definition, an intensive property. It is the volume of unit mass, so even if I have used half the paint, the volume per unit mass of the remainder has not changed.

In SI, volume per unit mass is m^3/kg and, whatever the measuring system, the dimensions are $[L^3/M]$.

TUTORIAL QUESTIONS

For this chapter, where few numbers have been met, the tutorial questions are really ones of reviewing some points and principles. The detail answers are not necessarily in the text but the principles are. Go back and review them now if difficulties arise.

1.1 Decide the system boundaries for (a) a house and (b) a bicycle and its rider, then identify the inputs and outputs. Could either be a closed system for any purpose?

1.2 Take (a) a cooking recipe and (b) the dietary information from a lemonade bottle. Put both into self-consistent units and add the dimensions.

1.3 Find from any reference source the units of kinematic viscosity and dynamic viscosity. Write out the dimensions of each and deduce the relationship between them.

Ideal gases and the ideal gas laws

2

2.1 THE FUNDAMENTAL LAWS

In the previous chapter, the terms temperature, pressure and volume were defined in a reasonably precise way. This sort of definition is necessary in any scientific context to avoid misunderstanding, such as where terms may also be used in a casual way as part of everyday language. The words elastic and plastic are good examples of this double usage. They have precise scientific meanings, to do with the way that things deform when bent or stretched or pressed. They also have everyday meanings, to do with things around the home.

A next step in thermodynamics is to study the **interactions** of pressure, temperature and volume – that is, the way in which they may influence one another or the effects a change in one may have on another (Figure 2.1). If these interactions are understood, then they can be used properly – such as for design purposes, performance assessment or the prescribing of safety precautions. This understanding of properties and their interactions in real-life engineering is usually a

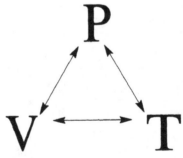

Figure 2.1 Pressure, volume and temperature interact.

mixture of theoretical study and analytical or empirical data gathering. It would be most unusual for a totally theoretical study to be used for anything of any significance – the experiments and trials need to be done. Here, however, we will be looking at the theoretical aspects.

● PQ 2.1 Explain why empirical information (that is, from tests or observations) should have any place in engineering calculations. Apart from straightforward errors, what could cause calculations to be wrong?

Any proper study of a real subject must be done in a logical fashion, normally starting with an easy case and progressing to more complex ones, or starting with known factors and replacing them gradually. The easiest starting point for looking at the thermodynamic behaviour of materials is to study gases. They are not perfect thermodynamic materials by any means but our view of them can be simplified to a considerable extent so that their elementary study is possible without becoming totally unrealistic. This simplification leads to the term 'ideal gases'.

In the previous chapter, the definition of pressure for a gas touched upon the 'kinetic theory of gases.' This is a long-standing elementary but important theory which gives a basic explanation to many points of gas behaviour. Even though later theories are far more sophisticated and yield more detailed information, the kinetic theory remains a good starting point for a first study of gas behaviour. Its important assumptions for the immediate purposes are:

● any gas will spread or expand to fill any space available to it;
● a gas consists of many rapidly moving, perfectly elastic, infinitely small molecules which take up negligible space;
● the gas is homogeneous.

Even with this simplified view of a gas, the theory also provides a foundation for the gas laws which we are to meet. I have not confused this part of the text by including any derivations of those laws from the theory here, but they appear in Appendix A.

● PQ 2.2 Look at this definition of a gas. What potential shortcomings are there? For instance, do the molecules always take up negligible space? If the molecules are not perfectly elastic, when would it matter?

The simplifications are just that, so there must be some inaccuracy somewhere in the results that the theory generates. Like any question of accuracy in any subject, we have to ask 'Does it matter for our present needs?' The answer obviously depends upon the circumstances. Think of a gas as a few molecules in a lot of otherwise empty space. If we consider the space occupied by the molecules themselves – not the whole of the gas volume, just the molecules – then this is likely to be more important at high pressures, when there is more gas molecule material present per unit volume, than at low pressures with less molecule material in a given volume. If the molecules are crowded together, their volume is going to be a more important part of the whole.

Similarly, the molecules are not perfectly elastic but that is only important when they hit containing walls or each other. Thus the actual operational conditions will dictate how important the deviations from ideal are, along with the quality of result we need. Notice the **need**, not **want**. It is always worthwhile deciding what is needed rather than what is wanted. They are not necessarily the same.

Any simplification is bound to bring with it some inaccuracy or a reduction of detail. When dealing with inaccuracies that simplification may bring, they must also be compared to other inaccuracies, such as in the use of instrumentation when measurements are taken. Looking specifically at gases in routine industrial use, it is fair to say that in the vast majority of cases, plant measurement inaccuracy – that is the accuracy of instrumentation and validity of sample – outweighs theoretical inaccuracy.

● **PQ 2.3** If I am taking a temperature of a gas, such as air in a room, why might the reading be inaccurate?

That latter point in the PQ answer – the so-called '*sample accuracy*' – is most important in any readings. Usually, a range of measurements may be taken to get a reliable average but if only one is taken it must be representative, not just some casual reading. The sample accuracy, the instrument accuracy and the theory accuracy have to be comparable, just like the use of 273, 273.2 or 273.16 in temperature conversion from Celsius to Kelvin has to be comparable to the accuracy of other data.

This does not mean that theories should not be improved but it does put it into context. For the so-called '*permanent gases*' – those such as oxygen, nitrogen and hydrogen which are gaseous at everyday temperatures and pressures – their behaviour obeys the kinetic theory to within about 5% for most of their usable range, especially when dealing with small changes of operational conditions. In fact, for most materials in the gaseous phase, well away from their liquefaction point (so that they really are gases) and not at high pressures (so that the molecules themselves are not taking up a lot of the space of the gas), this is often the case. I have avoided quite deliberately putting numbers to 'high pressures' because the values in question frequently vary with the material in question. Some illustrative graphical evidence is shown in Figure 2.2 and the whole is treated a little more analytically later. It is important not to lose sight of the fact that many simple theories in engineering can provide a lot of useful information.

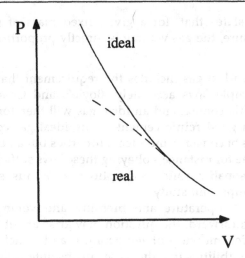

Figure 2.2 At constant temperature but near liquefaction.

With these conditions of use of the kinetic theory in mind, we can take an ideal gas – one which follows the ideas of the kinetic theory and the simple laws that derive from it – as the starting point for the study of the behaviour of thermodynamic materials. It does not matter at this stage that there is no such thing as an ideal gas. It is simply a starting point that is not totally unrealistic, perhaps best regarded as idealised forms of real gases – oxygen without imperfections, hydrogen without imperfections and so on. Thus an ideal gas based on nitrogen for instance would have the essential properties of nitrogen but fixed so that our *ideal nitrogen* obeys the simplifications of the kinetic theory and some simple laws which derive from it. The term 'ideal gas' thus does not imply a single possible or theoretical gas which is perfect in every way but the idealised forms of everyday gases.

The laws which deal with and define more closely such a gas are, naturally, called the *ideal gas laws* and they are really statements of or derivations from the laws determined by the observational work of Boyle and Charles around 200 years ago. Appendix A shows how they fit with the kinetic theory. Although both Boyle's and Charles' laws are commonly presented as simple mathematical equations and used as such, they are in fact quite precise statements. Since these laws arise fairly early in school science teaching, there is little need for anything other than the reminder statements:

- **Boyle's law** states that, for a given mass of gas held at a constant temperature, the gas pressure is inversely proportional to the gas volume;

- **Charles' law** states that, for a given fixed mass of gas held at a constant pressure, the gas volume is directly proportional to the gas temperature.

The idea of an ideal gas includes the requirement that an ideal gas follows some simple laws accurately. Boyle's and Charles' laws are simple laws in this context and an ideal gas will therefore obey them, thus giving us a good reinforcement of the ideal gas concept. Not a single perfect gas of constant and ideal properties but an everyday gas – nitrogen, methane for instance – obeying these laws sufficiently well for us to make reasonably valid assumptions and thus simplify their elementary but important study.

We have met temperature and pressure and volume but, while volume is straightforward, the question may arise about the other two properties – which measure of temperature and which of pressure? There are two possibilities in either case, an absolute value or a relative value. The absolute value means referring to an absolute zero – zero Kelvin for temperature or zero pascals (N/m^2) for pressure. Relative values mean referring to some local or handy value – atmospheric pressure or the freezing point of water, for instance.

- **PQ 2.4** Decide which pressure and which temperature are the ones to use in Boyle's and Charles' laws and identify their SI units.

Remember that everyday temperature and pressure scales – the tyre gauge, the greenhouse thermometer – are scales of convenience. If, for instance, the freezing point of water had been different from its present earthly value, then the Celsius scale would have been different.

So, logically, the *absolute values* are the ones to use. Here is a simple safety net, though. If there is ever any doubt, use the absolute values anyway. Notice too that the *conditions* associated with the laws are most important: Boyle's law, a given mass at constant temperature; Charles' law, a given mass at constant pressure. If these conditions are ignored, the laws don't work. A couple of worked examples will illustrate the use of these laws but first a reminder of the laws in their common mathematical form.

In Boyle's Law, volume and pressure are inversely proportional if mass and temperature are kept constant. Thus

$$p \propto 1/V \text{ or } V \propto 1/p$$

which, in the normal way of removing the 'proportional to' sign, becomes

$$pV = \text{a constant.}$$

Therefore, if a gas following Boyle's law changes from pressure p_1 and volume V_1 to pressure p_2 and volume V_2, we could write

$$p_1 V_1 = p_2 V_2$$

For Charles' law, volume and temperature are directly proportional if mass and pressure are kept constant. Thus

$$V \propto T \text{ or } T \propto V$$

and again this reduces to

$$V/T = \text{a constant.}$$

So if a gas following Charles' law changes from volume V_1 and temperature T_1 to volume V_2 and temperature T_2, we could write

$$V_1/T_1 = V_2/T_2$$

Be quite clear that the two constants on the right hand side of the equals sign for pV and V/T are not the same value. Each is saying, of course 'some constant value for the conditions and law in question and for that purpose only.' Each is a constant value for that particular case or calculation alone. Neither has any special value outside the immediate calculation – they are quite casual. None of these sorts of constants are values that can be looked up in a book, like some of the constants that will be met later. So although we may write $p \times V = \text{constant}$ that is only a convenient abbreviation for 'a constant value on this occasion'.

Since we are going to meet several constant values in this text for various purposes, I am going to introduce a convention to help identify which are important and which are casual. If these constant values are only applicable to say a given worked example or given set of laboratory conditions, then I will use a small letter c, as in 'constant'. If, as we will meet later, any constant value is of widespread importance and really is a constant value irrespective of the worked example or whatever else, then I will use an upper case C, as in 'Constant'.

Example 2.1 An ideal gas is held in a sealed balloon at 150 kN/m^2 and occupies 2 m^3. The temperature is kept steady as the pressure is raised to 600 kN/m^2. What happens to the gas volume?

The balloon is sealed, so the mass of gas is constant; the temperature is steady and we are told that the gas is an ideal gas, so Boyle's law applies, $pV = \text{constant}$. Notice, small c for this constant. Under these fixed conditions, whatever is done to the balloon of gas the product of absolute pressure times volume will remain the same. If the gas is initially at p_1 and V_1 and then changes to p_2 and V_2, we can write with confidence

$$p_1 \times V_1 = p_2 \times V_2$$

and putting in the numbers

$$150 \text{ kN/m}^2 \times 2\text{m}^3 = 600 \text{ kN/m}^2 \times V_2 \text{ m}^3$$

which tells us that the volume V_2 is decreased to 0.5 m³. Notice that the units and dimensions match on either side of the equals sign:

$$\text{kN/m}^2 \times \text{m}^3 = \text{kN/m}^2 \times \text{m}^3$$
$$[ML/T^2L^2] \times [L^3] = [ML/T^2L^2] \times [L^3].$$

Make quite clear in your own mind that this calculation is only valid because the conditions of Boyle's law – constant mass, constant temperature – apply.

Example 2.2 For the same gas in the same balloon, the gas temperature is raised from 10°C to 80°C whilst the pressure is held steady at 1.5 bar. What happens now to the gas volume?

We now have fixed mass and fixed pressure for our ideal gas, so this time it is Charles' law that applies. We aren't told a starting volume for the calculation and it doesn't follow from the previous example because we were not told the previous temperature – only that it stayed steady. However, notice carefully what we were asked – what happened to the volume? How did it change? We were not asked to state what was the initial or final volume, just what happened to the volume. Always answer the question that you have been asked – not the one that you think you have been asked!

Charles' law, for stated conditions (fixed mass, fixed pressure) tells us that the ratio of volume to temperature remains constant:

$$V/T = \text{constant.}$$

Once again, a *small-c* constant very particular to this calculation alone, not the same as the previous constant or any other that you are likely to meet. It means that V/T has a constant numerical value for the present purposes but for no other reason.

If the initial volume for this case was V_3 for the starting temperature of (10 + 273) – absolute temperature of course but using round numbers for comparability of data accuracy – and the final volume V_4 for the final temperature of (80 + 273) K, then

$$V_3/283 = V_4/353$$

which gives

$$V_4 = 1.247 V_3,$$

so the volume increases by just under 25%. Check that the units and dimensions fit. Again, make sure that you are quite clear that the conditions attaching to Charles' law – constant mass, constant pressure – **must** apply before this type of calculation can be done. (Figure 2.3).

Moving on from the two laws on their own, it is fair to say that there must be limits to the number of real occasions when two of the possible variables pressure, volume, temperature and mass, will be constant. The next step is to stay with the idea of a fixed mass of gas but look at a *joint variation* of pressure, temperature and volume. Everyday examples include the way that a car tyre pressure changes if the car is left in the sunshine or the way that a party balloon will expand if it is first inflated in a cold room and then moved to a warmer one.

It is apparent from Boyle's and Charles' laws that the fundamental relationships between pressure, volume and temperature for an ideal gas are fairly simple. A reasonable consequence of Boyle's law (pressure and volume) and Charles' law (volume and temperature) is that an equally simple relationship between temperature and pressure is also likely to exist. The relationship can be proposed from everyday experience but there must be conditions applied just like those attaching to the present laws.

● PQ 2.5 Propose a law, with conditions, which links pressure and temperature for an ideal gas.

Experience tells us that if a rigid can of gas is heated up, then the pressure in the can will rise. There have been enough gas cylinders and soup cans ruptured by rising temperature for this to be common

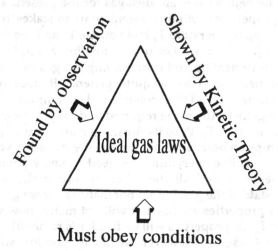

Figure 2.3 The ideal gas laws.

knowledge. Contents overheat – pressure rises – can bursts.

2.2 COMBINING THE LAWS

Boyle's law, Charles' law and the one derived in PQ 2.5 all had the conditions of constant mass and constant something else – the other gas property, whichever law is being quoted. The holding constant of one of the variables allows a clear statement of the relationship between the other two variables. It is reasonably obvious though, that even without one of the variables held constant, combined effects of change will always be possible for a given mass of gas. For instance, if the gas is heated up, it will either try to expand or to increase its pressure or some mixture of both perhaps.

The three laws already considered (Boyle's, Charles' and our new one) offer very simple relationships between two variables so it seems fair to expect that anything connecting all three variables of pressure, volume and temperature will also be simple.

● PQ 2.6 Without arguing it out too much, write down a likely relationship for a constant mass of gas where pressure, volume and temperature are able to change. That is, write down a combination of p, V and T in the style of Boyle's or Charles' law so that it reads:
$$\text{some ratio of } p, V \text{ and } T = \text{constant.}$$

Since a combined expression for pressure, volume and temperature is of **fundamental** importance – it gets us away from the need to specify a constant third property – it must be looked at analytically.

Suppose there is some gas in a closed flexible container, such as a strong balloon. Remember it is an ideal gas for the present and we are using absolute properties to eliminate elementary mistakes. Initially the gas is at pressure p_1, temperature T_1 and volume V_1 and the intention is to change it to p_2, T_2 and V_2. Let us say for this example that these changes will involve heating and compressing the gas in some combination. For our needs and to be quite general, it does not matter whether the gas is heated and then compressed or compressed and then heated or both together. All we are required to do is get the gas from its starting conditions of p_1, T_1, V_1 to its finishing conditions of p_2, T_2, V_2.

A quick interruption here. Since for our gas the pressure, volume and temperature really define everything we need to know about the gas, they also define or control all the other important thermodynamic properties – the **state** of the gas. It does not matter how we got the gas to p_1, V_1 and T_1, its properties are fixed. It will not matter how we get the gas to p_2, V_2 and T_2, its properties will be fixed. It's rather like travelling from London to Glasgow. However we do it, London will still be London and Glasgow will still be Glasgow!

Back to the gas. Take the change in a series of steps. However ideal our approach, the gas will not suddenly and magically change from one condition to another. As it is heated, for instance, it will change progressively. As it is compressed, it will change progressively. There will be many intermediate values of p, T and V that we could measure as the changes take place. Identify some of these intermediate values as p_a and p_b, V_a and T_a, then lay down some steps for the required change. These are quite arbitrary intermediate values and steps, so the outcome will be the same for any other other set of values and steps laid out logically.

(a) Hold T_1 steady and change p_1 and V_1 to p_a and V_a. As this step is at constant temperature T_1 (Boyle), then

$$p_1V_1 = p_aV_a$$

(b) Now change the temperature to T_2, holding V_a steady. As this step is at constant volume (our new law), the pressure changes from p_a to p_b:

$$\frac{p_a}{T_1} = \frac{p_b}{T_2}$$

(c) Hold T_2 steady and adjust the pressure from p_b to p_2. The volume must now change to V_2, since this is the stated volume for p_2 and T_2 – there is no choice. Thus (Boyle again)

$$p_bV_a = p_2V_2$$

But, from step 2,

$$p_b = p_a \frac{T_2}{T_1}$$

and from step 1,

$$V_a = \frac{p_1V_1}{p_a}$$

Therefore, substituting these values for p_bV_a we get

$$\frac{p_a T_2 p_1 V_1}{T_1 p_a} = p_2V_2$$

or

$$\frac{p_1V_1}{T_1} = \frac{p_2V_2}{T_2}$$

So, for a fixed mass of gas, the relationship between the gas pressure, temperature and volume must remain unchanged. For the given fixed mass of gas then, we can write

$$\frac{\text{gas absolute pressure} \times \text{gas volume}}{\text{gas absolute temperature}} = \text{a constant value,}$$

or

$$\frac{pV}{T} = \text{constant.}$$

Since this has been derived through a series of quite arbitrary steps or changes, it is a general conclusion and must hold good. Note that this is another *small-c* constant. Change the conditions, change the law being quoted and the constant changes also.

Example 2.3 In the previous balloon of gas (see Examples 2.1 and 2.2), a set of conditions was measured as 2 m³, 1.5 bar, 10°C. If the pressure and temperature change to 6 bar and 80°C, what will be the new gas volume?

Without going through all the steps, since they are similar to the previous examples, we can write

$$\frac{1.5 \text{ bar} \times 2 \text{ m}^3}{(10 + 273) \text{ K}} = \frac{6 \text{ bar} \times V}{(80 + 273)\text{K}}$$

which gives the new volume V of 0.624 m³.

Notice that the fundamental units of pressure, N/m², have been replaced by the bar, which is 100,000 N/m². This is a matter of convenience and I have done it because I have confidence in my ability to handle these convenience units. If there was any doubt whatsoever in my mind, then the safe route would have been to replace the bar with the fundamental unit of N/m².

The bar is a handy round number of fundamental units (100 000) and 1 bar is about everyday atmospheric pressure – notice about, not exactly! – which makes it doubly handy. It is often far more convenient to talk in these sorts of derived units than putting big numbers in front of the primary units. Thus atmospheric pressure is around one bar, car tyre pressures are usually around one and a half to two bars, mains tap water pressure is around one to one and a half bars and so on – far handier than saying 150 000 N/m² or whatever.

If in doubt, always use absolute values and fundamental units. It may take a few seconds longer in the calculation but it will help to avoid elementary mistakes. Only use the derived or convenience units if you are quite sure.

Example 2.4 If the gas now cooled down to 60°C and the volume reduced to 0.6 m³, what would be the new gas pressure?

Without going through all the steps with which you are now familiar, we can write

$$\frac{6 \text{ bar} \times 0.624 \text{ m}^3}{(80 + 273)\text{K}} = \frac{p \text{ bar} \times 0.6 \text{ m}^3,}{(60 + 273)\text{K}}$$

which gives the new pressure of 5.89 bar.

● **PQ 2.7 What temperature would now be needed to change the gas conditions to 7 bar, 0.7 m³?**

Since the mass of gas is unchanged – recall that this is an important condition for being able to write pV/T = constant – then the same answer should arise if we now go right back to the original conditions of 2 m³, 1.5 bar, 10°C, i.e.

$$\frac{1.5 \text{ bar} \times 2 \text{ m}^3}{(10 + 273)\text{K}} = \frac{7 \text{ bar} \times 0.7 \text{ m}^3,}{T\text{K}}$$

which gives a temperature for these latest conditions of 462K. If you follow these calculations through on a pocket calculator, then you will certainly have decimal place differences in the two answers. Remember though that all calculations can only be to the accuracy of the supplied data. In these calculations, the temperature conversion from Celsius to Kelvin of 273 was being used so, strictly, that rules the accuracy of the decimal place answers. If we had used 273.2 or 273.16, then more decimal places could have been shown in the answer – if the rest of the data were that accurate, of course.

So, for a fixed mass of gas, this kind of example shows that the relationship between pressure, temperature and volume stays steady. It does not matter which set of pressure, volume and temperature conditions are used in the calculations (provided they are a set and not a random mixture of values!) the answers will be consistent (Figure 2.4).

Up to now, the mass of the gas has been identified only as being fixed for each law or example and there has been no need to give any value to the mass. Carrying on from here, there must be a clear limit to the number of real life occasions or technical circumstances where a fixed mass of gas is met. The next step then is to expand the laws or relationships already derived to take account of changes of mass, such as might be met in a manufacturing process, for instance.

Take the simplest amount of gas, 1 kg. If 1 kg of gas is considered, then the volume V in the pV/T equation will have a special value, the specific volume, conventionally written as v_s. Remember that **specific** in the scientific context normally means '*per unit mass*'.

If the pV/T equation is now written for the specific volume

$$\frac{p \times v_s}{T} = \text{Constant}$$

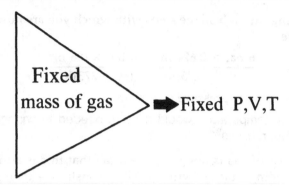

Figure 2.4 For a fixed mass of gas, the relation between pressure, volume and temperature is fixed.

then this new Constant (*large C, notice*) will have a special value, since it will refer only to 1 kg of gas. This special-value Constant, applying to unit mass of whichever ideal gas is being used, is called the **ideal constant** for that gas. Clearly, different ideal gases will have different values of their ideal constant, since they will have different values of specific volume.

● **PQ 2.8 Why will they have different values of specific volume?**

In looking at the gas laws of Boyle and Charles and all that followed them, several constants have been met but these were all quite anonymous. They were only 'temporary' constants. This latest Constant is, however, **very** much of consequence, so much so that it is tradition-ally given a particular letter to identify it, usually the capital letter *R*. It is a real Constant whose value can be calculated and whose value is recorded in data tables.

Take a little time to make absolutely sure that you understand the important difference between constants which arise almost casually, being constant values for a particular set of conditions and Constants which are truly Constant, always being applicable.

The *pV/T* equation thus becomes, with this true Constant,

$$\frac{p \times v_s}{T} = R,$$

written more commonly as

$$pv_s = RT.$$

There are two points to note carefully about this special Constant *R*.

● It will have a different value for a different ideal gas. Its value for 'ideal gas' oxygen is different to its value for 'ideal gas' carbon dioxide, for example.

- it is the ideal Constant for a particular named gas, so its value in real life may vary a little from the ideal value.

For the present, the second point is for information only. We are not, in this chapter, concerned with any variations from the ideal, other than noting that they will arise sometime.

- **PQ 2.9 What are the units of R?**

Since the value of volume used in the last $p \times V/T$ equation was the specific volume v_s, the equation was written as

$$\frac{p \times v_s}{T} = R$$

any that applies to unit mass of gas, 1 kg. That is a valuable equation but it really now needs modifying to accommodate any mass of gas. The chances of always dealing with one kilogram of anything are fairly remote!

For any mass of gas m, its total volume is the mass of gas multiplied by the volume per unit mass, $m \times v_s$, so if both sides of this last equation are multiplied by m

$$\frac{p \times v_s \times m}{T} = m \times R$$

and replacing $m \times v_s$, mass $\times$ specific volume, by V to represent the total volume of mass m, the equation becomes

$$\frac{p \times V}{T} = m \times R$$

and now cross-multiplying,

$$p \times V = m \times R \times T$$

or, in the usual fashion,

$$p V = m R T$$

and this is called the **characteristic ideal gas equation** or the *ideal gas equation of state*. A worked example will illustrate its use.

Example 2.5 The gas in the balloon of the previous worked examples is returned to its original conditions of 2 m³, 1.5 bar, 10°C. The gas is nitrogen and the ideal gas constant for that gas is 297 J/kg K. These gas constant values are available widely from standard tables and, as we shall see later, can also be calculated very simply. What is the mass of nitrogen in the balloon?

The first step is to make sure that the units are consistent and that absolute values, where appropriate or if there is any doubt whatsoever, are used. So 1.5 bar becomes 1.5×10^5 N/m². The bar is a convenient unit, remember but here we need all the values in fundamental units because a fundamental unit – mass m – is the outcome. Similarly, temperature becomes (10 + 273) K:

$$p \quad \times V \; = \; m \; \times \quad R \quad \times \quad T$$
$$1.5 \times 10^5 \text{ N/m}^2 \times 2 \text{ m}^2 = m \text{ kg} \times 297 \text{ J/kg K} \times 283\text{K}$$

so

$$m = 3.57 \text{ kg}$$

and the units (and the dimensions of course) agree, as the joule, J and the newton-metre Nm reduce to the same value.

Just as a reminder, while that value of R for nitrogen was the ideal value, many calculations based on the ideal values for everyday gases are going to be within 5% of the truth, especially for smaller changes in gas conditions.

Example 2.6 If more nitrogen was injected into the balloon to raise the mass of gas to 5 kg whilst the pressure and temperature were held steady, what would be the new balloon volume?

The unknown is now volume V for the new mass 5 kg. The other values do not change so we can write

$$1.5 \times 10^5 \text{ N/m}^2 \times V \text{ m}^3 = 5 \text{ kg} \times 297 \text{ J/kg K} \times 283 \text{ K}$$

which gives the new volume of 2.8 m³.

There is of course a simple short cut which can be used here since so many values are held steady. Only the volume and the mass are changed, so they must be directly related – the more the mass, the more the volume and so the new volume must be

$$V \text{ m}^3 = 2 \text{ m}^3 \times \frac{5 \text{ kg}}{3.57 \text{ kg}}$$
$$= 2.8 \text{ m}^3$$

Only use short cuts when you are really familiar with the calculations!

2.3 A UNIVERSAL EQUATION

The characteristic equation $p V = m R T$ is of the same form for any gas but the value of R is individual to a particular gas. The equation applies equally well to nitrogen, oxygen, hydrogen or any other 'ideal' gas but the actual numerical value of R depends upon which gas we are using.

At first, this seems like a step forward (we can now accommodate different masses of different gases) but with a distinct limitation – we can only use it for gases where we have to hand the value of R. Fortunately, by combining the characteristic equation with the outcome of Avogadro's hypothesis, a universal equation is derived.

Avogadro's hypothesis – sometimes called Avogadro's law, states that for given values of temperature and pressure, 1 mol of any gas occupies the same volume. As with Boyle and Charles, this law is repeated here as a reminder since you will have met it already in school science. As with those other laws also, if you have any doubts, now is the time to go back to a physics book and refresh your memory.

Like the other laws we have met, it is a little simplification of the truth but it is quite accurate enough for many industrial applications and for our present purposes (Figure 2.5).

● **PQ 2.10** Remind yourself – what is the 'mol'? You may have met variations of its name – the kmol, kilogram mole or something similar.

Therefore, according to Avogadro, 1 mol of hydrogen occupies the same volume as 1 mol of nitrogen or 1 mol of methane, for example, provided both gases are at the same temperature and pressure. Like the previous laws, this has of course to be read in the ideal gas context and there are deviations, as before, for real gases. Equally though, Avogadro's hypothesis gives a very close approximation to the truth for a wide range of practical applications, just as Boyle's and Charles' laws do.

Since the mol will be used in various places in this text, it may be useful to list a few **molar masses** now for reference. The values given are round-number values and thus any calculations using them must take

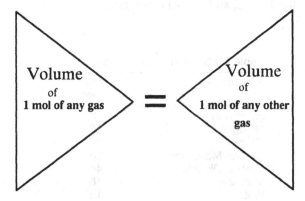

Figure 2.5 Avogadro's hypothesis.

account of their accuracy, just as with converting temperatures from Celsius to Kelvin earlier. Table 2.1 gives some values and others are available in many data or reference books.

As the mol (kmol or similar) is effectively the molar mass expressed in kilograms, then to round numbers, the volume occupied by 28 kg of nitrogen (molar mass 28) is the same as the volume occupied by 16 kg of methane (molar mass 16) under the same temperature and pressure conditions and so on. Recall that all this is under the 'ideal gas' conditions at all times, even though the laws may have very useful and practical applications.

Going back to the characteristic equation $pV = mRT$, the mass m in the equation is **any** mass under consideration. The values of pressure, volume and temperature will vary to accommodate different masses but m can be whatever value is necessary for the particular calculation. Any value can be chosen for any purpose provided the values of pressure, volume and temperature are compatible. Change the mass m and one or more of the values p, V and T must also change.

On this occasion then, let us choose as the undefined mass m the **molar mass**, the mass of 1 mol of whichever gas we are using. Identify this mass as MM and use it in the equation, which becomes

$$p V = MM \, R \, T.$$

This is a particular value of the equation chosen to suit our needs. The equation can be used for any mass we choose and we have chosen the numerical value (in kilograms, for whichever gas is to be studied) of the molar mass. Other than that, it is not special in any way. Note, though, that V in this special instance is the volume of the molar mass under the particular or chosen conditions of pressure p and temperature T. Since this is the **molar volume**, the volume associated with the molar mass and not just any casual volume and any casual mass, it is useful to give it a special identity, V_m so that we can write

Table 2.1 Some molar masses

Substance	Molar mass
Hydrogen	2
Nitrogen	28
Oxygen	32
Carbon Monoxide	28
Carbon Dioxide	44
Methane	16
Water	18

$$p V_m = MM R T.$$

Be quite clear that the equation is still the same characteristic equation as before. It has not changed in any way. It's just that we have chosen a particular value of mass m (MM) to put in it and we have identified the volume V (V_m) very carefully.

- PQ 2.11 What relationship can be deduced for R and MM, whichever ideal gas is being considered? Recall that the volume V_m of the molar mass MM of any ideal gas is Constant for given conditions of temperature T and pressure p.

Thus the product of the molar mass and the individual gas constant for any gas gives a Constant value. Since molar mass is fixed for any gas and we have seen that the gas constant for any gas is an individually fixed value, then the product ($MM \times R$) will not vary for ideal gases whatever the operational conditions. Say there were three gases, a, b and c, then

$$MM_a R_a = MM_b R_b = MM_c R_c$$

Thus for any gas (but still subject to the ideal demands) the product of its molar mass and its ideal gas constant gives yet another important and truly (ideally) Constant value. This one is called the **universal gas constant** and is usually designated R_m or $\tilde{R}$ or something similar. (Some texts use an ornate designation called 'curly R' which is a descriptive name.)

So we can write

$$R \times MM = R_m$$

or for the three gases a, b and c

$$MM_a R_a = MM_b R_b = MM_c R_c = R_m$$

and exactly the same for any other ideal gas.

Most data books and many text books carry values of common molar masses and they are, in any case, easy to calculate from molecular formulae and atomic masses. Another example will illustrate.

Example 2.7 If R for nitrogen is 297 J/kg K, what is the value of R for hydrogen, what is the value of the universal gas constant and what are its units? Take the molar mass of nitrogen as 28 kg/mol and that of hydrogen as 2 kg/mol.

For ease, let us identify nitrogen properties by the subscript N and hydrogen properties by the subscript H. Thus

$$MM_N \times R_N = MM_H \times R_H = R_m.$$

First, R_H

$$28 \text{ kg/mol} \times 297 \text{ J/kg K} = 2 \text{ kg/mol} \times R_H$$
$$R_H = 4158 \text{ J/kg K}.$$

Now R_m

$$28 \text{ kg/mol} \times 297 \text{ J/kg K} = R_m$$
$$R_m = 8316 \text{ J/mol K}.$$

Individual calculations of this nature for the value of the universal gas constant are likely to vary a little bit depending upon the quality of the data – how many decimal places in the molar mass – but a commonly quoted rounded value is 8314 J/kmol K and a yet more accurate one is 8314.5 J/kmol K. Recall that the accuracy of an answer must reflect the accuracy of the starting data.

● PQ 2.12 Now estimate the values of R for oxygen and for carbon dioxide. You already have the necessary information for these calculations.

So there is available a series of relationships which, while strictly for ideal gases, can be used with common sense for real gases in real circumstances. The only question to satisfy then is what does common sense mean here? Perhaps the most important point is to ask for what purpose calculations are being done. If for general guidance on a design or operation or estimating the effect or magnitude of a change of conditions, then the ideal gas relationships are adequate. If the work is for a purpose where extreme accuracy is paramount, then consideration **must** be given to the use of improved versions of the laws, such as we will meet in a later chapter.

It is, however, fair to say that, providing extremes of conditions are avoided and the gas in question is well away from its liquefaction point so that it is truly acting as a gas, then the laws are widely applicable to real gas demands. The closeness of applicability varies from one gas to another, though. For instance, at room temperature nitrogen obeys the ideal gas laws to half a per cent or so up to a pressure of 100 bar; oxygen has deviated by about 5% at that pressure. If temperatures rise, then the deviation is likely to increase. Again by way of example, nitrogen at 200°C and 100 bar deviates by around 4%.

There is plenty of information in the reference texts on how much real gases vary from the ideal values and it is commonplace to apply a known deviation factor to ideal gas law calculations for better accuracy when really necessary. The use of correction factors in thermodynamics is quite legitimate and you don't always have to calculate everything from first principles!! Go back to PQ 2.1 to remind yourself.

As with any other calculations or data, half the battle is knowing what is needed for the task in hand rather than what would be the perfect outcome. Summarizing these derived ideal gas relationships, we have:

- for a fixed mass of a given gas $\dfrac{p_1V_1}{T_1} = \dfrac{p_2V_2}{T_2} = \dfrac{p_3V_3}{T_3}$ etc.;
- for any given gas $pV = mRT$;
- for all gases $MMR = R_m$.

These are fundamental relationships and it is important that you understand them at this stage. If you have any doubts, go back now to make sure.

2.4 MIXTURES OF GASES

While these laws have been derived for and, up to now, applied to single gases, there are occasions when mixtures of gases either occur or are used in industry and commerce. Manufactured fuel gases, exhaust products of engines, air itself of course, are common examples. If gas mixtures are important, then it is equally important to know something of their properties and their behaviour.

For the whole of this chapter, we are only concerned with **unreacting gases** so for this part of the chapter our mixture of gases is also unreacting – i.e. they are inert, they do not burn or go bang or interact in any chemical fashion. Later we shall look at reacting mixtures, though. For these inert mixtures, there are two basic questions which need to be asked immediately because the answers could make life simple when studying gas mixtures. The questions are:

- are the properties of the mixture simply the averaged properties of the constituents of the mixture?
- does the fact that it is a mixture invalidate any of the laws which we have met so far?

In short, the answer to the first is, broadly, yes. The properties of an unreacting mixture of gases are more or less the average properties of the constituents, obviously taking into account the amounts of each present.

● PQ 2.13 If the answer to the first question is yes, what is the likely answer to the second, and why?

So, the gas laws which we have met and derived can be applied to unreacting mixtures by using the properties of the mixture. This example will illustrate.

Example 2.8 Air is made up approximately of 76.8% nitrogen and

23.2% oxygen by mass. What volume will be occupied by 2 kg of air at 30°C, 1.5 bar?

If this was a single gas, we could apply $pV = mRT$ directly and get the answer. We can still do much the same because we know p to be 1.5 bar (150 000 N/m^2) and T to be 30°C (303K) and we are asked V. What we need to assess for the 2 kg of mixture (air) is mR. We can do this by adding together the product of $m \times R$ for nitrogen and $m \times R$ for oxygen:

for the nitrogen, mass of nitrogen is 2 kg $\times$ 76.8%= 1.536 kg, for the oxygen, mass of oxygen is 2 kg $\times$ 23.2%= 0.464 kg.

Now look back through the chapter. For nitrogen, R is 297 J/kg K and for oxygen 260 J/kg K. Therefore, for the mixture

$$mR = (1.536 \times 297) + (0.464 \times 260)$$
$$= 576.8 \text{ J/K}$$

so that

$$150\ 000 \text{ N/m}^2 \times V\text{m}^3 = 576.8 \text{ J/K} \times 303\text{K}$$

and

$$V = 1.165 \text{ m}^3$$

Notice that the rounding off of answers is left until late in the calculation, even though rounded numbers are being used throughout. This is a simple way of making sure that rounding-off errors don't accumulate at each stage of the calculation.

Although it was included in the calculation, a value of R for air was not identified on its own because it was not needed on its own. As we can treat mixtures of unreacting gases in the same way as individual gases, we can get a value of R for mixtures of known composition.

- **PQ 2.14** Calculate a value of R for air.

Example 2.9 I estimate the volume of one of the tyres on my car to be 3.5 litres. The tyre pressure gauge records 2 bars for an atmospheric pressure of 1.1 bar and the car has been standing long enough for the air in the tyre to be at the atmospheric temperature of 21°C. What is the mass of air in the tyre?

First it is necessary to get all the values into absolute values – the litres into m^3, the pressure into N/m^2 (absolute, not gauge!) and the temperature into K.

Thus, V becomes 0.0035 m^3, pressure becomes 310000 N/m^2 (2 + 1.1 bar absolute, of course) and the temperature 294 K. The value of R

for air has just been estimated at 288 J/kgK:

$$310000 \text{ N/m}^2 \times 0.0035 \text{ m}^3 = m \times 288 \text{ J/kgK} \times 294 \text{ K}$$
$$\text{mass} = 0.0128 \text{ kg or 12.8 grams.}$$

Example 2.10 Assuming the tyres to be fairly rigid containers (which they are in reality), what happens to the tyre pressure if I leave my car in the sun and the air temperature in the tyre rises by 10 K?

The volume and mass of the air are unchanged as is the value of R for air, so the pressure must change in line with the temperature. We can thus write (check back if you are not sure)

$$p_1/T_1 = p_2/T_2$$
$$3.1/294 = p_2/304$$

which gives

$$p_2 = 3.2 \text{ bar.}$$

This does not seem too important but recall that tyre pressures are measured by gauge, not absolute, pressure normally. The gauge pressure is (absolute pressure − atmospheric pressure) so the tyre gauge would read

$$3.2 - 1.1 = 2.1 \text{ bar,}$$

compared to the recommended 2-bar pressure. That explains why tyre pressures should always be read with the tyres cool, not after standing in the sun or immediately after a journey, when the tyres will be warm.

Since the mixtures being studied are not reacting, it would still be possible to identify the individual gases in the mixture if we had the proper analytical equipment. It is relatively simple, for example, to separate the oxygen and the nitrogen in air. As the gas laws are exactly that – laws which govern the behaviour of gases – it is reasonable to say then that the laws will apply to the individual gases in the mixture at the same time as they apply to the mixture as a whole. It is rather like having a can of gas and saying that the laws apply equally to the left hand side of the can, to the right hand side and to the whole can (Figure 2.6).

For the present, only the relationship between pressure, volume and temperature will be considered and much can be illustrated by looking at the characteristic gas equation. Suppose we have a mixture of three gases, A, B and C, held in a container. As the gas laws apply equally well to mixtures and to individual gases, then $pV = mRT$ can be used in some form or other for the individual constituents of the mixture and

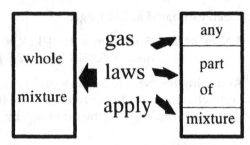

Figure 2.6 Gas laws apply to the whole and to the parts.

for the mixture as a whole. Identify the individual properties of the three gases A, B and C with the subscripts a, b and c.

● PQ 2.15 If I wish to apply the equation $pV = mRT$ to each of those three gases in the mixture, which of the five factors p, V, m, R and T are special to the individual gases?

The values of volume, mass and ideal gas constant therefore apply to the **individual** gases, even though we can measure a total mass, a total volume and determine an average gas constant for the mixture. The values of pressure and temperature apply to the mixture **and** to each individual component *equally*. The characteristic equation for each of the gases A B and C individually would thus be written as

$$p V_a = m_a R_a T,$$
$$p V_b = m_b R_b T,$$
$$p V_c = m_c R_c T.$$

The word *'partial'* is often used to identify the individual mass and volume contributions to a mixture. For instance, V_a is the **partial volume** of gas A in the mixture and it follows that the total volume V is the sum of the partial volumes

$$V = V_a + V_b + V_c,$$

and similarly for the **partial masses** of the gases A, B and C

$$m = m_a + m_b + m_c.$$

If an average value R_{mix} for the gas constant of the mixture was worked out, as was done for air in PQ 2.12, then the characteristic equation could also be used with the total mass m and the total volume V as

$$p V = m R_{mix} T$$

● PQ 2.16 This latest application of the characteristic equation has important practical uses, so summarize what has been said, in your own words, as briefly as possible.

Gas mixture analyses are important in many areas of industrial activity, such as in manufacturing processes and environmental protection. Since gas mixture analysis is normally on the **volume** basis (mostly because of the type of equipment used, as was noted earlier) a typical gas mixture analysis then would show the partial volume, usually as a percentage but it is relatively simple to convert this to a mass basis if necessary, either by using the gas densities or by a simple gas law technique.

Example 2.11 In this example we are going to do a calculation which illustrates some of the points just met and it will also give us a reminder about data accuracy and about what we are told and what we need to know. The answer takes some space because of the coverage given to these items.

An industrial process gas mixture is analysed as 20% oxygen, O_2; 50% nitrogen, N_2 and 30% carbon dioxide, CO_2. What is its analysis on a mass basis?

There is apparently very little information in the question but most such problems can be simplified by making sure of what information we need and by knowing where routine information can be found. There is some of both in this question.

As we are seeking to change from something in volume to something in mass, then an equation linking the two is necessary. The particular one which we have met is $p\,V = m\,R\,T$ and writing this for each of the three constituents reminds us that p and T are common whilst V for each is given – at least as a percentage. m is what we are to find – also at least as a percentage – which leaves R for each.

Take the latter first. Either we can find the individual value of R from standard tables or we can recall that $R \times MM = R_m$. The value of the universal gas constant appears in most data tables even if individual gas values don't. The molar masses of gases are also well established so, say we have from some everyday data table:

$$\text{universal gas constant} = 8314 \text{ J/kmol K};$$

$$\text{oxygen, molar mass 32, so } R_o = \frac{8314}{32} = 259.8 \text{ J/kg K.}$$

● PQ 2.17 Calculate the values of the gas constant for nitrogen and carbon dioxide in this way.

Notice that the accuracy of the answer depends upon the accuracy of the initial data. If we had used a value of 8316 J/kmol K for R_m, then the value of R for oxygen would have become about 260 J/kg K. I have done this quite deliberately to remind you that data table values for many factors do vary a little bit and that you **must** take this into account when performing any calculations. Well established figures like R_m are usually reported consistently, of course, and the common value is 8314 or maybe 8314.5 J/kmol K for a little more accuracy.

If we are going to report results to two decimal places, as for R_c, then the initial information must be accurate to two decimal places also. We could ask, for instance, is the molar mass of carbon dioxide exactly 44? If the answer is 'no', then we have either to:

● find more accurate figures to give a more accurate answer or;
● round off the answer in a sensible manner.

Whilst there is no substitute for accuracy, there is also no substitute for common sense. Both have their place in thermodynamics! My choice here would be to report R_c as 189 J/kg K, R_o as 260 J/kg K and R_N as 297 J/kg K. These roundings-off represent differences of 0.03%, 0.08% and 0.035% respectively compared to the calculated values, which are very small changes compared to everyday industrial instrumentation accuracy. As another aside, if calculations involve many steps, the safe way with rounding-off is to leave it all to the end, thus avoiding cumulative rounding errors. (Figure 2.7).

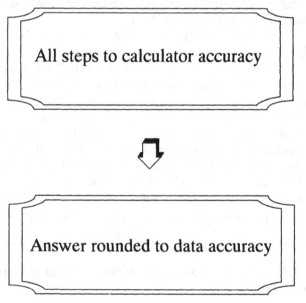

Figure 2.7 Keep the calculation as accurate as possible, then round off at the end.

This reminds us then that the accuracies of initial reference or tabular data and ultimate answer have to align but it also adds that the information provided by instrumentation has to be of the same order. It's no use having the molar mass of carbon dioxide to three decimal places if the analytical equipment is plus or minus two per cent!! As a little test, even with the most accurate instruments, try reading one which has a needle or pointer on a dial gauge. How accurately can you really read it? Now back to the calculation.

We are not told any absolute volume for the gas, so work on some convenient basis such as 1 m^3. There is thus 0.2 m^3 oxygen, 0.5 m^3 nitrogen and 0.3 m^3 carbon dioxide. We can write then for each 1 m^3 of the gas mixture:

$$\text{oxygen} \qquad p \times 0.2 = m_o \times 260 \times T;$$
$$\text{nitrogen} \qquad p \times 0.5 = m_N \times 297 \times T;$$
$$\text{carbon dioxide } p \times 0.3 = m_c \times 189 \times T.$$

At first sight, it may seem that we come to a stop because we do not know p or T – but we are asked the mass ratio or percentage, not the absolute amounts. As we continue through the calculation, you will see that p and T do not matter.

Thus

$$m_o = \frac{p \times 0.2}{T \times 260} = 0.00077 \, \frac{p}{T}$$

$$m_N = \frac{p \times 0.5}{T \times 297} = 0.00168 \, \frac{p}{T}$$

$$m_c = \frac{p \times 0.3}{T \times 189} = 0.00159 \, \frac{p}{T}.$$

This makes the total mass

$$m = m_o + m_N + m_c = 0.00404 \, \frac{p}{T}$$

these all being kg/m^3. Therefore, the mass ratios are

$$\frac{m_o}{m} = \frac{0.00077}{0.00404} = 0.1906$$

$$\frac{m_N}{m} = \frac{0.00168}{0.00404} = 0.4158$$

$$\frac{m_c}{m} = \frac{0.00159}{0.00404} = 0.3936$$

these now being kg of constituent per kg of mixture. Make sure that you are quite clear on the steps taken so far because to answer the question asked, we are now to convert these into percentages. This is the easiest step. We have found the mass of each gas per unit mass of mixture and percentage is amount per hundred units so the analysis by mass is

oxygen 19.06%, nitrogen 41.58%, carbon dioxide 39.36%.

Here's something to think about. Bearing in mind what has been said about accuracies and so on, would you report the mass analysis as it reads here or would you round off these numbers? Incidentally, if we had done the calculation by use of tabulated densities for the real gases, the answer would have been

oxygen 19.05%, nitrogen 41.67%, carbon dioxide 39.26%,

so the use of the ideal gas laws when actual real gas values are not to hand gives an acceptably accurate result in this case.

That was quite a lengthy answer but it covered some very important points. As a model of the mixture, it can be imagined that each gas in it occupies its own share of the total volume of the mixture. This share is commonly called the **partial volume**, which we have met and used successfully. Since in theory and to a very great extent in practice gases are mutually miscible, then there is normally no question of the gases actually each occupying a particular part of the total container, such as the left hand side or the right hand side. The gases are thoroughly mixed but each will still have its share, its partial volume, occupied by its own individual molecules spread throughout the total volume.

By the same token then, it can be seen that each gas in the mixture is contributing to the total pressure in the container. Use the same gases which we have just met in Example 2.9 and, for the sake of simplicity, pretend that we are dealing with a big rigid container, 100 m^3 and that the pressure inside is 2 bar. Although the gases are very well mixed, the nitrogen will take up 50 m^3 (albeit spread throughout the mixture) because it is 50% of the volume of the mixture. The oxygen will take up 20 m^3 and the carbon dioxide 30 m^3 for the same reason.

● PQ 2.18 If the oxygen and the carbon dioxide are now removed but care is taken not to change the temperature, what will now be the pressure inside the rigid 100 m^3 container?

If we do the same for the oxygen now, its pressure on occupying the whole box will be 0.4 bar and if the same was done for the carbon dioxide, its pressure on occupying the whole box would be 0.6 bar. This pressure of the individual constituents if given the whole volume to occupy is called the **partial pressure** and it is the contribution which the individual gas is making to the total pressure of the mixture. Note that the individual pressures just calculated add up to the total original pressure – 1 bar + 0.4 bar + 0.6 bar = 2 bar – but note also that the temperature must remain constant. There are always conditions to calculations of this nature. Incidentally, work out for yourself that there is no equivalent property of partial temperature.

This reference to partial pressures is an example of another long-standing thermodynamic law called **Dalton's law** which was formulated by experimental observation, just as Boyle's and Charles' laws were. Dalton's original statement said that each gas in a mixture acts as if it alone filled the vessel which contains the mixture. For engineering purposes, another investigator called Gibbs expanded the idea to give the **Gibbs–Dalton law**, which covers several thermodynamic properties of gas mixtures. Some we will meet later but the part of interest here has just been proved, that is:

the pressure of a gaseous mixture is the sum of the pressures which each gas alone would exert in the vessel at the mixture temperature.

For any gas mixture therefore, we meet total and partial pressures exactly as we have met total and partial volumes. Each is important in its own right and they must not be confused.

In summary, partial volumes add up to total volume and both the partial volumes and the total volume are subject to the total pressure. Partial pressures add up to the total pressure and both act on the total volume. We can write, using the identifiers already met,

$$\text{for the total mixture } pV = mRT,$$
$$\text{and for each component } pV_a = m_a R_a T_a \text{ (and so on)}.$$

● **PQ 2.19** Now write the parallel equations for partial pressure and total pressure acting on the mixture and the components.

Looking at these two partial pressure and partial volume equations for gas A,

$$pV_a = m_a R_a T \text{ and } p_a V = m_a R_a T$$

they show that $pV_a = p_a V$.

Example 2.12 Go back to the gas mixture in Example 2.11. If the total pressure at the point of measurement of that gas composition was 2 bar, what are the partial pressures of the three constituents?

Since $p_a V = p V_a$, then

$$\frac{p_a}{p} = \frac{V_a}{V}$$

and so on. The volume ratio (partial volume to total volume) is another way of expressing the volume per cent for each constituent, so the pressure ratio (partial pressure to total pressure) must also reflect the volume per cent. Thus 20% by volume oxygen must provide 20% of the total pressure and the partial pressure of the oxygen is 20% of the total, 2 bar, or 0.4 bar. Similarly, the nitrogen partial pressure is 50% of the total, or 1 bar, and the carbon dioxide 30% or 0.6 bar.

Up to this point and to enable the development of some fundamental relationships, the gas has been held in a closed container – the flexible balloon or the rigid can of the examples, for instance. The tutorial examples at the end of this chapter also refer to contained quantities. While the example conditions have spoken of pressure or temperature or volume change, no thought has been given to how these changes are wrought. The next chapter then deals with energy transfer to and from masses of gas and their consequent expansion or contraction, that is, the agents of change of condition.

PROGRESS QUESTION ANSWERS

PQ 2.1

There's a basic rule in engineering – if theory and practice do not agree, then it's the theory that's wrong! Most theoretical work has to have some simplification or generalization in it somewhere, or it would be far too intricate for everyday use. Any very detailed theory is also likely to be very specific in its applicability, so general theories or design guides are likely to be far more useful. Thus when something is designed for the first time, it is necessary to see how well the design information or theory applies. Any test or empirical information helps to strengthen the theory or design calculation.

Beyond that, there are often factors in engineering that defy any precise calculation – for example, how long will a car last before the bottom rusts away? Experiment is needed to allow for these usage factors and thus to strengthen again the designer's hand. There are many examples and empirical information is a vital part of the subject.

PQ 2.2

Without delving too deeply at this stage, we may question these points.

- Will gases fill any space? They may do in a balloon but if we look at the earth's atmosphere, the air stays near the earth and does not disappear into space.

- Are the molecules infinitely small? No, they are not. It is possible to measure the size of a molecule nowadays, so they can't be infinitely small, and different materials will also have different molecule sizes.
- Are they perfectly elastic? Highly unlikely. Even if we know nothing of gases, we do know that there is no perfect anything in terms of real life and molecules are no exception.
- Are they homogeneous? Yes if it's a single gas, but there are plenty of examples of gas mixtures which are not.

PQ 2.3

Two reasons may be:

- the accuracy to which the thermometer might be read;
- the thermometer not being in a representative position – is the temperature in the middle of the room the same as at the ceiling or at the wall?

PQ 2.4

We can do this by inspection. Take Charles' law first. If I have some gas at 10°C and I measure its volume as 1 m^3, then I reduce the temperature to 0°C, will its volume reduce in the ratio 0°C:10°C? That is, will its volume fall to zero, which is what should happen if the law is correct and if Celsius is the temperature to use? No, it will not, so clearly Celsius is not used for this purpose. Do the same exercise with Boyle's law.

The SI units for these laws are the same as in any other use of temperature (often represented as T in equations), pressure (p), volume (V) and mass (m) – K, N/m^2, m^3 and kg respectively.

PQ 2.5

Since $V \propto 1/p$ for constant mass and temperature and $V \propto T$ for constant mass and pressure, then it is reasonable to deduce that $p \propto T$ for a constant mass of gas held at constant volume.

Thus, where values are of course in absolute units, the law is of the form 'For a given mass of an ideal gas held at constant volume, the pressure is directly proportional to the temperature'.

PQ 2.6

There is nothing wrong with using common sense in thermodynamics as well as the analytical approach. Sitting and thinking about it can put you on the right lines, even if you then have to analyse to get the details correct. Since you are asked not to argue it out too much, you may say something like:

'pV is constant; V/T is constant; p/T is constant. Each has its own prescribed conditions of fixed mass and fixed third property, T, p and V respectively. Whatever I do to the gas, the likelihood is that pressure will try to rise as temperature rises, volume will try to fall as pressure increases, volume will try to rise as temperature rises. The gas might not achieve each of these changes but it is likely to try.'

The next step then is to put the bits of the argument together and offer

$$\text{'For a given mass of an ideal gas, } \frac{pV}{T} = \text{constant.'}$$

PQ 2.7

Again, using pV/T = constant, we can write

$$\frac{5.89 \text{ bar} \times 0.6 \text{ m}^3}{(60 + 273)\text{K}} = \frac{7 \text{ bar} \times 0.7 \text{ m}^3}{T\text{K}}$$

which gives the required temperature of 462K or 189°C.

PQ 2.8

Different gases have different densities – methane has a higher density than hydrogen but lower than nitrogen, for example. Most data books will give values. Density is mass per unit volume, specific volume is volume per unit mass – the inverse of density. If gas densities are different, then so will be their specific volumes.

PQ 2.9

The units for R can be derived directly from those of the rest of the equation. Writing the equation and putting the units of each component in also

$$p \text{ (N/m}^2) \times v_s \text{ (m}^3/\text{kg)} = R(\text{what?}) \times T(\text{K}).$$

The units must be consistent so, cross multiplying just the units,

$$\frac{(\text{N/m}^2) \times (\text{m}^3/\text{kg})}{(\text{K})} = \text{what?}$$

Thus the units of R must be

$$\frac{N \times m^3}{m^2 \times kg \times k} \text{ or Nm/kg K}$$

As the newton-metre, Nm, is an equal unit to the joule, J (check a school physics book if you have any doubt) and for a reason which will be seen later, the usual units for R are written as J/kg K. We will meet numerical values of R for different gases shortly.

PQ 2.10

For our purposes and, indeed, for most practical purposes, the mol is simply represented as the molar mass expressed in kilograms for any material. Thus as carbon dioxide has a molar mass of 44, a mol of carbon dioxide means 44 kg of carbon dioxide. Similarly, a mol of oxygen (molar mass 32) means 32 kg of oxygen.

PQ 2.11

Think of several different gases in different containers but all at the same pressure p and temperature T. We know that the molar volume V_m for all these gases must be the same because Avogadro's hypothesis tells us just that.

Thus if the characteristic equation is applied to all the gases in their own containers, p, V_m and T are all the same. The individual values of MM and R will be different but the product $(MM \times R)$ must also be constant. That is for any ideal gas, the product of its molar mass and its gas constant is a Constant value.

PQ 2.12

Table 2.1 gives the molar mass of oxygen as 32 kg/mol and 44 kg/mol for carbon dioxide. The value of R_m has just been found, so dividing R_m by the individual molar mass gives R for oxygen as 260 J/kg K and R for carbon dioxide as 189 J/kg K. Notice the rounding off, where appropriate and recall the reasons.

PQ 2.13

Since the properties of an inert gas mixture are the averaged properties of the constituents, then the answer to the second question is no, provided of course that the same conditions are satisfied as for single gases. The gas laws apply to these unreacting mixtures.

PQ 2.14

The contribution which each gas in the mixture makes to the overall value must depend upon the mass proportion of each gas in the mixture, since the units of R are J/kg K.

So, for the nitrogen the contribution is 297 × 76.8% and for the oxygen the contribution is 260 × 23.2%. Thus

$$R(\text{air}) = 297 \times 0.768 + 260 \times 0.232 \text{ J/kg K}$$
$$= 288.4 \text{ J/kg K}$$

It would be sensible to round this value to 288 J/kg K since the original example said 'approximately' when giving the air composition.

PQ 2.15

If I fix a thermometer and a pressure gauge to the container, they will measure the temperature and the pressure of the mixture. They will have no way of distinguishing between gases A, B and C which are all mixed up in the container and must therefore all be at the same temperature and pressure. So p and T are not individual to each gas. Each gas will experience the same pressure and the same temperature.

The mass of each gas is individual – it depends upon how much is present and the same applies to the proportion of the total volume taken up by each gas, so m and V are individual. Even though the three gases are thoroughly mixed, each must make its individual contribution to the total mass and the total volume. As a point of interest, when gas mixtures are analysed, they are usually analysed on a volume basis because of the type of equipment normally used for that purpose. Finally, each gas of course has its own value of R, the ideal gas constant for that gas.

PQ 2.16

One way of saying this may be – 'whatever the gas mixture, the characteristic equation can be used for the individual gases and for the gas mixture equally well, provided that the correct equation components are used'.

PQ 2.17

Similarly then,

$$R_N = \frac{8314}{28} = 296.9 \text{ J/kg K}$$

$$R_c = \frac{8314}{44} = 188.95 \text{ J/kg K,}$$

where R_o, R_N, and R_c are used to identify the Constants for oxygen, nitrogen and carbon dioxide respectively. These abbreviations are my choice – there's nothing official about them at all and you are free to use your own identities, provided you state them clearly.

PQ 2.18

As nothing else has changed, the nitrogen now occupies 100 m³ instead of 50 m³ at the same temperature. Recall Boyle's law – the pressure must fall from 2 bar to 1 bar as the volume occupied by the nitrogen has doubled at the same temperature.

PQ 2.19

The overall equation for the mixture is, of course, exactly the same, $pV = mRT$, so it is a matter of how to use the partial pressures for each constituent. Recall that the partial pressure is the individual gas's contribution to the total, so the individual partial pressure must be exerted on the total volume. Its influence may be because of the presence of gas A but that bit of the total pressure will also be felt by gas B and gas C – that is, the total mixture volume. We can therefore write

$$p_a V = m_a R_a T$$
$$p_b V = m_b R_b T$$
$$p_c V = m_c R_c T$$

TUTORIAL QUESTIONS

You are given the answers, rounded sensibly, to these numerical questions but not the workings. All the workings follow those already met in the chapter and you may have to look back for some data. Assume ideal gas properties unless told otherwise.

2.1 0.5 m³ of a gas at 1.7 bar undergoes an isothermal change to 2.3 bar. What is its new volume?   [0.37 m³]

2.2 A gas at 30°C expands isobarically (constant pressure) to 125°C. What is the ratio of final to original volume?   [1.3:1]

2.3 If 3 kg of nitrogen occupy 1.5 m³ at 18°C, what is its pressure? If atmospheric pressure is 1.05 bar, what would a conventional pressure gauge read if attached to the gas container?　　　　[1.73 bar, 0.68 bar]

2.4 What is the specific volume of air at 0.97 bar, 12°C?　　　[0.84 m³]

2.5 Given Table 2.1 and then looking back for the gas constant for nitrogen, what is the value of the gas constant for methane?[520 J/kg K]

2.6 By finding the relevant air analysis, determine the partial pressure of nitrogen in the atmosphere at 1.00 bar. If a sample of air at this pressure and 10°C was put into a rigid container and then its temperature raised to 91°C, what would be the partial pressure of oxygen in the container?　　　　　　　　　　　　　　[0.79 bar, 0.27 bar]

Energy and contained gases

3.1 THE NON-FLOW ENERGY EQUATION

When looking at the pressure–volume–temperature relationships for an ideal gas, the simplest circumstance is that of a contained, fixed mass – some gas perhaps in a sealed can or in a balloon or maybe trapped by a piston in a cylinder. For changes to occur in the pressure, volume or temperature, singly or in combination, of this gas, it has to expand or be compressed; it may be heated or cooled, again singly or in combination. The words compression and expansion, heating and cooling are easy to use but what is needed for these processes to take place? Something has to cause the change.

● PQ 3.1 Thinking of simple practical ways of heating, compressing and so on, what common factor links them? There is a clue in the chapter title.

At this stage, accept the everyday meaning of **energy** in its various forms. It is one of those items which are easier to recognize than to define. As an exercise, try to define it and each subsequent term which you use. It becomes interesting! Start for instance with one common definition, a capacity for doing work. Work is done when a force moves through a distance. A force is . . . ? and so on.

Where energy is used in a thermodynamic operation, it is usual to talk about **energy transfer**, whatever the process. So the cooking of food would involve an energy transfer, blowing up a balloon would need a transfer of energy and so on. Energy transfer takes place from an energy source – a cooker, your lungs or whatever – to an energy sink – the food, the balloon (Figure 3.1). In this respect, we will meet the terms heat transfer and work transfer shortly.

Even in the simplest of changes to a gas, there is an energy involvement and for our immediate purposes we will stay with a fixed mass of gas in a closed container. The forms or types of energy which

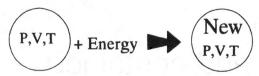

Figure 3.1 Energy transfer causes change.

may be used in the simplest changes are quite limited but even this simple study can yield some important thermodynamic relationships.

• **PQ 3.2** I have a sealed flexible balloon of gas. First I heat it a little and then I compress the balloon. In the most general terms, what forms of energy have I used?

Thus, **heat transfer** and **work transfer** are the agents of change in this case. It doesn't matter how it is done or what equipment is used, those words and concepts apply. So for the balloon of gas, where does this energy transfer go, both the work and the heat transfer? A little must go into the balloon skin if it stretches (storing energy like an elastic band) or if it gets warm (storing energy like hot water) but this is a very small, almost negligible amount. The vast majority of the transferred energy goes into the gas, changing pressure or volume or temperature or, far more likely in real life, a mixture of two or three of these gas properties.

We will look at these changes and see what they mean in energy terms, but you can imagine some of the effects. If the gas pressure rises, then it may be storing energy to do work later, just as air from a compressor does work when driving a pneumatic drill. If the gas temperature rises, then it could be used as a heat source for some other purpose, rather like warm air from a hair drier. Volume changes may be associated with both pressure and temperature changes (recall $pV = mRT$) so maybe the temperature and pressure effects cannot be separated simply every time into saying pressure for doing work and temperature for heat. For a start, however, stay with simple cases and simple energy changes.

The gas is undergoing change – being warmed, being compressed – so its energy content is changing because things are being done to its pressure and/or its volume and/or its temperature. Since the gas is effectively some very mobile molecules in a lot of otherwise empty space – the kinetic theory – any gas energy changes have to be associated with the molecules rather than the empty space. The energy of the molecules, their kinetic energy which is of course an indication of gas temperature, is called the **internal energy** of the gas. It is a reasonably descriptive term, as you can see. The gas will **always** have some internal energy if its molecules can move, so the internal energy

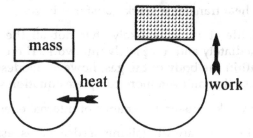

Figure 3.2 Heat transfer can give work transfer.

can increase or decrease (temperature rise or fall) depending upon what is done to the gas.

As an aside, it is strictly incorrect to talk of energy being added to a body or a material in the way that milk may be added to tea. It is more to do with a change of energy state but, for clarity and simplicity, terms like 'adding heat' or similar sorts of words will be used.

The law of conservation of energy says, at its simplest, that energy can be neither created nor destroyed but only transformed. So any changes to the gas in the balloon must end up with the energy forms (heat transfer, work transfer, internal energy changes) balancing each other out before and after. There must be an **energy balance**

If the balloon is heated by pouring hot water over it, the energy transferred from the hot water has to reappear in the gas. The only energy sink within the gas body is the molecules themselves, so their internal energy must increase and consequently the gas molecules must be moving quicker. Depending upon if and how the balloon expands, the pressure will change because the molecules are hitting the walls quicker and more often because their kinetic energy has risen – yet more of the kinetic theory.

● PQ 3.3 If instead I just compress the balloon – with no heating or cooling – where does the work transfer go?

These are single changes – either heating or compressing. Of course, the heated balloon is likely to expand and it would be a simple matter to use this expansion to do some work, such as by raising a weight which was resting on the balloon. (Figure 3.2).

● PQ 3.4 Heat is being added to the balloon of gas and the expanding balloon is being used to do some work. Write down, in words not symbols, an equation to represent the energy balance for this.

In any work or heat transfer operation in real life, there will be some inevitable **energy losses** – friction in machine surfaces, heat escaping from hot surfaces – so the equation in answer to PQ 3.4 should read something like heat transfer = work transfer + losses.

Equally in real life it is quite likely that not all the heat will be transferred immediately and completely into work. There is likely to be some change within the body of the gas, however modest, a change of internal energy, so that an even more complete equation would read

heat transfer = work transfer + a change of internal energy + losses.

For the moment, as we are still talking of ideal gases, assume that we have an ideal process so that incidental losses are negligible. The changes here then are those of **heat transfer**, **work transfer** and **internal energy change**. Notice that the equation says heat transfer and work transfer, meaning finite amounts. No heat was being transferred previously and no work was being done previously. Everything was in a steady state. However, the earlier reference to internal energy when some hot water was poured over the balloon spoke of an increase of internal energy, as did the more complete equation. The gas had internal energy before the event because the molecules were moving about. They receive some heat, so they move about a bit quicker – their kinetic energy rises. The internal energy has changed from one value to another, whereas the other two factors (heat and work transfer) are discrete doses.

This example of a balloon being heated and the heat being used to do work is a demonstration of the so-called **First Law of Thermodynamics**. This law says that heat and work are mutually interchangeable. There are formal statements of the law, as there are for most scientific or natural laws but that straightforward interpretation will do all that is necessary for our purposes. In fact, we can go further and say that, given the right apparatus and the right operational circumstances, all forms of energy are interchangeable. Like the gas laws, though, the First Law was formulated when some other forms of energy were not recognized as such, hence the reference to heat and work. The law, though perhaps obvious to us now, enables some important relationships to be derived (Figure 3.3).

Since this chapter is to do with energy usage in gas processes, it will

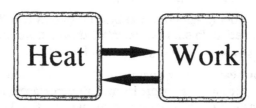

Figure 3.3 The First Law.

be useful to have the verbal equation of a couple of paragraphs back represented in a conventional mathematical form. As we are dealing still with gases held in sealed containers, there is no flow of the gas into or out of the container. The sensible thing therefore is to identify any relevant equation quite clearly as having no flow component (because there is no flow) and the common mathematical representation of that verbal equation is thus called the **non-flow energy equation**. It is in fact an example of the First Law of Thermodynamics and it is handy to use the initials NFEE for the equation.

Instead of writing the words of the equation, the letters Q, W and U are used commonly for heat transfer, work transfer and internal energy respectively. There are a few qualifications and conventions to observe. First, the work being done, whether it is the expanding balloon pushing a weight or a machine compressing the balloon, is termed 'external work'. The reason is logical in that the use or provision of work is ultimately outside the gas system boundary – the balloon skin in this case. The energy that is seen as work is transferred across the boundary – the external weight is pushed up or the external compressing machine presses the balloon skin. The work is not actually done inside the gas. It is supplied to or from the outside, hence external.

Second, as noted carefully earlier, any heat or work transfer may cause a change of internal energy of the gas, so in a symbols equation, ΔU is used to emphasize the fact that it is a **change** rather than something new. Third, the upper case (CAPITAL LETTERS) is used to represent **total** amounts, with lower case (small letters) being used for **specific** amounts – amounts per unit mass.

● PQ 3.5 Using the letters Q, q, W, w, ΔU, Δu as just indicated and m for mass, write down the relations between total and specific amounts of heat, of work and of internal energy change.

Fourth, there is a logical convention in respect of the use of plus or minus signs to denote the direction of flow of work and heat transfer or the change of internal energy.

● PQ 3.6 Still using the balloon of gas as a good system example, if heat is transferred to the gas and its internal energy rises, what would be a logical sign convention for these two factors?

So, there is a reasonably obvious convention for heat transfer and internal energy change but work transfer may require a little extra thought. If the balloon and its contents are doing work – external work transfer in lifting the weight – then it is a positive change, $+W$. It is something positive that the gas is doing. With the heat transfer, $+Q$ is something positive being done on the gas; the increase of internal energy is something positive being done within the gas; the work transfer – raising the weight – is something positive being done by the gas. Listing these then:

- heat transfer into the system (the gas) is positive $+Q$
- heat transfer out of the system (the gas) is negative $-Q$
- increase of internal energy of the gas is positive $+\Delta U$
- decrease of internal energy of the gas is negative $-\Delta U$
- external work done by the system (the gas) is positive $+W$
- external work done on the system (the gas) is negative $-W$.

It is worth spending some time on understanding this convention as it is important in many areas of thermodynamics. The plus and minus signs can be illustrated by writing down in equation form the relations between Q, W and ΔU when any two of them are involved. For instance, if the gas is heated by external heat transfer without doing any work, the heat transfer and the rise of internal energy must balance. Thus

$$\begin{array}{ccc}\text{heat transfer into gas} & = & \text{change of internal energy} \\ \text{(added to)} & & \text{(increase of)} \\ +Q & = & +\Delta U \end{array}$$

If instead the heat transfer to the gas is used immediately to do external work, that is, the gas receives heat and uses it promptly to raise a weight, then

$$\begin{array}{ccc}\text{heat transfer into gas} & = & \text{external work done by gas} \\ \text{(added to)} & & \text{(done by)} \\ +Q & = & +W \end{array}$$

- PQ 3.7 Now write down an equation which says 'Heat transferred into the gas is shared between some external work being done and a rise of internal energy.'

The answer to PQ 3.7 gives a common representation of the non-flow energy equation. As is usual in mathematical equations, the + sign is omitted unless absolutely necessary so the equation becomes

$$Q = W + \Delta U$$

and there are variations such as for specific amounts

$$q = w + \Delta u$$

and for calculus purposes

$$dq = dw + du.$$

While the lead into that NFEE equation was by use of positive amounts – heat added, work done, internal energy rise – the equation must be equally applicable for negative values of any of the parameters Q, W or ΔU. It is saying, in effect:

whatever the heat transfer into or out of the system, it must be balanced by combined changes in work transfer and internal energy level.

● **PQ 3.8** If there is no heat transfer but the gas balloon is used to raise the weight – that is, to do external work – then write down the NFEE in a form which deals with this.

This representation of the NFEE is then valid whether the individual amounts or changes are negative or positive.

These examples will illustrate the use of the equation.

Example 3.1 The contents of the balloon which we have already met are heated by the addition of 1.2 kJ. The subsequent expansion of the balloon is used to do 800 Nm of external work. What is the net change of internal energy of the gas? Heat transfer adds heat (strictly heat energy but it is commonplace to use the word heat alone), so we have $Q = +1.2$ kJ. External work is done by the gas, so we have $W = +800$ Nm. Note that though heat is usually expressed in joules and work in newton-metres these two (J and Nm) are the **same** in quantity and dimensions. Make sure though that you are not about to compare kJ to Nm! There's a thousand-fold difference!

$$Q = W + \Delta U$$
$$+ 1200 \text{ J} = +800 \text{ Nm} + \Delta U$$

whence

$$\Delta U = +400 \text{ J}.$$

The joule is used commonly for internal energy changes.

Example 3.2 The same balloon contents are compressed using 300 Nm of external work but the balloon is insulated so that there is no external heat transfer. What is the energy balance for this operation?

In this case, the external work is done on the gas, so it is a negative value according to the convention which we are using:

$$Q = W + \Delta U$$
$$O = -300 \text{ Nm} + \Delta U$$

whence

$$\Delta U = 300 \text{ J}.$$

Just check through the equations to make sure that the numbers and signs have been transferred correctly either side of the equals sign.

Example 3.3 In a more complicated operation, the gas now does 350 Nm of external work whilst the balloon is also cooled, losing 200 J by external heat transfer. What is the new change of internal energy?

The gas is doing work, so $W = +350$ Nm and at the same time the gas is being cooled so $Q = -200$ J, thus

$$-200 \text{ J} = +350 \text{ Nm} + \Delta U$$

whence

$$\Delta U = -200 \text{ J} - 350 \text{ Nm}$$
$$= -550 \text{ J}.$$

Again, make sure that you have followed the movement of the numbers about the equals sign to give this answer. Always check back if you have any doubts.

The value of the non-flow energy equation, NFEE, is not limited to the ability to do simple calculations about gases in balloons or similar sorts of containers, fortunately. It can be used to derive some important relationships in thermodynamics. For this next purpose, suppose that some ideal gas is trapped in a cylinder by a perfectly sealing piston. This is highly unlikely in real life but, like the ideal gas concept, it gets rid of imperfections so that we can look at the fundamentals. The piston can move up and down the cylinder to let the gas expand or contract but no gas can escape past it.

Figure 3.4 shows the idea with the piston area being A and the initial length of the cylinder of gas being L_1. The starting volume, say V_1, of gas is thus $A \times L_1$, and suppose the gas pressure in the cylinder is p, all values being in consistent units. Now suppose the gas is heated in some way so that it expands a little but the heat input is just sufficient also to hold the pressure at p. As it expands then, the gas will push the piston a little way up the cylinder so that the new cylinder length holding the gas is L_2 and thus the volume, now V_2, has increased to $A \times$

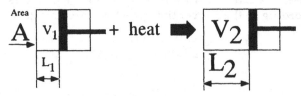

Figure 3.4 Some gas trapped in a cylinder by a piston.

L_2. Nothing else has been changed, just some heat added so that the gas can expand a bit while holding its initial pressure.

● **PQ 3.9** Define carefully the term 'work'.

Pressure is force per unit area so force is pressure times area. The force acting on the piston is then the gas pressure within the cylinder (which of course acts equally in all directions for a contained gas) times the piston area. Now the piston has moved through a distance $(L_2 - L_1)$ so that the external work done, force × distance, is

$$W = \text{force} \times (L_2 - L_1)$$
$$= (p \times A) \times (L_2 - L_1)$$
$$= p \times A \times (L_2 - L_1),$$

but

$$A \times (L_2 - L_1) = A \times L_2 - A \times L_1$$
$$= V_2 - V_1.$$

That is, the change in volume of the gas so that the external work W is

$$W = p \times (V_2 - V_1)$$

or

external work done = gas pressure × change of volume
$$W = p \times \Delta V.$$

Notice, incidentally, that the piston in this example does not necessarily have to be connected to a machine to do work. If the piston moves, then it must be pushing back the surrounding atmosphere, and that is doing work. The work required to push air out of the way is very important in motor vehicle aerodynamics – different circumstances, same principle!

While this has been derived very simply for a carefully contrived operation, it does generate a fundamental expression for the calculation of **external non-flow work transfer**. Note carefully that the expression relates to non-flow work specifically. Since the conditions for the derivation were simple and idealized, then the expression must be fundamentally correct. That is the basis of most fundamental observations – a simple circumstance with no sophistication to qualify its application.

Although, as far as possible, calculus workings are held in Appendix B, it is worth mentioning the one for non-flow external work here, for completeness. In its calculus form, an infinitessimally small change of volume, dV, would be considered so that, whatever the set of conditions, any change of pressure p would be negligible and the resulting expression would be, as derived in Appendix B,

$$dW = p \, dV.$$

You see that our simply derived expression is only a particular example of that general calculus version and we can therefore accept it with added assurance. For the present, the expression $W = p \times \Delta V$ for non-flow external work transfer is adequate in its non-calculus form.

● PQ 3.10 Define specific heat capacity and write down an equation which relates specific heat capacity, mass, temperature change and heat supply.

Some texts, especially older ones, miss out the word capacity. If you look up a definition, then you may only find specific heat mentioned rather than **specific heat capacity**. The change is a technical one but there is no change in its application. The letter c, in various forms, is commonly used as the abbreviation for specific heat and specific heat capacity in old and newer texts.

In the definition used in the answer to PQ 3.10, the vague terms 'unit quantity' and 'unit temperature change' appeared. This was quite deliberate for two reasons:

● the same general definition applies to measuring systems other than SI and the real world still uses systems other than SI;
● occasionally, for very particular reasons which we will not meet here, the unit quantity may be volume rather than mass.

As far as we are concerned, specific heat capacity will be in SI units, the unit quantity will be mass 1 kg and unit temperature change will be 1 K. A little warning – the joule is a rather small unit for many specific heat capacities, so tables of values which are available in data books or reference texts often report specific heat capacity in kilojoules, kJ. Do take careful note when using values in calculations.

Example 3.4 By way of a simple illustration to remind you about specific heat capacity, suppose a piece of steel, mass 2 kg, is to be raised from 20°C to 55°C. Its specific heat capacity is 0.48 kJ/kg K. What is the heat (strictly, heat energy, recall) requirement for this, assuming no losses?

$$Q = m \times c \times \Delta T \, \text{J}$$
$$= 2 \text{ kg} \times 480 \text{ J/kg K} \times (55\text{–}20) \text{ K}$$
$$= 33\ 600 \text{ J}.$$

Note that the specific heat capacity was given in kJ but was changed to J for the calculation and that the temperature change, though given in °C, is numerically identical when used as K. The rise from 20°C to 55°C is identical at 35°C or 35 K. If the specific heat capacity had been used in the equation as kJ, then the value of Q would also be in kJ, of course. Do take particular care – it is easy to make a trivial error when using numbers like this.

Continuing with the cylinder and piston arrangement which was used to derive $W = p \, \Delta V$, you recall that if a contained gas is heated by an external supply of heat, the gas can use this heat to increase its internal energy or to do external work or some combination of both. Let us take one extreme – the heat transfer is used just to heat up the gas. That is, all the heat transfer is used to increase gas internal energy which means, of course, to raise the gas temperature since temperature is a measure of internal energy.

We have to identify some parameters so let the gas mass be m, the heat supplied be Q, the temperature change be ΔT and the gas specific heat capacity be c, in the self-consistent units which we have just determined. If you have any doubts whatsoever, though, you must write down the full equation to include all the units and then check that the units balance. We can now write the equation

$$Q = m \times c \times \Delta T,$$

and the specific heat capacity is thus

$$c = Q/(m \times \Delta T).$$

So there is what appears to be a sensible definition of specific heat for a gas – it is related to mass, heat supply and a temperature change, just as if we were dealing with the solid steel of Example 3.4.

However, if the other extreme is now considered – all the heat transfer is used to do external work, there is no temperature change because there is no change of internal energy of the gas since $Q = W$ and $\Delta U = 0$. There is still heat being supplied to the gas though. It is the same gas using the same heat supply – so does the same value of specific heat apply?

If there is no temperature change, then the definition appears to fall down at this extreme of all the heat going into work. The reason is simply that the gas in this case is an **agent of energy transfer**. The heat is changed into work but the gas itself, in effect, undergoes no change. In this very particular case then, there is no reasonable way in which we may derive a specific heat capacity. However, this is an extreme and it is easy to think up ways of using supplied heat between the two extremes, to share the heat transfer between doing external work and increasing the gas internal energy.

● **PQ 3.11** What if the heat transfer was used for a combination of doing some external work and raising the internal energy a little?

If in a series of experiments the same gas of mass m received the same amount of heat transfer Q but the proportions of the heat transfer used for external work varied, then the gas temperature rises would also vary. This is because the amount of heat transfer available for internal energy rise varies accordingly. Suppose the temperature changes are ΔT_1, ΔT_2 and ΔT_3 for three such tests. Writing down the equation $Q = m \times c \times \Delta T$ for each test thus gives

$$Q = m \times c_1 \times \Delta T_1$$
$$Q = m \times c_2 \times \Delta T_2$$
$$Q = m \times c_3 \times \Delta T_3$$

Each equation is accurate, yet each gives a different value of c because each has a different but accurately measured value of ΔT (Figure 3.5).

With any given gas, then, there can be an infinite number of specific heat capacities. This is because an infinite number of combinations of external work and internal energy changes can be thought up, to give an infinite number of temperature changes for the same quantity of heat transfer. This might be a fascinating thought but it is quite useless, so it has to be rationalized to give some worthwhile and usable values.

The way that it is done is to define two values for very special conditions – the heating of a gas held at constant volume and the heating of a gas where the pressure is kept constant. These values of specific heat capacity for a gas are called, logically, the **specific heat capacity at constant volume** and the **specific heat capacity at constant pressure**. Conventionally they are designated c_v and c_p and are some-

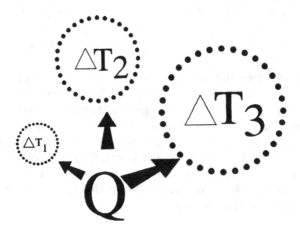

Figure 3.5 Same heat supply, different temperature rises.

times called the **principal specific heat capacities** of a gas. The actual values are individual to different gases, of course, just like specific heat capacities of metals are individual to those metals. Let me repeat that these are two very special values but they are by no means the only values that could be generated, as PQ 3.11 indicated. The reasons for them being both special and very useful will soon become apparent – they are not randomly chosen values for academic purposes!

We have already met and derived one value.

● PQ 3.12 Which? How? As a clue, write down the equation for non-flow external work. Now decide which parameters were zero for one of the extreme cases already discussed for the gas trapped by a piston in a cylinder.

If, instead of being held at constant volume so that no external work transfer took place, the gas had been allowed to expand steadily so that the initial pressure p was maintained in the cylinder throughout the operation, then this would be called a constant pressure operation and the associated value of the gas specific heat would be c_p.

Example 3.5 A rigid can holds 100 g of nitrogen. The can is well insulated and is fitted internally with an electric heater. The heater supplies 370 J to the nitrogen and the gas temperature rises by 5 K. What is the specific heat capacity of nitrogen and which value has been determined?

$$Q = m \times c \times \Delta T$$

thus

$$370 \text{ J} = 0.1 \text{ kg} \times c \times 5 \text{ K}$$

and

$$c = 740 \text{ J/kg K}.$$

As the can is rigid, the volume of the gas cannot change and the value thus determined is c_v. Since the volume has not changed, there is no external work done ($p\Delta V = 0$), so this value is for all the heat transfer being supplied to raise the gas internal energy.

Note that the mass was given in grams but was carefully converted to kilograms for the equation, so that the units were self-consistent. While the answer therefore came out in J/kgK, it is quite common-place in reference tables to give specific heat capacities in kJ/kgK, so this value would normally be reported as 0.74 kJ/kg K.

Example 3.6 A well insulated piston and cylinder is now fitted with an electric element to supply 520 J of heat to 100 g of nitrogen. The piston is allowed to move down the cylinder to hold the pressure steady at its initial value and the temperature rise is again 5 K. Which value of specific heat capacity has been determined this time and what is it?

Again we can write quite accurately, even for different conditions,

$$Q = m \times c \times \Delta T$$
$$520 \text{ J} = 0.1 \text{ kg} \times c \times 5 \text{ K},$$

and

$$c = 1040 \text{ J/kg K or } 1.04 \text{ kJ/kg K},$$

As the piston was allowed to move along the cylinder to maintain a steady pressure p, this new value is the specific heat capacity at constant pressure, c_p.

● PQ 3.13 Why is this value bigger than c_v? After all, the gas temperature has risen by the same amount, 5 K. Here is a clue – is the piston doing work? On what?

Notice that it is a quite different numerical value to c_v. Be clear that whilst c_v is the value for 'all heat going to internal energy', c_p is not the value for 'all heat going to work'. It is a very special value of constant pressure, where some of the heat transfer goes towards external work and some goes towards an increase of internal energy.

● PQ 3.14 In Example 3.6, how much of the heat transfer was used to do work and how much was used to raise the internal energy?

Although the different operations (constant volume and constant pressure in the two Examples 3.5 and 3.6) gave different values of specific heat capacity, this was entirely due to the fact that the heat transfer was used for different purposes. Whatever the combination of external work and internal energy change, there will always be a certain amount of the supplied heat transfer required for any given gas temperature rise. It is most important to appreciate that the specific heat capacities of any gas, ideal or real, are controlled by the way in which the heat transfer to the gas is being used.

The non-flow energy equation can be used to find out a few things about these specific heats and, in fact, other points of importance in thermodynamics. That is the general aim of the next chapter.

PROGRESS QUESTION ANSWERS

PQ 3.1

Each of the processes mentioned earlier – heating, cooling, expansion, compression – is accompanied by some energy usage. If you use a foot pump to blow up a bicycle tyre, you use energy. A fridge uses energy to

cool food – there are plenty of everyday examples. The common factor
then is energy usage.

PQ 3.2

I have used some heat at first and then done some work. Both are
examples of energy transfer – heat transfer, then work transfer. In both
cases the energy has been transferred from a supply, often called the
source, to the user of the energy, often called the sink.

PQ 3.3

Again there must be a change in the gas molecules energy picture. Since
there is no externally supplied heating or cooling, this compression
(work transfer) must benefit the gas internal energy. I have done work
on the gas and the energy used has to reappear somewhere. We will
take a closer look at that soon.

PQ 3.4

If the heat added is then all used to do work, a simple equation would
be

$$\text{heat transfer} = \text{work transfer.}$$

PQ 3.5

For mass of gas m, the total heat transfer Q is m times the specific (per
unit mass) heat transfer q. Therefore, $Q = mq$ and similarly for work
transfer and internal energy change – $W = mw$ and $\Delta U = m\Delta u$. Total
amounts upper case, specific amounts, lower case.

PQ 3.6

Think of the system boundary (the balloon skin) and the system
contents, the gas.
  If the gas internal energy rises, then internal energy has been added,
so it is logical to write $+\Delta U$. If the internal energy falls, then we would
write $-\Delta U$. If heat is transferred into the system, then $+Q$ is sensible
and if heat is transferred from the system, then we would write $-Q$.

PQ 3.7

The heat transferred into the gas is $+Q$; some external work done is $+W$ and the rise of internal energy is $+\Delta U$, so the equation is

$$+Q \qquad = \qquad +W \qquad + \qquad +\Delta U$$
(added to the gas) (done by the gas) (increase of, for the gas)

PQ 3.8

Since $Q = 0$, then the NFEE becomes

$$0 = W + \Delta U$$

which tells us that

$$W = -\Delta U.$$

That is, the work can only be done by a fall of gas internal energy.

PQ 3.9

The straightforward definition is that work is done when a force moves through a distance.

PQ 3.10

Specific heat capacity is the quantity of heat required to raise unit quantity of a substance through unit temperature change. In SI, the unit quantity of mass is 1kg. The unit temperature change is 1 K and the heat supply is measured in joules, J, so the units of specific heat capacity are J/kgK.

Thus for m kg of material of specific heat capacity c J/kg K to be raised through $\Delta T K$, the heat supply is

$$Q = m \times c \times \Delta T \, J$$

PQ 3.11

Where there is a mixture of work done and internal energy change, then there will be a relationship between heat supplied and temperature change but it will vary with the proportion of the supplied heat which is used for changing the internal energy.

The basic definition of specific heat capacity still applies but, clearly, for a gas it must be accompanied by some sort of qualification which defines how much of the supplied heat transfer is for internal energy change such as 'when half the supplied heat is used to do external work', or something similar.

PQ 3.12

Just after defining specific heat capacity in PQ 3.10, we looked at one extreme of heat transfer to a gas. In that one, no work was done and all the heat was used to raise the gas internal energy. Since no external work was done, we can write $W = p\Delta V = 0$. The gas certainly has a pressure which we called p so, for $p\Delta V$ to be zero, then it must be ΔV that is zero. Since ΔV is zero, the volume has remained constant, so the value which was derived was the specific heat capacity at constant volume, c_v.

PQ 3.13

Two things are happening here – the gas is getting warmer and the piston is moving. Even if the piston is not connected to anything, as it moves down the cylinder, it must do work by pushing the surrounding atmosphere away.

   Thus the heat supply is used for both changing gas internal energy (getting warm) and doing work (moving the piston). The gas is then both using the heat directly to raise its internal energy and as an agent of change in doing work. The heat supply, as indicated by the specific heat capacity, must be higher for this case than for the single case of raising the internal energy alone.

PQ 3.14

We saw in Example 3.5 that 370 J would raise the gas temperature by 5K. All the heat transfer was used for that single purpose but it seems reasonable to suggest that, however we go about it, we will always need 370 J to raise the temperature of 100 g nitrogen by 5K. In Example 3.6, there was both temperature raising and work doing. Therefore, the difference between the heat transfer in the two cases must represent the demand on the heat transfer for the external work done, that is (520 − 370) = 150 Nm.

TUTORIAL QUESTIONS

All these questions refer to contained, fixed masses of ideal gases, with no losses.

3.1 How much work must be done on a gas at 3 bar to reduce its volume from 2.7 m³ to 1.8 m³?   [270 kNm]

3.2 If 50 kJ of heat energy are transferred at constant volume to 5 kg of a gas whose specific heat capacity c_v is 740 J/kg K, what would be the change of temperature?   [13.5 K]

3.3 In a non-flow process, the heat transfer to gas in an engine is 100 kJ and 78 kNm of work are extracted. What is the change of gas internal energy?   [+22 kJ]

3.4 In a constant pressure process at 5 bar, 3.93 kg of gas expand from 0.7 m³, 27°C and 300 kNm of external work are done. What is the new gas volume and temperature?   [1.3 m³, 284°C]

3.5 If the specific heat capacity for this gas c_p is 1.04 kJ/kg K, what is the necessary heat transfer to the gas?   [1050 kJ]

3.6 If 184 kJ of heat is transferred to 1 kg of oxygen at 10°C and no work is extracted, what is the change of volume? If the specific heat capacity at constant volume, c_v, for the oxygen is 650 J/kg K and its pressure doubles during this heating, what is its new temperature and what is the change of internal energy?   [nil, 293°C, +184 kJ/kg]

Derivations from the non-flow energy equation

4

4.1 INTERNAL ENERGY AND SPECIFIC HEAT CAPACITY

We saw in the last chapter that there could be numerous values of gas specific heat capacity, all quite legitimate and all depending upon the conditions under which the specific heat was measured. Two particular values – that when the gas is heated and held at a constant volume c_v, and that when a gas is heated and allowed to expand carefully so that the pressure is held constant c_p. For this chapter, we are staying with the non-flow energy equation (NFEE) so we are still dealing with contained and constant masses of ideal gas.

If the gas is heated and held at **constant volume**, then there is no external work done ($W = 0$) because the value of ΔV, the change of volume, is zero in the expression $W = p\Delta V$. All the heat transfer to the gas Q is used in changing the internal energy of the gas ΔU. The NFEE, $Q = W + \Delta U$, can therefore be written as

$$Q = \Delta U$$

for this particular condition.

The general SI definition of specific heat capacity is the amount of heat required to raise the temperature of 1 kg of a substance through 1 K. If the mass of gas here is m and it increases in temperature by an amount ΔT when Q is fed to it, we can write in the usual fashion

$$Q = mc_v\Delta T.$$

Note carefully that the very particular value of gas specific heat capacity c_v is used because this is a constant volume heating process. If it was some other sort of process, then c_v would not be appropriate. So, as this is the same value of Q as in $Q = \Delta U$, we can relate the change of internal energy of an ideal gas ΔU to a temperature rise ΔT.

● **PQ 4.1** Write down, from the two expressions containing Q, an equation which relates ΔU to ΔT for this constant volume operation.

Example 4.1 The specific heat at constant volume c_v of oxygen around room temperature is 650 J/kg K. If oxygen at room temperature is heated through 20 K, what is the internal energy change?

No value of the gas mass is given, so no total change of internal energy can be derived. What can be determined, though, is the change per unit mass and this will have to suffice in the absence of any other information.

Notice too, that reference is made to room temperature. Later, we will investigate the effect of conditions on some real gas thermodynamic properties but, for the present, we must assume that c_v is constant. If you refer back to Chapter 2, you will recall that the ideal gas laws can be applied with confidence for small changes to real gases. 20 K is a small change in this context.

We therefore have

$$\Delta U = mc_v\Delta T$$

and $m = 1$ kg for this calculation, so that

$$\Delta U = 1 \text{ kg} \times 650 \text{ J/kg K} \times 20 \text{ K} = 13\,000 \text{ J or 13 kJ.}$$

The above equation, which gives the value of ΔU for a temperature change of ΔT has been derived from a very particular operation – a *constant volume* one. We are, though, dealing with a thermodynamic property of a material, the internal energy of a gas. However any material property is measured – density or viscosity for instance – then its value is just that, whatever acceptable measuring technique is used. The factors which determine the value or a change in value of that property are consistent and are **independent** of the measuring technique.

In the present case, the change of internal energy has been shown to be related to three factors – mass m, specific heat capacity at constant volume c_v and the change of temperature ΔT. For a given amount of gas m is constant; if the temperature change is stated then ΔT is fixed; c_v is also a thermodynamic property. These three factors will then determine the change of internal energy of an ideal gas whatever the operation and however the change is measured.

Thus for an ideal gas undergoing **any** operation or process, the change of internal energy will **always** be given by

$$\Delta U = m \times c_v \times \Delta T,$$

more often written as

$$\Delta U = mc_v \Delta T.$$

The fact that was derived from a constant volume process is convenient but otherwise immaterial. This relationship thus holds for all other ideal gas processes and we can use it quite freely.

● PQ 4.2 Recall the definition of an ideal gas.

Up to now we have been dealing with ideal gases and will continue to do so but their definition can now be refined or made more precise since various laws which control their behaviour have been introduced or derived. One definition which is often quoted is that ideal gases obey the equation $pV = mRT$ and that the specific heats c_p and c_v are constant for a given ideal gas. Soon, we shall be able to look at that more critically but, for now, we will use it as a **quantifiable** (meaning that we can put numbers to it) representation of ideal gases.

Turning now to specific heat capacity at constant pressure, this will apply to the gas quite simply when the heating (or cooling, of course) process is a **constant pressure** one. For simplicity and clarity, stay with heating alone and think of some ideal gas held by a piston in a cylinder (Figure 4.1).

Some ideal gas is held by a piston in a cylinder. The piston can slide up and down the cylinder but is leakproof, so no gas can escape, just as before. It is a non-flow process. Suppose that heat is transferred to the gas and the whole process controlled so that, although the piston moves along the cylinder as the gas warms up and expands, it does so at a rate that keeps the gas pressure constant. It is a so-called **constant pressure** process – a descriptive name, as you can see.

● PQ 4.3 If this is a constant pressure process, the specific heat capacity is c_p. Heat Q is transferred to the gas and its temperature rises through an amount ΔT. If the mass of gas held in the cylinder is m, write down an expression connecting these four items c_p, Q, m and ΔT.

If any gas warms up because of a heat supply to the gas, in any way whatsoever, there must be a relationship between the heat supplied and the temperature change. There must be some value of specific heat

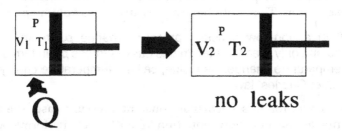

Figure 4.1 The piston and cylinder again.

capacity involved also, and this value will depend upon the operational conditions. Recall that in a non-flow process, heat supply may be shared between external work done and gas internal energy change. The constant pressure condition change is a particular example which is easily identified.

For our immediate purposes, that is sufficient justification to talk about it and it is not special in any other way. The piston movement in the cylinder could just as easily have been controlled so that the pressure rose to twice or three times its original value instead of moving to keep a constant pressure. This constant pressure process is just one example of what could be done.

● **PQ 4.4** If the change of volume of the gas, as it expands and pushes the controlled piston up the cylinder at constant pressure, is $\triangle V$, write down an expression for the external work done. If this piston is simply a leakproof fit but otherwise free to move, on what is this external work done? Will that make it constant pressure?

4.2 THE IDEAL GAS CONSTANT

The heat transfer to the gas in the previous constant pressure process is used in two ways.

● **PQ 4.5** What are they?

Some of the heat supply is therefore used to do external work and some to increase the gas temperature or gas internal energy. The constant pressure process is not the only one to split the heat usage in this way of course. It is possible to think of many other ways, each of which will separate the heat supply in different proportions, maybe one quarter going to do work or maybe one half, for instance. The general non-flow energy equation $Q = W + \triangle U$ applies to **any** such process but it is valuable to concentrate on the constant pressure one for now.

The external work part of this is represented by $p\triangle V$ and, at all times for our ideal gas, the expression $pV = mRT$ holds good. Looking at that equation, it should be possible to combine it with the external work $W = p\triangle V$ expression. They both have a pressure times volume part.

● **PQ 4.6** In a non-flow constant pressure change p, m and R are unchanging and only V and T alter. In this case, express pV in terms of m, R and a temperature change. As a clue, call the temperature change $\triangle T$ and remember Charles' law.

We now have, for this particular constant pressure case, the three components of the non-flow equation (Q, W, $\triangle U$) in forms which include $\triangle T$ so the NFEE which is

$$Q = W + \Delta U,$$

can be written

$$mc_p \, \Delta T = mR\Delta T + mc_v\Delta T.$$

Check back on each of those factors if you have doubts.

We can cancel $m\Delta T$ from each component of the equation to give

$$c_p = R + c_v$$

and this rearranges to give the important relationship

$$R = c_p - c_v.$$

That is, for any ideal gas (and, therefore, fairly closely for real gases, you recall) the ideal gas constant is measured as the **difference** between the specific heat capacities at constant pressure and constant volume. As with the internal energy change derivation $\Delta U = mc_v\Delta T$, this measure of R is the measure of a **property**. So the fact that this latest expression has been derived by reference to a special process – constant pressure – is almost immaterial. It's just like measuring density by use of a special apparatus. The fact that it is a special apparatus does not matter; it is the property value that is important. This relationship between R, c_p and c_v applies to all ideal gases because, at the outset, no particular gas was mentioned – it was a general approach.

Example 4.2 A gas has a molar mass of 28. If its specific heat capacity at constant pressure is 1040 J/kg K and at constant volume is 743 J/kg K, what is the value its gas constant and of the universal gas constant? Assume the gas to be ideal.

$$R = c_p - c_v = 1040 \text{ J/kg K} - 743 \text{ J/kg K}$$
$$= 297 \text{ J/kg K}.$$

Note that the units of R are the same as the units of the specific heat capacities.

The universal gas constant is given by

$$R_m = (MM) \times R = 28 \text{ kg/mol} \times 297 \text{ J/kg K}$$
$$= 8314 \text{ J/mol K}$$

Note the minor difference between the value calculated here and the value calculated in the last chapter – it depends upon the accuracy of the starting data.

4.3 MORE ON EXTERNAL WORK AND IDEAL GASES

When a contained ideal gas – a fixed mass of gas held in a sealed can or balloon or by a piston in a cylinder perhaps – is subject to change by work or heat transfer, the ideal gas laws and the NFEE **always** apply.

The special fixed mass changes at constant volume and constant pressure have already been met but they are just that – very special examples which lead to some important conclusions. There are, of course, many more possibilities for change than just these two processes, as many as you care to think of and thus too many to analyse easily or in any worthwhile fashion. It would be useful then to have some sort of general expression which covered these other conditions of change, as the chances of meeting processes other than constant pressure or constant volume are obviously higher than just those two (Figure 4.2).

Since pressure, volume and temperature are closely connected, it is possible to describe changes in terms of any two of these parameters because the third must follow – $pV = mRT$ and so on. One conventional way of identifying a general change, that is, with no special conditions and not necessarily a constant for anything other than the fixed mass, is to represent the change in the form

$$pV^n = \text{constant.}$$

Yes, another constant but one of those *small-c* constants that have no special value. It means simply that for whatever heat or work transfer process is being practised, the product of pressure and volume raised to some unspecified index n is constant. For another transfer process with a different index n, the constant value would be a different one. Let me repeat that it is important to understand which are **real** Constants – always the same value, such as the universal gas constant – and which are casual constants for some stated conditions. Add this latest to your list of casual ones – not the important, fixed ones.

Put another way, if the equation applied to gas being heated or compressed or something and it changed from p_1, V_1 to p_2, V_2 and then to p_3, V_3, these pressures and volumes could be related by writing

$$p_1V_1^n = p_2V_2^n = p_3V_3^n.$$

Example 4.3 An ideal gas has an initial volume of 0.5 m³ but is now

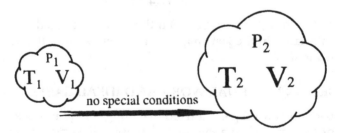

Figure 4.2 A general change.

compressed from its initial pressure of 2 bar to 3 bar. If the change follows $pV^2 = $ constant, what is the new volume?

We can write in this case

$$p_1V_1^2 = p_2V_2^2$$

thus

$$2 \times 10^5 \times 0.5^2 = 3 \times 10^5 \times V_2^2 \text{ N/m}^2 \times \text{m}^6.$$

Note the units – N/m² for pressure but the volume is m³ raised to the index $n = 2$.

If you are quite sure, we can take a short cut by putting the pressures in bars as it appears quite simply on both sides – but only do that if you are sure

$$2 \times 0.5^2 = 3 \times V_2^2 \text{ bar} \times \text{m}^6.$$

In either case, this gives

$$V_2 = 0.408 \text{ m}^3.$$

Example 4.4 If the same volume of the same gas was compressed from 2 to 3 bar according to $pV^{1.4} = $ constant, what would be the new gas volume?

As before, taking the short cut and without repeating everything,

$$2 \times 0.5^{1.4} = 3 \times V_2^{1.4} \text{ bar} \times m^{4.2}$$

which gives, for the new case

$$V_2 = 0.374 \text{ m}^3.$$

● PQ 4.7 What would be the similar general equation involving V and T? p and T? Remember $pV = mRT$ still applies.

Note that this expression $pV^n = $ constant (and its pressure–temperature and volume–temperature equivalents) is a general equation which covers some changes to the temperature, pressure or volume of a fixed mass of gas because of work or heat transfer. It is **not** an exclusive equation which claims to cover every possibility and it does not mean that there can be no other way of showing the changes. It does, though, fit quite a few everyday possibilities in real life thermo-dynamics, which is why it appears here.

Looking at the external work transfer for a start, the external work which may be done on or by the gas under a change of this general form $pV^n = c$ is readily determined. The easiest way is to use some calculus but, as we are trying to avoid that as sensibly as possible, this is consigned to Appendix B. The expression or equation (whose derivation is quite straightforward as you can check in the appendix) which yields work transfer for a change of gas volume and pressure from p_1 and V_1 to p_2 and V_2 under the conditions of $pV^n = $ constant is

$$W = \frac{p_1V_1 - p_2V_2}{n - 1}.$$

Since $pV = mRT$ also applies, then we can also write

$$W = \frac{mRT_1 - mRT_2}{n - 1}$$

Example 4.5 A gas undergoes a change according to $pV^n = $ constant. If the pressure rises from 10 bar to 20 bar and the initial volume was 8 m³ with an index n of 3, what work is done on or by the gas?

We can write $p_1V_1^3 = p_2V_2^3$ which gives $V_2 = 6.35$ m³. I have, quite deliberately, omitted the working so that you should really check this answer.

Now the pressure in bars has to be changed to N/m² since the answer is to be in Nm, not some other unit. 1 bar $= 10^5$ N/m² The external work transfer is given by

$$W = \frac{p_1V_1 - p_2V_2}{n - 1} = \frac{10 \times 10^5 \times 8 - 20 \times 10^5 \times 6.35}{3 - 1}$$
$$= -23.5 \times 10^5 \text{ Nm or } -2.35 \text{ MNm}$$

• PQ 4.8 Explain what all the numbers are, what the units are and whether it is work done on or by the gas.

Example 4.6 If the gas in Example 4.5 was 90 kg of nitrogen, what would be the temperature change for the stated volume and pressure change?

In PQ 4.7 it was shown that $TV^{n-1} = $ constant applied to the general non-flow change. We know also that $pV = mRT$ also applies so, for the initial conditions in fundamental units, we can write

$$\begin{array}{cccc} 10 \times 10^5 \times 8 & = 90 & \times 297 & \times T_1 \\ \text{N/m}^2 & \text{m}^3 \quad \text{kg} & \text{J/kgK} & \text{K} \end{array}$$
$$T_1 = 299.3\text{K}.$$

Thus

$$\begin{array}{cc} \dfrac{299.3 \times 8^{3-1}}{\text{K} \quad (\text{m}^3)^{n-1}} & = \dfrac{T_2 \times 6.35^{3-1}}{\text{K} \quad (\text{m}^3)^{n-1}} \end{array}$$

and

$$T_2 = 475K.$$

So the temperature change, to the nearest degree, is 176K. Note that the rounding off was left to the end again.

Earlier, a process of no external work transfer (the constant volume process) was examined and it yielded the useful information on internal energy. If now the other option of no heat transfer is inspected, this yields yet another useful relationship.

Where a process takes place with no heat transfer, whether it is in the heating of ideal gases or in real processes or in anything else, it is called an **adiabatic process**. No heat enters or leaves the system. Applying this to the non-flow energy equation would mean simply putting $Q = 0$, so for this very particular circumstance of no heat transfer

$$0 = W + \Delta U.$$

For any ideal gas or no losses theoretical process, the term usually used in this case is **reversible adiabatic**. It is a descriptive name, and it can be interpreted as saying that because we are dealing theoretically with some system which has no imperfections, then the whole process could be reversed back to the original condition with no losses. The word reversible is omitted occasionally in discussion but it really should be included if the discussion is about ideal processes. Any reference to adiabatic in this text, unless otherwise stated, implies reversible adiabatic. Although some would argue, it may help to read reversible as meaning 'perfect', at this stage (Figure 4.3).

Since, at present, we know nothing of the pressure–volume equation for a reversible adiabatic change, let us suppose that it is possible to use the general equation $pV^n = $ constant. This is a fairly commonplace approach in many technical disciplines. If you do not have a *dedicated* (meaning applied solely to a particular need) equation, then try a relevant general equation and put the particular conditions in it to see what happens. That is what we are doing here.

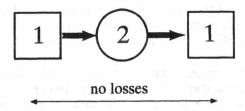

Figure 4.3 Reversible change.

We already have expressions for W and ΔU:

$$W = (p_1V_1 - p_2V_2)/(n - 1) = mR(T_1 - T_2)/(n - 1),$$
$$\Delta U = mc_v(T_2 - T_1).$$

Notice which way round T_1 and T_2 are in each of these – check back if you are unsure, because it is important.

These can be put directly into $0 = W + \Delta U$ to give

$$0 = (p_1V_1 - p_2V_2)/(n - 1) + mc_v(T_2 - T_1).$$

Using $pV = mRT$

$$0 = mR(T_1 - T_2)/(n - 1) + mc_v(T_2 - T_1)$$

and rearranging this equation then gives

$$mR(T_1 - T_2)/(n - 1) = mc_v(T_1 - T_2)$$

so that

$$R/(n - 1) = c_v.$$

As we have also proved earlier that $R = c_p - c_v$, we can change to

$$(c_p - c_v)/(n - 1) = c_v$$

and this rearranges to

$$(c_p - c_v)/c_v = n - 1$$

which gives

$$n = c_p/c_v.$$

Now be quite clear that this value of n, the ratio of the specific heat capacities at constant pressure and at constant volume, c_p/c_v, is **only** for a reversible adiabatic change. It is a very particular value of n, not the only value or anything else.

● PQ 4.9 For nitrogen, c_p = 1.04 kJ/kg K and R = 297 J/kg K. What is the value of the index n for a reversible adiabatic process using nitrogen?

Adiabatic processes do have real engineering applications, such as in the study of the design of engines, power stations and fridges, so this value of n, the **ratio of the principal specific heat capacities**, is given its own identifying letter, the Greek letter gamma, γ. So for reversible adiabatic changes, this particular case of the general law can thus be written pV^γ = constant.

As an aside at this stage, γ does have a wider importance since it has a fairly steady numerical value for various groups of gases. For instance, with diatomic gases such as O_2, N_2, CO and H_2, the value of γ stays around 1.4. It has a different numerical value for other groups of gases but again the value holds fairly steady and is thus a useful piece of information for design purposes.

The use of a general expression (pV^n = constant) in the absence of a dedicated one has worked out well in this case of a reversible adiabatic change. Sometimes it doesn't work out quite so completely but it is often worth a try.

• **PQ 4.10** Have a look at the general equation pV^n = constant and write it down for an **isothermal** (no temperature change) non-flow process. Now write down the external work expression for this isothermal change.

The general expression still holds, but just for this special case it yields zero divided by zero, which cannot be evaluated! Whilst the isothermal expression pV = constant is a particular case ($n = 1$) of pV^n = constant, just as pV^γ is a particular case, we clearly cannot continue that approach to get an expression for external work in an isothermal change.

Just as deriving $W = (p_1V_1 - p_2V_2)/(n - 1)$ was made easy by the use of calculus, so this little difficulty can be overcome and, for the same reason as before, the derivation is consigned to Appendix B. The derived expression for work transfer during an isothermal process is, for changes from p_1, V_1 to p_2, V_2,

$$W = pV \log_e (V_2/V_1).$$

Note that the product pV in that equation has no identifying subscripts since we are here talking about a Boyle's law change where pV is constant, so any compatible values can be used whether they are p_1 and V_1, p_2 and V_2 or any intermediate values. For instance,

$$W = pV \log_e (V_2/V_1) = p_1V_1 \log_e (V_2/V_1) = p_2V_2 \log_e (V_2/V_1)$$

all give the same answer. Incidentally, recall from school mathematics that logarithms to base e (natural logarithms) are written both as $\log_e$ and ln.

Example 4.7 For the change of pressure and volume in Example 4.3, what is the external work done?

$$W = \frac{p_1V_1 - p_2V_2}{n - 1} = \frac{(2 \times 10^5 \times 0.5 - 3 \times 10^5 \times 0.408)}{2 - 1} \text{ Nm}$$
$$= -0.224 \times 10^5 \text{ Nm}.$$

Note that the short cut of just using bars cannot be used here because we want an answer in Nm, not some other unit. The answer is negative, so the work is being done on the gas.

Example 4.8 What is the external work transfer for the conditions of Example 4.4?

$$W = \frac{(2 \times 0.5 - 3 \times 0.374)}{1.4 - 1} \times 10^5 = -30.5 \text{ kNm.}$$

We cannot use the bars shortcut but we can, of course, transfer the 10^5 to the outside of the bracket as both terms inside the bracket involve the pressure in bars. Note the answer uses kNm which is fairly commonplace.

Example 4.9 For a constant temperature change, what would be the final volume and the external work done? The initial volume, the initial pressure and the final pressure are as the previous examples.

For a constant temperature (isothermal) change, $p_1V_1 = p_2V_2$ thus

$$2 \text{ bar} \times 0.5 \text{ m}^3 = 3 \text{ bar} \times V_2$$
$$V_2 = 0.333 \text{ m}^3.$$

For the work

$$W = pV \log_e (V_2/V_1)$$
$$= 2 \times 10^5 \times 0.5 \log_e 0.667$$
$$= -40.55 \text{ kNm.}$$

While we could use bars directly in the determination of V_2, because bars appeared on either side of the equation in their own right, the pressure had to be converted to N/m^2 to find the external work as the answer is, of course, in Nm – or, as presented here, in kNm.

4.4 ENTHALPY

Many industrial processes involving gases or vapours operate at a constant pressure, or very nearly so. Steam generation in power station boilers or in process units in the plastics industry; warm air flowing through paint drying ovens; flames and hot gases in a brick- or glass-making furnace are all everyday examples (Figure 4.4).

● **PQ 4.11** If an ideal gas is raised from temperature T_1 to T_2 at constant pressure, write down the equation which shows how much heat has been transferred to the gas.

T_2 is higher than T_1 because we are told that the temperature is raised. The heat transfer is positive, $+Q$. If the temperature had fallen because the gas was being cooled, then T_2 would have been lower than T_1 and the heat transfer would have been negative, $-Q$. In general, that

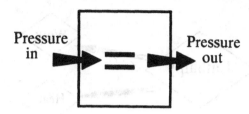

Figure 4.4 Constant pressure processes.

equation for heat transfer is abbreviated to

$$Q = m\, c_p\, \Delta T,$$

where ΔT is the temperature change. Care has to be taken to get the right ΔT, that is, making sure whether it is positive (temperature rising) or negative (temperature falling). However, if you write

$$\Delta T = (\text{final temp}) - (\text{initial temp})$$

that is automatically covered.

In real life processes where the gas pressure (warm air, combustion gases, whatever) is substantially constant, as in the examples above, then the specific heat capacity at constant pressure c_p is especially valuable. This grouping of m, c_p and ΔT is then important in those cases and a particular word, **enthalpy**, is used for that product of mass, specific heat at constant pressure and temperature. Thus $m\, c_p\, \Delta T$ is the **enthalpy change** associated with a temperature change ΔT.

Just as with internal energy ($\Delta U = m\, c_v\, \Delta T$), enthalpy change is usually the important measure. Both, though, are referred conventionally and conveniently to 0°C as a datum, meaning that data tables show these as having zero value at 0°C.

- **PQ 4.12 Will a gas at 0°C really have zero enthalpy?**

If we wished to be very precise, then enthalpy should be referred to zero absolute but, for real life engineering, zero Celsius is far more convenient and manageable. So it is correct to speak of 'enthalpy above 0°C' but the last bit is usually assumed or taken as read and the word enthalpy is used alone. As with any other abbreviations or arbitrary datum levels, if you have any doubts then state it fully.

The enthalpy of a gas above the zero Celsius datum is then

$$Q = m\, c_p\, T,$$

where T is the gas temperature, T°C. While the term enthalpy is quite common on its own, the term *'enthalpy content'* is also met. Rather like the heat content mentioned earlier, enthalpy cannot really be seen as

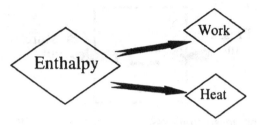

Figure 4.5 Enthalpy – working ability and heating ability.

something added in the way that milk is added to tea but it sometimes helps to understand changes by viewing it as something which can be added or taken away.

To make it quite clear, real materials will have internal energy and enthalpy in a measurable quantity if their temperatures are above zero Kelvin. For convenience, zero Celsius is the common reference point for these properties of internal energy and enthalpy so, unless it is clearly stated otherwise, normal measures of internal energy and enthalpy are amounts above zero Celsius. Should any calculations involve temperatures below 0°C, then negative values are normally recorded.

Great care has to be taken when referring to any datum levels which are arbitrary. It is really a matter of experience and usage bringing in to line the academic and the practical interpretations. If there is any doubt, always refer everything to the proper or absolute datum.

Example 4.10 2 kg hydrogen, c_p = 14.4 kJ/kg K, is held at 150°C. What is its enthalpy? If its temperature now falls to 125°C, what is the enthalpy change? If its temperature now falls to −30°C, what is this further enthalpy change? Referring to 0°C, what is the specific enthalpy of hydrogen at −30°C? Assume for this calculation that the specific heat capacity remains constant.

The enthalpy or enthalpy content above 0°C is

$$Q = m\, c_p\, T = 2 \text{ kg} \times 14.4 \text{ kJ/kg K} \times 150\text{K}$$
$$= 4320 \text{ kJ}.$$

Note that the gas temperature of 150°C translates to 150K in the calculation – because it is a temperature above the 0°C datum and the numerical difference is the same whether it is degrees C or K. It is not a case of changing a temperature in degrees Celsius to a temperature in degrees Kelvin – it is the temperature above the chosen datum.

For the temperature falling to 125°C, the gas temperature has changed by 25K. 150°C − 125°C = 25°C = 25K. The change of enthalpy is

$$Q = m \, c_p \, (T_2 - T_1)$$
$$= 2 \times 14.4 \times (125 - 150) \text{ kJ}$$
$$= -720 \text{ kJ}.$$

Now when the temperature falls to $-30°C$, the next temperature change is from $125°C$ to $-30°C$ so ΔT is $(-30 - (+125)) = -155K$. This enthalpy change is then

$$Q = 2 \times 14.4 \times -155 \text{ kJ}$$
$$= -4464 \text{ kJ}$$

Specific enthalpy, like specific anything, is the enthalpy per unit mass. The specific enthalpy of hydrogen at $-30°C$ (referred to $0°C$, that is the same as 30 degrees Kelvin below the datum)

$$q = 14.4 \text{ kJ/kg K} \times -30K$$
$$= -432 \text{ kJ/kg}.$$

Since the factors m, c_p and T involved in enthalpy are all true thermodynamic properties of the gas, then enthalpy also must be a true **thermodynamic property**.

To examine enthalpy a bit further, apply the First Law non-flow energy equation to a non-flow constant pressure process. Remember that in non-flow processes, external work is related to $p\Delta V$ and heating is related to ΔU:

$$Q \quad = \quad W \quad + \quad \Delta U$$
enthalpy = external work + internal energy change.

or

$$Q \quad = \quad p \, \Delta V \quad + \quad \Delta U$$
enthalpy = work ability + temperature ability.

Enthalpy then can be viewed as the **total energy available** from the gas – its external work capability component and its internal energy changing ability component, the combined capability of the gas to do some work and to raise a temperature, Figure 4.5.

Since enthalpy is useful in real life constant pressure manufacturing operations, it is given its own designatory letter. The small letter h is usually used for specific enthalpy (per unit mass) or the capital letter H for total enthalpy. This is of course in line with other uses of small and capital letters for other properties. A total enthalpy change would be ΔH and the specific enthalpy change Δh, which is $\Delta H/m$.

Thus in the non-flow example, recalling that $p \, \Delta V = m \, R \, \Delta T$ and that $R = c_p - c_v$

$$\Delta H = p \, \Delta V + \Delta U$$

$$= mR\Delta T + mc_v\Delta T$$
$$= mc_p\Delta T.$$

Enthalpy has been introduced here by reference to its industrial significance in constant pressure processes and examined by reference to non-flow processes. The reason for this approach is simple – it sets a proper scene instead of looking at some vague theoretical approach. However, that is not the sole reason for its existence. Whatever the actual operational circumstances, the grouping of heating or temperature changing ability and work doing ability in its various forms is still enthalpy.

In the next chapter, we will move on to flowing gases rather than just non-flow operations. Here we will see enthalpy again. If the enthalpy of a gas flowing into and then out of a system is measured, this change of enthalpy will show the total energy yielded by the gas during the process.

PROGRESS QUESTION ANSWERS

PQ 4.1

Since for this operation $Q = \Delta U$ and $Q = mc_v\Delta T$, then the required equation is

$$\Delta U = mc_v\Delta T$$

PQ 4.2

For our purposes, an ideal gas is in effect any real gas (such as hydrogen or methane or nitrogen) with any detail imperfections taken out and behaving such that it obeys the simple laws like Boyle's and Charles' laws. It is a consequence of this that ideal gas properties are assumed constant.

PQ 4.3

$$Q = mc_p\Delta T.$$

PQ 4.4

$$W = p\Delta V.$$

If the piston was connected to something – a pump, maybe – then the external work is done on the pump mechanism. If the piston was not connected to anything, then the work has to be done by pushing back the surrounding atmosphere outside the piston. It is thus likely to be a constant pressure operation because the piston will have to be in equilibrium with the atmosphere to be stationary. It will start at atmospheric pressure and will finish at atmospheric pressure.

PQ 4.5

Some is used to do external work ($W = p\triangle V$) and some is used to raise the gas internal energy ($\triangle U = mc_v \triangle T$). Make sure that you understand why the internal energy change involves c_v even though we have a constant pressure process.

PQ 4.6

Suppose that the constant pressure process starts with gas volume V_1 at temperature T_1 and ends with V_2 and T_2, so that $pV_1 = mRT_1$ and $pV_2 = mRT_2$. Then

$$pV_2 - pV_1 = mRT_2 - mRT_1$$

or

$$p(V_2 - V_1) = mR(T_2 - T_1)$$

thus

$$p\triangle V = mR\triangle T$$

This is just one way of doing it, so you may well have found others.

PQ 4.7

What needs to be done is to get rid of p in the pV^n equation. Since $pV = mRT$, then we can put $p = mRT/V$ and so

$$pV^n = \frac{mRT}{V} V^n = mRTV^{n-1}.$$

Since m and R are constant values (fixed mass of a certain gas), this can be simplified to

$$TV^{n-1} = \text{constant}.$$

Work through the same way to

$$T^n/p^{n-1} = \text{constant}.$$

This also emphasises the fact you have to be quite clear on which are real Constants, with unchanging values and which are just casual constant values for a particular purpose. As always, if you have any doubts or are not quite clear about this, check back now!!

PQ 4.8

The numbers are the values of pressure and volume in SI fundamental units, so that trivial mistakes are avoided, i.e. 10 bar becomes 10×10^5 N/m², the initial volume is 8 m³ and so on. Index n is only a number, so it has no units. One mega newton-metre is 1 000 000 Nm.

Look back to the table in Chapter 3. If W is negative, then it shows that work is being done on the gas. The sign convention is thus worthwhile – it tells us immediately whether work is being done on or by the gas.

PQ 4.9

$R = c_p - c_v$ thus 297 J/kgK = 1040 J/kgK $- c_v$ which gives $c_v = 743$ J/kg K. Take care to convert the given data to common units.

$$n = c_p/c_v = 1040/743 = 1.4$$

and the answer is a number, as the units cancel out.

PQ 4.10

Boyle's law is the fundamental one for an isothermal change, as a constant temperature is one condition for that law to apply. For a constant temperature, it tells us that $pV =$ constant. If this an example of $pV^n =$ constant, the index n for an isothermal change must be $n = 1$.

Boyle's law also tells us that $p_1V_1 = p_2V_2$ of course, and with $pV =$ constant, the external work expression becomes

$$W = \frac{p_1V_1 - p_2V_2}{n - 1} = \frac{0}{0} = \text{incalculable.}$$

So here is an example where we cannot use the general law approach.

PQ 4.11

The equation, where gas mass is m, the heat transferred is Q and the specific heat capacity at constant pressure is c_p, is

$$Q = mc_p (T_2 - T_1).$$

PQ 4.12

No. Zero Celsius is a handy real-life datum for practical use, not an absolute measure. If the freezing point of water had been at another temperature, then the datum would have been different. The only likelihood of anything having zero enthalpy or zero internal energy is at absolute zero, OK?

TUTORIAL QUESTIONS

Note the units as well as the numbers when checking the answers, which are rounded. You may have to seek data but assume ideal gas properties. Recall that $R = c_p - c_v$, $\gamma = c_p/c_v$ and $R_m = (MM) \times R$.

4.1 If, for nitrogen, the principal specific heat capacities are 1.040 kJ/kg K and 0.740 kJ/kg K, what are the specific enthalpy change and the specific internal energy change for a temperature rise of 30 K?

[+31.2 kJ/kg, +22.2 kJ/kg]

4.2 If the changes in specific enthalpy and specific internal energy for oxygen are 45.5 and 32.5 kJ/kg for a temperature change of 50 K, what are the values of the Ideal Gas Constant and the ratio of principal specific heat capacities for oxygen?   [260 J/kg K, 1.4]

4.3 By finding a value for the universal gas constant and knowing that the hydrogen molecule is diatomic, determine the values of its principal specific heat capacities.   [c_v = 10.39 kJ/kg K, c_p = 14.56 kJ/kg K]

4.4 Nitrogen (γ = 1.4) undergoes a reversible adiabatic change from 1 bar, 3 m^3, 300 K to 2 bar. What are the new volume and temperature?

[1.83 m^3, 366 K]

4.5 What is the external work transfer for an ideal gas change according to pV^2 = constant, with initial conditions p_1 = 3.5 bar, V_1 = 0.7 m^3 and final condition p_2 = 5.1 bar?   [−50.8 kN m]

4.6 In an isothermal change, 4 kg of nitrogen initially at 2.5 bar, 100°C is compressed to 6 bar. Find the work transfer and the changes of enthalpy and internal energy.   [−388 kNm, nil, nil]

4.7 Two kg of gas is compressed according to pV^3 = constant from 5 bar, 30°C to 150°C and it then expands isothermally to its original pressure. What are the work transfers of each stage and the net work transfer? The values of the specific heat capacities c_p and c_v for the gas are 1.05 and 0.75 kJ/kg K respectively.

$$[-36 \text{ kN m}, +127 \text{ kN m}, +91 \text{ kN m}]$$

4.8 Carbon dioxide is heated from 1 bar, 15°C to 250°C according to $pV^{1.3}$ = constant. Assuming that its specific heat capacity at constant pressure is constant at 0.90 kJ/kg K, determine

(a) the change of enthalpy [211.5 kJ/kg]
(b) the change of internal energy [167 kJ/kg]
(c) its new pressure and specific volume [13.27 bar, 0.0745 m³/kg]
(d) the specific work transfer [−148 kN m/kg]

Steady flow energy equation 5

5.1 STEADY FLOW AT LARGE

There are fairly obvious limits to the number of different circumstances in real life when gases are expanded or contracted, heated or cooled in closed containers without any of the gas flowing in or out. Sooner or later it is necessary to look at what happens when there is a **flow** of gas through a system – an air compressor or a fan heater, for instance – and the simplest condition to examine first is when the mass flow rate is **steady**. Steady here has the usual meaning of regular or unchanging. In that case, the flow rate does not change and there is no storing, pulsing or anything to interfere with the steady flow. The mass flow rate of gas entering the system equals the rate of gas leaving the system (Figure 5.1).

This equality of mass flow rate entering or leaving a system can be represented mathematically in a very simple fashion by the so-called **continuity equation**. The last sentence of the last paragraph said, in effect,

$$\text{Mass flow rate in} = \text{Mass flow rate out.}$$

Since we are, at present, dealing with a gas, then the density and

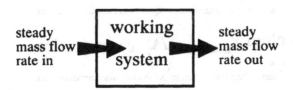

Figure 5.1 Steady flows, mass in = mass out.

volume can be used to express mass flow rate since mass = density × volume.

● PQ 5.1 Rewrite the mass flow rate equation to use volume and density. Do temperature and pressure matter?

Now if we look at the volume flow rate of a gas down a pipe, as in Figure 5.2, there is a velocity c m/s through the pipe of cross-sectional area A m².

If the gas density at the prevailing temperature and pressure is ρ kg/m³, then the mass flow rate is $\rho \times c \times A$ kg/s. Conventionally, c is one of the letters used for velocity but as some texts also use u and v, take care to identify velocity clearly when writing equations or doing calculations. Take even greater care that you do not mix up the letter c when used for velocity and the letter c when used for specific heat capacity. Fortunately, when dealing with gases, the latter will have a subscript as in c_v, but still take care. The Greek letter ρ is also used conventionally for density in most texts.

That pipe was a straight one but what happens when the pipe changes cross-section, as is common in many manufacturing plants? For the irregular shaped pipe of Figure 5.3, look at three places along the pipe.

Identify them as planes 1, 2 and 3. As we are dealing with steady mass flow rate we can write

> mass flow rate at plane 1 = mass flow rate at plane 2
> = mass flow rate at plane 3
> = mass flow rate at **any other plane**.

Since the mass flow rate has to be steady – what goes in must come out – the local gas densities and velocities are likely to change with the changing cross-section to keep the mass flow rate steady. If the pipe narrows, the gas has to travel faster, for instance.

● PQ 5.2 Use subscripts 1, 2 and 3 to identify local values, such as ρ_1, A_2 and so on. Now write that mass flow equation in terms of velocity, density and cross-sectional area.

These two equations, the mass flow words and the density, velocity and area symbols, are examples of the continuity equation. While this has been derived by reference to a piece of pipe, the equation applies

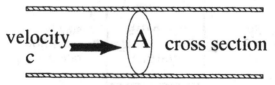

velocity $\longrightarrow$ A cross section
c

Figure 5.2 Volume flow rate down a straight pipe.

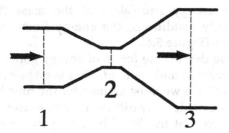

Figure 5.3 Flow down an irregular pipe.

widely. You would come to the same conclusion if we looked at air flow into a house through a door and out through a window, for example. Thus the continuity equation will also apply to the flow of air through a compressor or a fan heater or any relevant industrial equipment.

Similarly, whilst this continuity equation is aimed at gases for our immediate purposes, it is a simple extension to say that some form of the continuity equation must apply to anything which flows steadily – water through a river mill is a good example. Since, as we will see, the continuity equation and the steady flow energy equation of this chapter's title can be applied to flows other than just gases, then it is worthwhile introducing the term 'working fluid.'

At their simplest, the words *'working fluid'* can be taken to mean gas or vapour or liquid, whichever is appropriate, each time some general statement is made. The derivation is straightforward – it refers to anything fluid which may be used in a working system as an agent of energy transfer. Thus the wind turning a windmill, steam driving a power station turbine and water powering a waterwheel are all examples of working fluids. Incidentally, fluid means not solid. The words fluid and liquid are often confused in everyday speech. A liquid is a fluid but not all fluids are liquids – gas (not liquid!) is a fluid and vapour (also not liquid!) is a fluid.

Most real life steady flow (steady mass flow rate is often abbreviated to steady flow) operations are associated with energy usage. The examples of steady flow already mentioned have something to do with energy usage.

● PQ 5.3 Pick a couple of steady flow operations and identify the energy usage. Keep the words simple.

If the steady flow of the working fluid is also accompanied by a steady flow of energy, even though the energy may be changed from one form to another in the process, then these steady operations can also be examined and described by the so called **steady flow energy equation**. This is commonly identified by the initials SFEE (remember

NFEE) and is the energy equivalent of the mass flow continuity equation – for steady conditions, the energy flows in and out of a system must balance (Figure 5.4).

It means of course that **all** the forms of energy going in and coming out have to be recognized and counted. That is often easier to say than to do in real life, but here we must assume for the time being that such an assessment can be made. Any difficulties come later!

We are going to look at the SFEE by taking some examples of its application first, then by developing the lessons learnt into a generally applicable steady flow energy equation. As a start, go back to the waterwheel of the type found in some old mills. The flow of water over a waterfall strikes the wheel paddles and drives the wheel round which in turn drives some machinery in the mill. The energy supply or input to the system is the **gravitational potential energy** of the water due to the mass flow rate of water, the height of the waterfall and the influence of gravity. This term is often abbreviated, rightly or wrongly, to simply potential energy, omitting the word gravitational.

The main ultimate or intended energy output from the system is the **work** done inside the mill by the rotating machinery.

These are the main energy features. However, there will be other energy transformations to measure. Friction in the mill wheel and machinery moving parts must be overcome. The work done in overcoming friction is revealed as **heat**, just as in the brakes of a car. The water approaching the head of the waterfall, albeit often very slowly, will have velocity hence **kinetic energy**. The effect of the water hitting the wheel paddles will generate heat – a very small amount maybe but still an **energy transformation**.

While the main energy features – the waterfall height and the work done – can be measured quite readily, as can water velocities, the other transformations like the friction losses will be somewhere between difficult and impossible to measure. Note the use of the word 'losses' where energy is dissipated in an unproductive way.

● PQ 5.4 Is the inability to measure these losses easily and accurately really very important?

Example 5.1 A waterfall is 10 m high and the water flow rate over it is 1 tonne/s. It feeds a mill and produces 90 kW of useful work in the mill. What is the efficiency of this system and how big are the energy losses? Take the acceleration due to gravity g as 9.81 m/s^2 and ignore

$$\boxed{\text{Energy into system}} \; = \; \boxed{\text{Energy out of system}}$$

Figure 5.4 Steady flow energy equation.

any water velocity at the head of the fall.

Gravitational potential energy is the product of mass $\times$ g $\times$ height for a fixed mass, or mass flow rate $\times$ g $\times$ height when dealing with a continuous operation. If you have any doubts about this, please refer back to a school physics book because we will continue to meet potential energy.

So for the waterfall, the potential energy of the water flow – a flow rate in this example rather than a single fixed amount – is

$$1 \text{ tonne/s} \times 9.81 \text{ m/s}^2 \times 10 \text{ m} = 98\,100 \text{ Nm/s} = 98.1 \text{ kW}.$$

Again, make sure that you follow through the conversion of units. Any doubts – revise now. The approach velocity of the water can be ignored, so this potential energy is the only incoming energy source.

The rate of energy supply from the waterfall is thus 98.1 kW and the mill develops 90 kW of useful work, so,

$$\text{energy conversion efficiency} = \frac{90 \text{ kW}}{98.1 \text{ kW}} \times 100\% = 91.74\%.$$

The losses are the difference between the energy supplied and that used for the intended purpose or $(98.1 - 90)$ kW, i.e. 8.1 kW.

- **PQ 5.5 Where do these losses end up, eventually?**

Example 5.2 If the approach velocity of the water at the head of the fall is 1 m/s and the velocity of the water leaving the wheel is 3.5 m/s, what is the value of the unaccounted losses?

Kinetic energy is (mass $\times$ velocity^2)/2 or (mass flow rate $\times$ velocity^2)/2 as appropriate. For the water supply to the mill, this represents an extra amount of energy and for the water leaving the mill, this is one of the unproductive losses.

Thus the additional supply is $(1000 \times 1^2)/2 = 500$ W and the measured kinetic energy loss is $(1000 \times 3.5^2)/2 = 6125$ W. The overall equation then becomes

potential energy + kinetic energy of water at head of fall =
useful work + kinetic energy of water leaving mill + other losses,

so that

$$98.1 \text{ kW} + 0.5 \text{ kW} = 90 \text{ kW} + 6.125 \text{ kW} + \text{other losses}$$
$$\text{other losses} = 2.475 \text{ kW or } 2.5\% \text{ of ingoing energy.}$$

These water mill calculations are also good examples of two other points. First, the fluid flow is *steady* as the rate of flow of water over the fall is intentionally constant and this illustrates a practical application of the continuity equation. In this case, the rate of flow of water over the fall, round the wheel and out of the mill stream must all be the same if the operation is a steady flow one.

Secondly, it highlights the fact that in most real operations – car engines, power stations, washing machines for instance – there are some **important** features and some very **minor** ones. For the water mill, assuming that the mill is in generally good working order, the height of the waterfall, the amount of water available and the work done are important and they **dominate** the numbers in the calculation as they would in a real mill's working. The rate at which the mill can work is what we need to know. The other energy losses may be assessed for completeness but, unless there is some malfunction that needs investigating, that is about the limit of their value.

Looking at the numbers, the unaccounted losses at 2.5% are approaching measurement accuracy in many real installations. It is thus common to group small factors as 'errors and unaccounted losses'. It is vital in real-life engineering to appreciate:

- what we need to know for the assessment of a process or operation;
- how accurate or reliable measurements are, so that minor measurements are put into perspective;
- whether it is worth measuring minor items or whether they should be grouped together.

This does not mean never attempt to make a complete assessment nor ignore small items. It means that good scientists and engineers use a lot of common sense!

So what about the general usefulness of the steady flow energy equation? Many engineering processes involve unsteady flows of both fluid and energy. The motor car engine is a good example even when it is working a steady rate, such as cruising along a motorway. We take a look at engines later, so this is just a summary for our immediate purposes.

Fuel and air mixture enters a cylinder of the engine through the carburettor or the fuel injection system but only when an inlet valve is open – Figure 5.5, stage 1. The valve closes, the mixture flow stops and the trapped fuel + air mixture is compressed and ignited and does work on the piston – stage 2. Finally, the exhaust valve opens and the burnt or exhaust gases are ejected – stage 3. Overall, with that series of events being repeated for each cylinder of the engine, this is a very discontinuous operation yet the steady flow energy equation can still be applied.

- PQ 5.6 How can this be so? Go right back to system boundaries.

Figure 5.6, which follows from PQ 5.6, is a demonstration of the value of choosing the **correct system boundaries** when trying to define or

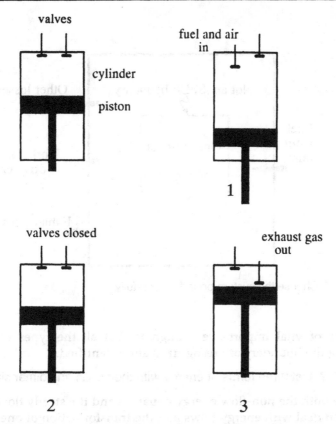

Figure 5.5 Some components and stages of a car engine working.

study any process or operation. A system boundary around bits of the engine will have unsteady flows across it but a boundary a little further out than just the engine will have steady flows across it. In this case, the SFEE can be used comfortably and conclusions could be drawn from engine tests to say how much work is done by how much fuel or, if the engine was in a car, what the average fuel consumption may be, for example, with no reference whatsoever to the components of the engine.

A little forethought about what is really to be studied and where to draw the system boundaries – which can be imaginary, of course – can simplify apparently complex problems. The SFEE can be used wherever a real or imaginary system boundary can be drawn such that there is a steady flow across that boundary. What happens inside the boundary, whatever mixture of events takes place, is in that respect of little concern. We can ignore the physics and chemistry of combustion inside the car engine because that all happens inside the car system boundary.

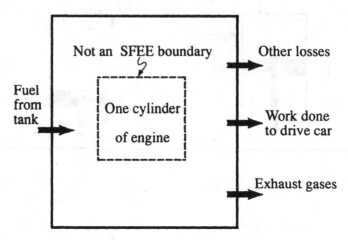

Figure 5.6 Choose the system boundary carefully.

What is of vital importance though is that all the types of energy crossing the boundary are recognized and quantified.

- **PQ 5.7** List three forms of energy with their units and dimensions.

Since both the non-flow energy equation and the steady flow energy equation deal with energy flows and the transformation of one form of energy into another, the units and dimensions of the various energy forms must always be consistent. That is always a useful check.

- **PQ 5.8** For a car travelling along a flat road at a steady speed, list some of the energy forms involved. Are these the same if the car climbs a hill steadily?

In the following example which uses the SFEE, various energy forms come together and their units need to be checked for compatibility. Also, note that the energy forms are written down as per second. There is no rigid single way of using the SFEE:

- it may be a balance of total amounts;
- it may be a balance on the basis of per unit mass of working fluid;
- it may be a balance on the basis of per unit time.

As long as the equation and the balance are generated for the particular process or operation, on a sensible basis with units and dimensions agreeing throughout, then the steady flow energy equation is valid.

Example 5.3 A car of mass 1 tonne is climbing a hill at a speed of

3 m/s. The hill is such that the car climbs through a height of 9 m in 30 s. At the speed of 3 m/s, the force required to overcome the car's resistance to movement and thus to maintain its kinetic energy is 1.76 kN. The fuel flow rate is 4.6×10^{-4} kg/s and its calorific value (energy value) is 44 MJ/kg. Estimate the proportion of the fuel's energy which is used to drive the car up the hill. Note that the rest of the fuel's energy is dissipated as exhaust heat, some incomplete combustion inside the engine and water radiator heat loss.

Some of the points in the question will be explored in later chapters but here is a good instance of the way in which something works being almost immaterial. We are asked a specific, clear question – what proportion? To reach an answer, it does not matter whether the car is petrol or diesel driven, new or old, big or small engine, working well or not. It is always important to recognize what is needed, not what it might be nice to know or anything else.

The car can thus be regarded as a simple system – a box, even – which is involved in various energy transformations. Figure 5.7 does just that.

Since the fuel flow is known in kilograms per second and the car

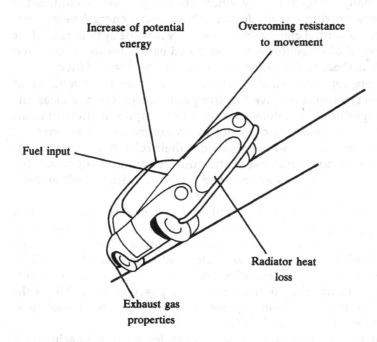

Figure 5.7 A box of energy changes.

speed is in metres per second, one easy way of using the SFEE is to do everything on a time basis.

$$\text{Energy in, fuel alone} = \text{fuel supply rate} \times \text{calorific value}$$
$$= 4.6 \times 10^{-4} \text{ kg/s} \times 44\,000\,000 \text{ J/kg}$$
$$= 20\,240 \text{ W.}$$

Note carefully that the units have been made self-consistent and that the end result is an energy supply rate, J/s or watts, W.

$$\text{Energy to drive uphill} = \text{increase of potential energy} + \text{overcoming}$$
$$\text{resistance to movement, both per second}$$
$$= \text{mass} \times g \times \text{height increase per second} +$$
$$\text{force} \times \text{distance per second}$$
$$= 1000 \text{ kg} \times 9.81 \text{ m/s}^2 \times (9/30) \text{ m/s} +$$
$$1760 \text{ N} \times 3 \text{ m/s}$$
$$= 2943 \text{ W} + 5280 \text{ W} = 8223 \text{ W.}$$

$$\text{Useful proportion} = \frac{8223}{20\,240} \times 100\% = 40.63\%.$$

5.2 THE GAS TURBINE AS AN EXAMPLE OF THE SFEE

There are many forms of energy which are likely to be encountered in everyday engineering – fuel calorific value, kinetic energy, heat, work, gravitational potential energy and so on. Any real-life system is likely to include several of these forms but, as noted earlier for the waterwheel and waterfall, there is usually some dominance by a few of them. While any complete representation of an energy balance must address all possible energy forms involved in the particular problem, a clear and careful recognition of the **dominant** ones can simplify matters in many cases. Note again that this does not mean ignore the lesser ones – it means recognize them and then consider their relative importance. As with real gases and ideal gases, the differences are important but sometimes the differences are masked by other things, such as measurement accuracy. This was met briefly earlier.

An examination of the principles of operation of a gas turbine engine (the aircraft jet engine, for instance) will illustrate a few of these points as well as giving a useful exercise in the numbers, units and energy forms involved. It will also give an insight into some simple modelling approaches for apparently complex operations. There is no substitute for taking an initial simple view, wherever possible, to highlight the important features. We've already met that philosophy with ideal gases and the NFEE and seen that it is true.

A gas turbine engine is a relatively complex piece of machinery – complex in the sense of having a large number of components from nuts and bolts to combustion chambers and so on, not in the sense of very difficult to understand. It is indeed an excellent example of very simple principles being recognized and applied with great benefit – using some work in conjunction with some fuel to increase the energy content of a working fluid and thus to provide a large amount of work when that energy content is recovered. The word 'set' is often used, as in a gas turbine set, to cover all the components.

A gas turbine set or system (Figure 5.8) has three main sections – the first uses the 'some work', the second uses the 'some fuel' and the third yields the large amount of work. The sections may be physically separate but commonly they may all be contained within an integrated structure – like an aircraft engine. A set will include a *compressor*, which forces air from the atmosphere into a *combustion chamber*. Here fuel, usually gas or a light fuel oil such as kerosine, is burnt using the compressed air. This generates hot, high-pressure combustion product gases which exhaust from the combustion chamber to drive a *turbine*. Some of the work available from this turbine is used to drive the compressor at the beginning of the sequence and the rest of the work is used for another purpose – as in an aeroplane engine or an electricity generator.

While the actual set contains many carefully manufactured and matched components, it can be represented quite simply and adequately – or **modelled** – as three connected boxes for study using the steady flow energy equation. Figure 5.8 illustrates this and the principal energy identities encountered, with Figure 5.9 being a cutaway sketch of a typical gas turbine set, to show how the boxes and the real components align.

● **PQ 5.9** You see the principal energy forms but can you name another which would contribute to a complete energy balance?

So, getting back to the set. The air has been compressed in the compressor section and heated in the combustion chamber and the resulting working fluid (hot exhaust gases) is passed through to the

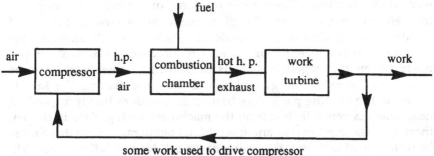

Figure 5.8 Essentials of a gas turbine set; h.p. = high-pressure.

Figure 5.9 A real gas turbine. (14 000 KW DR60 gas turbine having simple cycle thermal efficiency of 37.2%. Courtesy of Dresser-Rand.)

work-doing turbine. Here we can see some of the minor items in the total energy balance. Since the gases push the turbine blades round rapidly (several tens of thousands of revolutions per minute), the rotating components have kinetic energy. Because mechanical components are moving together – shafts in bearings – there will be some energy use in overcoming friction and thus maintaining the kinetic energy of the rotating parts. Any friction in machines finally appears as heat which is generally lost from the machinery casing. Also in the set, there are high combustion and operational temperatures, so there must be further heat loss from the casing of the set. No insulation is perfect.

By a large margin, though, the energy transformations that dominate in a well-operated set are those involved in the compression of the air, the combustion of the fuel and the driving of the turbine to yield work. A gas turbine set is a continuously running engine, with steady flows of air and fuel, steady rates of combustion and steady work output. The operational conditions can be adjusted – such as for an aircraft taking off, cruising and landing – but it is as good an example of steady operation as is likely to be met.

While the set has been described as complementary components – compressor, combustor and turbine – each of these three may be treated as a system in its own right and an appropriate system boundary drawn. The whole can be studied using the steady flow energy equation and so can each constituent.

Example 5.4 The turbine of a horizontally mounted jet engine receives a steady flow of hot gas from the combustion chamber at 7.25 bar, 870°C and 165 m/s, exhausting at 2.25 bar, 620°C and 305 m/s. If the casing is very well insulated, what is the work output? The specific heat capacity at constant pressure c_p of the working fluid gases may be assumed constant at 1100 J/kg K.

The turbine is that section of the set to which the hot, high-pressure gases flow from the combustion chamber and from which work is taken. Some has to be used to drive the compressor but most of it is used for the set's main task, such as driving an electricity generator. These gases have finished the burning process in the combustion zone, so we can assume that they are unlikely to change in any way, other than releasing some of their energy to drive the turbine, cooling and losing pressure as they do so.

The turbine is horizontal, so there is essentially no change of gravitational potential energy (the energy due to the height of something above a datum level) of the gases as they enter and leave. The unit is very well insulated so, while this cannot be a perfect heat retainer, it does mean that the heat loss is considered negligible compared to the other energy transfers. Therefore, the energy forms involved in any significant way are those related to the temperature, pressure and volume (the gas enthalpy) and velocity of the gases (kinetic energy) and the work delivered by the turbine as a result of the gas flow. Everything else is either too small to be of significance or is zero.

● PQ 5.10 Remind yourself of the definition and value of the property enthalpy.

Incidentally, take care to note the difference between external work as delivered in a non-flow process and the work delivered here in a flow process. The non-flow work is dependent on pV changes alone, as in $\int p\,dV$. In this flow process, the work delivered is related to all the energy changes of the gas, as we will see. This question of *flow* and *non-flow* work is explored a little further in Appendix B.

Since the gases do not change in any way other than by losing energy through the turbine, the outgoing gases have the same mass and specific heat capacity but their temperature is lower. The change of enthalpy can be expressed as the product of mass, specific heat capacity and change of temperature from turbine inlet to turbine outlet as

$$\Delta H = mc_p \Delta T$$

Notice that although the gas pressure is given, it does not enter into the calculation separately, as its effect is part of the concept of enthalpy – the pV bit if you like. As an aside, there would also be a pressure effect on the specific heat capacity which would need to be determined in a more accurate calculation. Specific heat capacities are not really constant, especially with changing temperatures (we look at this later), so this is a simplification for this illustrative calculation.

● **PQ 5.11** Can you think of any other reasons for wishing to know the gas pressure anywhere in the system? What if I was designing the turbine and what if I had not been told the velocities, just duct sizes?

No figure is given for mass, so the sensible approach is to calculate on some convenient basis such as 'per unit mass of gas'. This type of reporting is fairly commonplace, especially where different engines or machines are being compared in the same sort of task. So the enthalpy change becomes

$$m \times c_p \times \Delta T = 1 \text{ kg} \times 1100 \text{ J/kgK} \times (870 - 620) \text{ K}$$
$$= 275\,000 \text{ J}.$$

Similarly, the kinetic energy change of the gases is per unit mass:

$$m \times \frac{(c_1{}^2 - c_2{}^2)}{2} = \frac{(165^2 - 305^2)}{2} \text{ kg m}^2/\text{s}^2$$
$$= -32\,900 \text{ J}.$$

● **PQ 5.12** Why is the kinetic energy term negative? Are the units consistent?

The work delivered then is given by the change of energy of the gases, the source of the work, as

$$\text{work} = (275\,000 - 32\,900) \text{ J/kg of gas}$$
$$= 242\,100 \text{ J/kg or 242.1 kJ/kg}.$$

This example has mentioned various energy forms and it leads on to the development of a general equation for wide applicability. We will be returning to the gas turbine example a little later, however.

5.3   A GENERAL STEADY FLOW ENERGY BALANCE

The examples met up to now investigate and apply the simple statement

> For steady overall energy flows through a system, the energy coming into the system equals the energy going out of the system.

Each example – the turbine, the waterwheel – has featured some energy forms and a balance was prepared in each case. What is needed for wide use, of course, is an all-embracing equation. With such an equation, whatever the specific example, the relevant energy forms will feature and the irrelevant will be eliminated.

The First Law of Thermodynamics says that work and heat are mutually interchangeable and the non-flow energy equation puts this to good use in saying something like

$$Q = W + \triangle U$$

or one of its variations which we have already met. Any *imbalance* between work and heat is reflected in a change in internal energy.

● PQ 5.13 Using that as a pattern, write a similar expression for the steady flow energy equation, saying that the imbalance between heat and work is reflected in – what?

What is now needed is a thorough look at what energy forms may contribute to the **imbalance** $\triangle E$. For this, suppose that there is an imaginary steady flow engine or machine involving many possible energy components. Figure 5.10 shows the idea, with the machine represented by a box. As a reminder, it is quite acceptable to represent engines, machines, processes and so on by simple boxes in real-life engineering, if the structural features are not particularly important to the discussion or calculation in hand.

Some working fluid, say air, goes into the machine through a *port* (an opening of any kind may be called a port in engineering parlance) near its base at a pressure p_1, temperature T_1, velocity c_1 and mass flow rate $\dot{m}$. Where m is used for mass (e.g. 3 kg), it is conventional to use $\dot{m}$, usually called '*m dot*', for mass flow rate (e.g. 3 kg/s). The simple use of a dot above other parameters also indicates a rate rather than a single amount. Thus 'V dot' would be volume flow rate or flow per unit time.

The working fluid leaves the machine from an exhaust port near the top at pressure p_2, temperature T_2 and velocity c_2. Since it is a steady

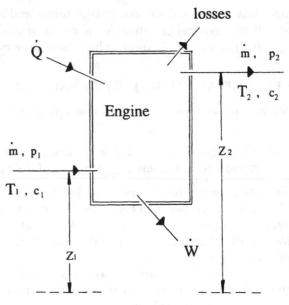

Figure 5.10 A steady flow engine.

flow machine, the rate of gas leaving is also $\dot{m}$. Take note that this is a flow **rate** (kg/s) in this case, not just an arbitrary total amount nor is there any the need to use 'per unit mass'.

Under the inlet conditions, the mean value of the air specific heat capacity at constant pressure is c_{p1} and at outlet c_{p2}. Up to now, we have assumed a constant value for specific heat capacities, as in the first formulation of ideal gases. They do change a little, however, and they vary with temperature and pressure. This is discussed in the next chapter, so please accept it and the term '*mean value*' for now so that the general expression which we are investigating presently can be as wide as is reasonably possible. Mean value is the expression commonly used for an average value between two conditions, so c_{p1} is the average value between the common datum of 0°C and T_1 and c_{p2} is the average value between datum and T_2.

The inlet is at a height z_1 above the machine base and the exhaust or outlet at height z_2. Some extra heat is fed to the machine at a rate $\dot{Q}$ ('Q dot', amount per unit time), say by an electric heating element inside the machine, to contribute to the overall energy balance. The machine delivers work, such as by driving some external device, at a rate $\dot{W}$.

The actual way in which the energy changes are done does not matter, since all we have is a quite imaginary engine or machine being supplied with some forms of energy and giving out some forms of energy. However, since it is a machine involving heat supplies and energy changes, there must be some energy losses, that is, energy which is not used in the intended fashion. These may arise, for instance, through heat losses to the surroundings from the warm surface of the machine.

● PQ 5.14 List the energy forms supplied to and leaving the machine with their self-consistent units. If the heat supply Q had been gasoline rather than electricity, would the list alter and, if so, how?

As the steady flow energy equation is exactly that, an equation, then the components of the equation have to balance. The essence of the SFEE is **what goes in must come out**. As before, the pressure, volume, temperature factors can all be accommodated in the enthalpy term. It is, of course, reasonable to ask 'Why quote pressure if the enthalpy term allows for it anyway?' Apart from the fact that it is built into the enthalpy term, pressure does influence specific heat capacities of real gases as does temperature. Thus a real study would need the pressure for determining, from available standard reference tables, specific heat capacity values for inclusion in the calculation.

Kinetic energy and potential energy have their standard formulation as do work and heat, so the equation can be laid down something like

(enthalpy + kinetic energy + potential energy) of gases going in +
heat added through the electric element =
(enthalpy + kinetic energy + potential energy) of gases going out +
work delivered + losses.

● PQ 5.15 Bearing in mind that this imaginary machine has a gas flow rate $\dot{m}$ (that is, readings would be in kg/s, not just kg) and the heat and work parts are also given as rates of supply, change that equation from its words into a mathematical form.

It is easy to say '*losses*' in the equation as though they were measured very simply. As a practical point and as mentioned briefly earlier, individual losses are often quite difficult to assess with any accuracy. They are diverse and, in a well run machine or process, they should be small anyway. Since in real measurements there are likely to be errors – errors, that is, of ultimate accuracy of instruments, not errors due to carelessness – it is commonplace to group together any losses with accumulated measurement errors. A term such as '*unaccounted losses*' is used in practice for this purpose. The important point is that all efforts should be made to minimize the numerical value of this 'sweepings-up' term.

A generalized form of the steady flow energy equation can thus be worked up. This particular one as covered here is not the only way to approach the question, nor does it claim to include every possibility. What it does illustrate, though, is that care has to be taken to consider every possible contributor to the picture. Only when this is done can the negligible terms be identified and called truly negligible.

Example 5.5 For this pretend machine, using air as the working fluid, the following data are measured or taken from reference tables for the temperature and pressure conditions under which the machine is running.

Air flow rate	30 kg/s
Ingoing temperature	325 K
Ingoing mean specific heat capacity, c_{p1}	1.017 kJ/kg K
Ingoing velocity	5 m/s
Height of inlet port above datum	1 m
Heat input via electric element	50 kW
Outgoing temperature	300 K
Outgoing mean specific heat capacity, c_{p2}	1.015 kJ/kg K
Outgoing velocity	2 m/s
Height of exhaust port above datum	4 m
Work recovered from machine	?
Losses	assessed from experience as 2% of ingoing heat flow Q.

I'm going to use this example as an opportunity to revise one or two points. If you are happy with your understanding, then just work through the calculation. If you have any doubts, go step by step and revise the points which arise. I'm also going to take some space to explore some points about the various factors involved.

The ingoing enthalpy is

$$\dot{m} \, c_{p1} \, T_1 = 30 \times 1.017 \times 325 \text{ kJ/s}.$$

Now this temperature T_1 is the air temperature above absolute zero and is quite correct when dealing with enthalpy absolutely. However, for practical convenience and by commonly accepted practical convention, you recall that enthalpy is usually measured above 0°C. The reasons are straightforward, in that 0°C is easy to imagine and easy to measure and most people have experienced 0°C. Other properties, not just enthalpy, are referred to 0°C for the same reason. Should any working fluid be below this arbitrary datum of 0°C, then negative values are recorded. For instance, air at −10°C has an enthalpy, referred to 0°C, of about −10.15 kJ/kg.

So the ingoing enthalpy in this pretend practical example is of the air at 52°C rather than 325 K. Recall also the use of the round number conversion of 273 K as equivalent to 0°C where this is in line with other accuracies. As an aside, since we have enthalpy changes on both sides of the equation, any minor inaccuracy of not using 273.16 K will more or less cancel out anyway.

Thus, referring the enthalpies to 0°C,

ingoing enthalpy $= \dot{m}\, c_{p1}\, T_1 = 30 \times 1.017 \times 52$ kJ/s $= 1586.52$ kW,
outgoing enthalpy $= \dot{m}\, c_{p2}\, T_2 = 30 \times 1.015 \times 27$ kJ/s $= 822.15$ kW,
units check: kg/s kJ/kgK K   kJ/s.

● **PQ 5.16 Why can I put temperature units in K when I have just discussed changing everything to degrees Celsius?**

Ingoing kinetic energy $= \dot{m}\, c_1{}^2/2 = 30 \times 5^2/2 = 375$ W,
outgoing kinetic energy $= \dot{m}\, c_2{}^2/2 = 30 \times 2^2/2 = 60$ W,
units check: kg/s m^2/s^2 Nm/s or W.

Note how small this particular energy input is compared to the heat flow or enthalpy input – watts compared to kilowatts, less than one tenth of one per cent of the air's ingoing enthalpy. Here is an example of something which can be measured quite easily and accurately – the mass flow rate and velocity – which then proves to be negligible.

Ingoing potential energy $= \dot{m}\, g\, z_1 = 30 \times 9.81 \times 1 = 294.3$ W,
outgoing potential energy $= \dot{m}\, g\, z_2 = 30 \times 9.81 \times 4 = 1177$ W,
units check: kg/s m/s^2 m   Nm/s or W.

A similar comment applies to the contribution of gravitational potential energy – a very small amount which, nevertheless, can be measured accurately.

Unaccounted losses are 2% of the ingoing heat energy; 2% of 50 kW = 1 kW. The estimated unaccounted losses are more than the change in the easily measured kinetic energy ke and potential energy pe contributions:

$$ke_1 + pe_1 - ke_2 - pe_2 = 375 + 294 - 60 - 1177 = -568 \text{ W}$$
$$\text{estimated unaccounted losses} \qquad = 1000 \text{ W}.$$

This again reinforces the point made earlier about the relative value of various measurements. Notice also that if the incoming enthalpy had been rounded off to 1587 kW, the rounding-off error would be of similar size to those small, measured changes. It is generally good practice to leave roundings-off as late as possible, when their real importance is known. It is important to know what is important! Take care though – this does not say that kinetic energy and potential energy inputs are negligible, only that this was so in this particular example.

Putting everything together in self-consistent units, say kW, the SFEE becomes

$$1586.52+0.375+0.294+50=822.15+0.060+1.177+1+W$$

thus

$$W = 812.8 \text{ kW.}$$

Example 5.6

(a) What is the change of specific enthalpy of a gas, specific heat capacity c_p 1.04 kJ/kg K, if its temperature falls through 10K?

(b) What would be the new velocity of a gas, initially travelling at 150 m/s, if the same amount of energy per unit mass had to be recovered from a change of gas kinetic energy?

(c) What is the minimum ingoing gas velocity which would yield this specific energy?

(d) What height of a waterfall would have the equivalent gravitational potential energy per unit mass?

Assume all properties to be constant and ignore all losses and imperfections.

(a) Specific enthalpy is the enthalpy per unit mass, h, so the change of specific enthalpy is Δh.

$$\Delta h = m \, c_p \Delta T = 1 \text{ kg} \times 1.04 \text{ kJ/kg K} \times 10K = 10.4 \text{ kJ.}$$

(b) The change of specific kinetic energy for a new velocity c m/s is

$$1 \times (150^2 - c^2)/2 \text{ J.}$$

The required energy recovery is 10 400 J (both have to be in the same units – do not mix J and kJ!). Thus, with both sides in J

$$1 \times (150^2 - c^2)/2 = 10\ 400$$
$$c = 41 \text{ m/s.}$$

(c) The minimum ingoing gas velocity to yield this 10 400 J must be for the condition of zero outgoing velocity. In real life, this is an unlikely condition because all working fluids have to be exhausted from a system sooner or later. There are, however, many examples where the ejected working fluid is at a very low velocity, for instance where exit ducts are far bigger in flow area than inlet ducts. These approach therefore an outlet velocity of zero. If the minimum inlet velocity is c_{min} and the outlet velocity is taken as zero, then again in J both sides

$$1 \times (c_{min}^2 - 0^2)/2 = 10\ 400$$
$$c_{min} = 144 \text{ m/s.}$$

Notice how close this is to the original 150 m/s, which allowed an exit velocity of 41 m/s. The 'velocity squared' factor is significant.

(d) For the same amount of energy to be recovered from gravitational effects, the expression is the potential energy one mgz, thus, again J

$$1 \times 9.81 \times z = 10\ 400$$
$$z = 1060 \text{ m}$$

Now that is an unlikely height for a waterfall, so this simple sum tells us that far more mass would be needed to recover the required amount of energy.

Each of these are simple uses of the steady flow energy equation – you don't have to be designing complex machines to put it to good use.

5.4 USING THE GAS TURBINE TO BRING A FEW POINTS TOGETHER

As a brief but interesting diversion from the steady flow energy equation, look at the fundamental processes in the gas turbine engine compressing air, adding energy by way of fuel combustion, then taking hot, high-pressure working fluid and extracting useful work. Remember that we are using terms like 'adding energy' as handy simplifications. You don't really add energy in the way that sugar is added to coffee but the effects are much the same!

The PQs in this section are intended to make you look to the fundamentals and not to get you bogged down in the details of gas turbines or, indeed, any other machinery.

● PQ 5.17 (a) What are the main ingoing and outgoing energy forms for the gas turbine? There's only one of each.
(b) So, in words, how can we express the efficiency of the gas turbine as a useful work producer?

It is always valuable to identify the main components, inputs and outputs, of any system. It helps you to see how well or otherwise the system is working and thus what shortcomings need attention. Never get into detail if the principles are not laid out properly.

Now think back to ideal gas processes – changes to do with heat, work and internal energy. In the compressor of the set, the aim is to do work on the air – compress it. Ideally, there would be no heat loss as the work is intended to go into the gas, not to be lost. If there is no heat loss, then the process is adiabatic. Recall the word 'reversible' in this context also.

● PQ 5.18 Imagine some air – a pretend package of air if you wish – going through the compressor. Will its pressure alone change?

Example 5.7 For illustration, a gas turbine compressor may run at up to 4 or 5 bar and the air temperature on exit may be around 200°C. If we look at this as a reversible adiabatic compression of air, for which $\gamma = 1.4$, from 300 K, 1 bar we have

$$\frac{T_1^{\gamma}}{p_1^{\gamma-1}} = \frac{T_2^{\gamma}}{p_2^{\gamma-1}}$$

and, rearranging,

$$T_2^{1.4} = 300^{1.4} \times \frac{5^{0.4}}{1^{0.4}}$$

thus

$$T_2 = 475 \text{ K or } 202°C.$$

This PQ and the example illustrate the elements of **mathematical modelling** of a process. As a mathematical model is refined then the details of what happens gets better but the principles established by the first 'bare essentials' look are still valid. Find the important features and see how they control everything before using time on detail. A parallel is the physical aerodynamic modelling of a modern car. It's the basic shape that controls the aerodynamic drag. Having the mirrors streamlined is only worthwhile when the big factors are correct!

● PQ 5.19 What would be the ideal operation for the combustion chamber? Remember that it takes in high-pressure gas (air) and puts out high-pressure hot gas (exhaust).

If the combustion chamber was part of a non-flow process, then we could write $Q = \Delta U$. This is, however, a flow process (see Appendix B as already mentioned) so some energy has to be used to generate the flow of working fluid through the system. Even so, $Q = \Delta U$ does show the maximum possible temperature change since $\Delta U = mc_v\Delta T$ and thus gives a measure of how much fuel energy is being used for the prime purpose – the actual temperature rise compared to the ideal. As a point of interest, the combustion conditions are such that the air temperature rises to 700–800°C or so, depending upon the design and other operational demands. As with the previous performance figures, these numbers are by way of indication and not intended to be rigid.

The work turbine can be considered similarly. The aims of the compressor and the combustion chamber have been to provide high-enthalpy working fluid to do useful work in the turbine. There is little point in wasting that enthalpy.

● PQ 5.20 So the turbine should be – what? Remember that, in effect, it's the compressor in reverse.

As an aside but leading on from the effects of expansion through a turbine on the gases, this explains why you often find frost forming on the valves or pipework of compressed-gas bottles if the gas escapes or is used quickly. Although it's not a true adiabatic process, the gases expand with a consequent loss of both pressure and temperature. If there is moisture in the surrounding air (quite normal) then the expanding gases make the valves or pipes cold, hence freezing out some of the air's moisture.

Example 5.8 Some air, specific heat capacity c_p 1.04 kJ/kg K, is at 5 bar, 600K and it expands reversibly and adiabatically down to 1 bar through a turbine. What is the specific enthalpy yield? Ignoring any other energy contributions such as the kinetic energy of the air, if the turbine was 93% efficient, what would be the work yield?

The enthalpy change is

$$\Delta H = mc_p\Delta T.$$

We are asked to find the specific change, Δh, so $m = 1$ kg and we are given c_p, so there remains only ΔT to determine.

Refer back to Example 5.7 since this is exactly the same process, a reversible adiabatic change also using air, so $\gamma = 1.4$. Thus the temperature after expansion is

$$T = 600 \times (1/5)^{0.4/1.4} = 378.8K.$$

Thus the specific enthalpy change Δh is

$$h = 1.04 \times (600 - 378.8) = 230 \text{ kJ/kg}.$$

If the turbine is 93% efficient, then 93% of this would be yielded as useful work, the rest being used to overcome various losses. Note that this example is for a turbine alone – it does not say that it is part of a gas turbine set so we are not talking about the remaining 7% of the work being used to drive the compressor. The 93% would be shared between useful external work output and that needed to drive the compressor.

The useful work yielded is therefore $0.93 \times 230 = 214$ kJ/kg which, as work is usually recorded in Nm rather than J, is 214 kNm of useful work per kg of working fluid.

So, as a quick summary, the steady flow energy equation applies to any system where there is a steady mass flow rate of working fluid accompanied by a steady energy pattern. Under these conditions, the equation says simply

what goes in . . . must come out,
energy going into the system = energy coming out of the system,

and there is a direct parallel in mathematical equation terms between the non-flow and the steady flow energy equations, for instance

$$\text{non-flow } Q = W + \Delta U$$
$$\text{steady flow } Q = W + \Delta E.$$

PROGRESS QUESTION ANSWERS

PQ 5.1

Since the equation refers to mass flow rate, then any subsequent equation must also refer to flow rate. The equation therefore becomes

volume flow rate in × density of gas going in =
volume flow rate of gas out × density of gas leaving.

Since gas volume and gas density are mentioned, then temperature and pressure need to be included, since gas density depends upon these.

PQ 5.2

$$\rho_1 \times A_1 \times c_1 = \rho_2 \times A_2 \times c_2 = \rho_3 \times A_3 \times c_3$$

PQ 5.3

There are plenty, so mine are a fan heater and a waterwheel. In the fan heater, the electric heating element gives a continuous flow of heat to the air, which in turn flows steadily because the fan turns steadily. In a waterwheel, the steady flow of water over a waterfall drives a paddle wheel, changing gravitational potential energy (waterfall height) into kinetic energy (wheel going round) in a steady manner.

PQ 5.4

While it would be good to measure everything accurately, the main items can be assessed quite well. Thus the mill operator can make a good estimate of how much milling can be done – which is the important thing.

PQ 5.5

Eventually, all these losses end up as heat being dissipated to the world at large.

PQ 5.6

Although each cylinder of the engine is working in a discontinuous fashion – mixture in, burnt, does work, residue out – the whole engine is working continuously. Fuel is always flowing from the fuel tank and the wheels of the cruising car are always going round. So there are steady flows of energy – fuel to the engine and work at the wheels. The clue to applying the SFEE here is to draw the correct system boundary.

PQ 5.7

Choosing, for instance, kinetic energy, gravitational potential energy and work, which we have already met.

kinetic energy	mass, velocity	$\dfrac{\text{kg m}^2}{\text{s}^2}$	$\dfrac{[\text{M L}^2]}{[\text{T}^2]}$
potential energy	mass, g, height	$\dfrac{\text{kg m m}}{\text{s}^2}$	$\dfrac{[\text{M L}^2]}{[\text{T}^2]}$
work	force, distance	Nm	$\dfrac{[\text{M L}^2]}{[\text{T}^2]}$

As usual, check any of these now if you aren't sure. Notice that all the energy forms here – and any others which you may have selected – all resolve to the same dimensions. If these were energy flow rates, then there would be of course another [T] in the denominator to make it $[\text{T}^3]$.

PQ 5.8

The energy forms include: the fuel supply; the work done in driving the car along the road (which means overcoming aerodynamic and rolling resistances); the heat in the exhaust gases; and the heat dissipated by the radiator.

The same forms will be met if the car climbs a hill but now the car's gravitational potential energy is also increasing, so that must be included.

PQ 5.9

The heat loss from the housing or casing of the set is one. As there are high temperatures involved, the casing must get warm and so must lose some heat, however well insulated it may be. Any friction within the moving components will also contribute ultimately to this heat loss also. The final exhaust gas, after the work has been extracted, is likely to have some kinetic energy, and if the set was not mounted horizontally there would be a potential energy change from the ingoing to the outgoing working fluid.

PQ 5.10

The property enthalpy is measured as $H = m\,c_p\,T$. It is the working fluid's energy by virtue of its combination of pressure, volume and temperature. It does not include the energy which may be available from the working fluid by virtue of its velocity (kinetic energy) or its location (gravitational potential energy). It is usually the change of enthalpy that matters in a working system, measured as $\triangle H = m\,c_p\,\triangle T$. Refer back to Chapter 4 if needed.

PQ 5.11

If I am designing anything, like a chair or a van or a turbine, then I need to know what loads may be applied – what size person the chair has to support, what weight the van has to carry, what gas pressures and temperatures the turbine has to withstand.

If I had only been told the duct sizes through which the inlet and exhaust gases flow and not their velocities, then both temperature and pressure would be needed to assess gas density, and thus velocity.

PQ 5.12

The gas has a higher velocity when leaving the turbine than when entering it, so the kinetic energy of the gas – the working fluid – has risen. This outgoing energy increase has to be supplied from the incoming energy, so what is used for that cannot be used for turbine work, hence negative. Check through to show that the units resolve to joules in either case, recalling that J and Nm are the same here.

PQ 5.13

The step forward for the steady flow energy equation is to include all the energy forms to do with through flow, so a simple expression may be

$$Q = W + \triangle E$$

where the imbalance between heat and work is reflected as an accumulation, $\triangle E$, of several other energy changes.

PQ 5.14

Draw a box like Figure 5.10 so that the various energies can be seen readily.

Going in ... enthalpy, $\dot{m}\, c_{p1}\, T_1$; kinetic energy, $\dot{m}\, c_1^2/2$; potential energy, $\dot{m}\, g\, z_1$; heat from the electric element, $\dot{Q}$.
Coming out ... enthalpy $\dot{m}\, c_{p2}\, T_2$; kinetic energy, $\dot{m}\, c_2^2/2$; potential energy, $\dot{m}\, g\, z_2$; work done by the machine $\dot{W}$ and losses.

If gasoline rather than electricity was the energy source, then $\dot{Q}$ would be the same but:

- the mass of the fuel would have to be included, along with any enthalpy or kinetic energy of the fuel;
- the mass of the exhaust gas would be increased in line with the mass of the fuel;
- the specific heat capacity of the exhaust gas would change a little because the exhaust would be combustion products and not just air.

PQ 5.15

$$\dot{m}\, c_{p1}\, T_1 + \dot{m}\, c_1^2/2 + \dot{m}\, g\, z_1 + \dot{Q} =$$
$$\dot{m}\, c_{p2}\, T_2 + \dot{m}\, c_2^2/2 + \dot{m}\, g\, z_2 + \dot{W} + \text{losses}.$$

PQ 5.16

The air is at 52°C and the enthalpy is measured above the 0°C datum. So the difference between the datum and the ingoing temperature is 52 degrees and a temperature difference of Celsius or Kelvin is numerically the same. In this case, then, the air is at 52°C and the air is also 52 K above the datum.

PQ 5.17

(a) Fuel energy (thermal energy) going in and useful work coming out. Since some of the work generated is used back in the compressor, this last term means the work available for another use or the net work output.

(b) One way is to find the ratio of net work output to fuel energy input. Since the fuel energy is realised as heat or thermal energy, the conventional term is thermal efficiency, thus

$$\text{thermal efficiency} = \frac{\text{net work output}}{\text{thermal input}}.$$

PQ 5.18

No. Its volume will decrease, of course. Additionally for a given mass of gas – our pretend package – undergoing a reversible adiabatic change, we not only have the usual $p\,V^\gamma$= constant (small c!) but we can also write $T^\gamma/p^{\gamma-1}$ = constant (another small c), so the temperature will also rise.

PQ 5.19

Ideally, all the fuel calorific value (the energy available on combustion) should be used to raise the working fluid temperature. So the fuel energy should not be used to do any work in the combustion chamber. In practice, since the working fluid has to flow as well as get hot, the combustion chamber operation is reasonably close to a constant pressure process.

PQ 5.20

It should be adiabatic – no heat losses. The gases will lose pressure and temperature as they expand through the turbine and thus yield useful work.

TUTORIAL QUESTIONS

Recall that the steady flow energy equation deals with steady flows of energy and working fluids. At the end, everything must balance – what goes in must come out!

Note that the calorific value of a fuel is the common way of recording its potentially usable energy content, the amount of energy which it may feed into the system to which it is supplied.

5.1 Gases enter a horizontal turbine at 15 kg/s, 25 m/s, 500°C and leave at 100°C, 10 m/s. If the specific heat capacity of the gases at constant pressure is 1.05 kJ/kg K and the turbine delivers 6 MW of work, what are the unaccounted losses?   [304 kW]

5.2 A decorative garden waterfall is used to drive a paddle wheel which in turn drives a small generator. The generator produces electricity at 24 W to power lamps which illuminate the waterfall. The energy conversion efficiencies of of the paddle wheel and generator are 55% and 75% respectively. The water flows over the fall at a rate of 2 kg/s, being near static at the head of the 3 m high fall. If there are no losses other than through the stated efficiencies, at what velocity does the

water leave the paddle wheel? Take the paddle wheel exit as the datum
for the waterfall height.   [Approx 0.53 m/s]

5.3 In a horizontal gas turbine set which is under development, liquid
fuel of calorific value 42 MJ/kg is burnt at a rate of 0.1 kg/s. The related
combustion air supply is 2 kg/s from near-static surroundings at 25°C
but the exhaust gases finally leave the system at 100 m/s, 600°C. The
mean specific heat capacity at constant pressure of the working fluid
between 25°C and 600°C is 1.1 kJ/kgK and the casing losses are 130 kW.
15% of the work developed in the turbine is used to drive the
compressor part of the set. What is the total and net work output rate of
the unit?   [2.73 MW; 2.32 MW]

5.4 A car of mass 1.5 tonnes consumes fuel of calorific value 45 MJ/kg at
a rate of 16 km/kg when travelling at 100 km/h on a horizontal road.
What is the change in fuel consumption when the car climbs a steady
gradient at a vertical climb rate of 0.5 m/s? Assume that all losses or
efficiencies are otherwise unchanged.   [Increases by 9.4%]

5.5 A power station boiler burns coal to make steam with a thermal
conversion efficiency of 90%. This steam then does work in a turbine, at
an efficiency to be found. In turn, the turbine drives an alternator to
generate electricity, the efficiency of this stage being 95%. If the
calorific value of the coal is 28 MJ/kg and it is burnt at a rate of 1000 t/h
to generate 2700 MW, what is the overall conversion efficiency of the
power station and of the steam turbine stage?   [34.7%, 40.6%]

Real gases 6

6.1 PERFECT AND SEMI-PERFECT GASES

To recap part of Chapter 2, the term *'ideal gas'* is not intended to mean that there is just one possible theoretical or ideal gas. They can be numerous, just as there are numerous real gases. It is perhaps convenient and clearer to think of ideal gases as idealized forms of real gases. As an example of the small but important differences between real gases and their ideal versions, look at the specific heat capacity of oxygen. This is just a random example. Any other common gas will show exactly the same sort of features, although the numbers will be different. Up to now, any derivation and most of the calculations have assumed or used constant specific heat capacities for the ideal gases.

In fact the specific heat capacity at constant pressure of real oxygen does vary – at 400°C it is about 1026 J/kg K and at 800°C it has risen to about 1101 J/kg K, a change of over 7%. There is this sort of variation for all real gases, although the actual values of the property and the percentage change do depend upon the particular gas in question. Specific heat capacities of gases (and many other materials) rise progressively with temperature, the variations for most gases (again, other materials also) being recorded in reference books or data tables such as those listed at the end of this text.

It was stated in Chapter 2 that real gases follow the ideal gas laws quite closely over much of their usable range, perhaps with no more than a 5% variation in many cases, dependent upon the size of range. Here, then, is a little dilemma – the real gas properties can vary more than their obedience of the laws and this can therefore raise a question mark over the value of ideal gas concepts. Let it be clear – for many everyday uses and for modest changes of operational conditions, the ideal gas approach does all that is necessary. However, there are bound to be times – research work, development of new processes for example – when more accurate theories and gas models are needed. It is important therefore to make the theory line up with real life needs as closely as possible. One first step for ideal gases is to sub-divide them into **perfect** and **semi-perfect** forms. The difference is straightforward and it concentrates on the effect of temperature on properties. In particular, it often concentrates on the specific heat capacity properties, so we too will do just that to illustrate (Figure 6.1).

A perfect gas will obey the ideal gas laws and will have **constant** values of properties, notably the principal specific heat capacities c_v and c_p. A semi-perfect gas will obey the laws but may have properties, again notably specific heat capacities, which **change** with temperature. In practice, this means that some property changes which may be important can be accommodated without developing complete new theories or adding undue complications. It provides a simple refinement of an existing valuable theoretical approach. The perfect gas half of the subdivision is already covered – the ideal gases which we have been using up to now – so now we should concentrate on the semi-perfect gases.

As with the first definition of ideal gases, semi-perfect gases are **simplified representations** of everyday gases. Whereas up to now the basis was that no properties changed and the gases obeyed the simple laws and their derivations, for the semi-perfect case we allow for property changes but the gases otherwise still obey the laws and so on. In no way does this introduction of semi-perfect gases invalidate for our purposes any of the laws or their consequences. It simply refines them. Thus, for instance, internal energy is still the product of mass, temperature and specific heat capacity at constant volume. The refinement is that the specific heat capacity is not constant and this has to be covered in any calculation.

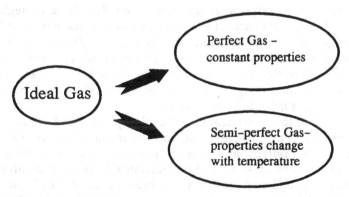

Figure 6.1 A simple refinement.

● PQ 6.1 If I read that the specific heat capacity of nitrogen at 30°C is 743 J/kg K and at 150°C it is 747 J/kg K, what precisely does that mean?

Thus values of specific heat capacities which refer to a particular temperature are just that – they apply to the *stated temperature*. If I wish to assess the internal energy or enthalpy of a gas at say 30°C and then at 150°C, (the usual zero degrees Celsius datum) I need specific heat capacity values which are not just applicable at those individual temperatures. I need values which deal with the **range** from the datum to the stated temperature.

● PQ 6.2 What are these values? Words, not numbers. You have already met them somewhere!

To assess values of or changes to internal energy or enthalpy for example, we have to use the **average** values of the properties between the temperatures in question. Average values used in this way are often called **mean values** and mean values of properties are usually identified by a line over the property symbol, such as in $\overline{c_p}$. This would usually be said as 'c_p bar'. For many properties, especially over modest ranges, the mean value is the arithmetic mean of the starting and finishing particular values. However, to be absolutely sure, tabulated values should be used wherever possible.

Example 6.1 A rigid can holds 2 kg of nitrogen at 30°C. The can is heated so that the nitrogen temperature rises to 150°C. The mean values of $\overline{c_v}$, the specific heat capacity at constant volume, between the usual 0°C datum and those temperatures are 743 (0°C to 30°C) and 745 (0°C to 150°C) J/kg K. The individual values of c_v, the ones particular to a given temperature, are 743 J/kg K at 0°C, 743 J/kg K at 30°C and 747 J/kg K at 150°C, to the accuracy of this calculation. What is the change in internal energy of the nitrogen?

The rigid can description is taken to mean that there is negligible volume change. Of course if you heat up any can, its volume is likely to change but the implication here is that the change is so small compared to other effects that it may be neglected. This is not just an academic simplification – there are good industrial examples too where the volume change of a container is modest for most purposes, commercial gas cylinders for instance.

Now look at the numerical values of the specific heat capacities. At the lower temperature, the actual and the mean values are identical. In fact, this is a reflection of data presentation accuracy – they are not identical because of some hidden technical reason. The difference between c_v at 0°C and 30°C is very small – maybe 0.1 J/kg K if the values had been presented to one decimal place – so that the mean value $\overline{c_v}$ rounded to calculation accuracy is the same as the two individual values.

As the temperature rises, so the c_v value differences increase and at 150°C, there is a significant difference (747 J/kg K compared to 743 J/kg K at 0°C) so that this is seen in the mean value between 0°C and 150°C. Notice that the mean value here can be taken as the arithmetic mean of values at 0°C and 150°C – (743 + 747)/2 = 745 J/kg K – in this case because they are still quite close. The safe way, though, is always to use data tables whenever they are available because virtually no thermodynamic property changes are arithmetic (equal steps) ones.

The internal energy of a gas at any time must be related to the conditions at that time so, taking the usual convenient datum of 0°C, the initial value (from zero at 0°C to whatever it is at 30°C, using the mean $\overline{c_v}$) of the nitrogen's internal energy is

$$m \times \overline{c_v} \times \Delta T = 2 \text{ kg} \times 743 \text{ J/kg K} \times 30\text{K}.$$

- PQ 6.3 Are you quite clear why K and °C can be mixed like this?

The final value (also from zero at 0°C to whatever it is at 150°C and using the relevant mean $\overline{c_v}$) is 2 kg × 745 J/kg K × 150K, so the change ΔU is

$$(2 \times 745 \times 150) \text{ J} - (2 \times 743 \times 30) \text{ J} = 178.92 \text{ kJ}.$$

It would, incidentally, be sensible to round the answer to the same accuracy as the starting data, say 179 kJ.

For this particular example, you can calculate that the difference between this value of internal energy change and the one given by using a constant value of c_v is about half of one per cent. If, instead, I had used the particular value of c_v at 150°C (747 J/kg K) rather than the mean value, the incorrect answer would have been 179.52 kJ. Neither are big differences in this calculation but that's just because of the numbers used. It could be very important indeed, so remember to use the mean values where necessary, not the particular or individual temperature values.

The ideal gas view of internal energy has still been used, quite reasonably, but the answer has been improved in its accuracy by including the fact that the specific heats are temperature dependent. The difference is not always so small.

• PQ 6.4 From a table of air properties, c_p and c_v at 300K are given as 1.005 and 0.718 kJ/kg K respectively at a particular pressure. At 1000K these have risen to 1.141 and 0.854 kJ/kg K respectively. What can you say about R and γ for this gas?

The sort of change shown by this PQ is reasonably typical of real gases. The numerical changes in c_p and c_v with temperature more or less keep pace with each other so that their difference is constant but their ratio changes. Notice carefully that we are dealing only with changes of temperature at this stage. If pressure is also changed, then other effects may be important. Again, where necessary data are freely available for all common gases.

To review briefly, the introduction of the concept of semi-perfect gases is a straightforward **refinement** of the initial ideal gas concept. For gases, many of their properties are influenced by *temperature* to a greater or lesser extent, and this refinement recognizes that. Including this in calculations means that the results of those calculations are just that bit more accurate. These increased accuracies would apply to heat transfer, work transfer, internal energy as we have seen, enthalpy and any other function or product or property where specific heat values play a part. Let it be stated yet again, though, that the introduction of this refinement does not invalidate the laws and relationships used so far. It simply **improves** them.

To put the importance into context, it may be useful to have a brief table of specific heat capacities at constant pressure, kJ/kgK, of some gases at different temperatures (Table 6.1). Note that these values relate to Celsius, not Kelvin as in the last PQ. It is sensible to use both temperature measures like this, then you get into the habit of checking the units every time.

Apart from showing the value of refining the ideal gas approach, by including the varying specific heat capacities and the need to use mean

Table 6.1 Some changes of c_p with temperature, at about 1 bar

°C	Oxygen	Nitrogen	Hydrogen	Air
0	0.913	1.038	14.24	1.005
200	0.963	1.051	14.53	1.026
400	1.026	1.084	14.61	1.068
600	1.068	1.135	14.78	1.114
800	1.101	1.177	15.12	1.156
1000	1.126	1.210	15.53	1.185
Overall increase % 0–1000°C	23.3	16.6	9.1	17.9

properties, one or two other points arise from this table.

First, provided temperature changes are modest, say within a 200K (same as 200°C) range, then most of the changes to the specific heat capacities are in the 5% or less range. Even the exceptions – oxygen from 200 to 400°C for instance – are but a little more.

Second, an average specific heat capacity value can be determined for a mixture from the individual values of its constituents. Recall what was said about mixture properties in Chapter 2.

● PQ 6.5 Confirm that statement by looking at the figures for air, which you can take as 23.3% oxygen and 76.7% nitrogen by mass. Note that there is a little rounding of values in the table for brevity, so don't expect absolute alignment!

Third, specific heat capacities of individual gases vary from gas to gas. Look for instance at the difference between oxygen and hydrogen. The latter is far higher than most but it does emphasize the point. Specific heat capacities are **individual properties** for individual materials, just as colours or densities change from one material to another.

● PQ 6.6 Will this variation of specific heat capacities make much difference to the value of the universal gas constant as calculated from the values of R and MM ($R_m = R \times MM$), when temperatures change?

Of course, this PQ conclusion must be read within the limits of measurement accuracy, but it is more than adequate for most common applications. Yet again, it underlines the fact that the ideal gas concept is of real practical value and all that is being done now is to refine the idea for greater accuracy where necessary. The introduction of the semi-perfect variation is a first step – it recognizes the influence of temperature on specific heat capacities in particular – but it is not the end of the story.

6.2 SOME OTHER PROPERTY DIFFERENCES

Much work has been done by researchers to record the influence of operational conditions on gas properties and the influence of temperature on gas specific heat capacities is a good example. There are other property changes that are worth noting when looking at the differences between real and ideal (especially perfect) gases and some of those are summarized here.

When we first met ideal gases, in Chapter 2, the point was made that the ideal gas concept can be used (within the indicated limits of accuracy) where gases are **well away** from their liquefaction points (the temperature and pressure conditions where a gas condenses into a liquid). The following comments also apply mainly to gases well away from their liquefaction points and, therefore, are to a great extent statements of fact rather than derivation. Few of the numbers are of special significance other than being property values. They are for example only, or to illustrate a point, not for recording.

The effect of pressure is an obvious one to consider. The influence of pressure is far *smaller* than the influence of temperature under normal circumstances insofar as temperature and pressure effects can be compared directly. Pressure has virtually no effect on the specific heat capacity at constant volume c_v of gases. The effect is negligible for most commonplace calculations and this has been measured well into the hundreds of bars pressure range. The specific heat capacity at constant pressure is, however, variable. For instance, c_p for nitrogen at about 300 bar is around 6% higher than at atmospheric pressure at room temperature. If the same readings were taken at around 800°C, then the difference would fall to less than 2%. Temperature is a very important influence on many thermodynamic properties (Figure 6.2).

For oxygen under the same conditions the values of the difference are about double, and for hydrogen are virtually negligible. The effects are individual to gases and note that these examples are at quite high pressures. Modest changes, say a few bar which therefore means most industrial and commercial processes, would be of little or no consequence.

Table 6.2 shows effects of pressure up to about 1000 bar, whereas the previous comments applied to about 300 bar. Notice again the temperature influence – it really is the dominant factor. This kind of information is commonplace, such as in those references listed at the end of this book.

Thinking of the two reference or principal specific heat capacities, c_p and c_v, it is fair to expect different temperature and pressure effects since they refer to the different operational conditions of constant pressure and constant volume. It is quite likely, therefore, that pressure and temperature effects will be different for the two values. Remember,

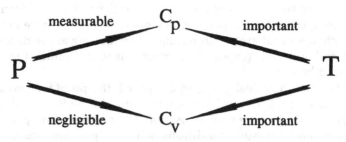

Figure 6.2 Temperature and pressure effects.

Table 6.2 Percentage increase of c_p of some gases for pressure rises from atmospheric to about 1000 bar

Gas	250°C	800°C
Oxygen	35	4
Nitrogen	14	3
Hydrogen	2	1
Carbon dioxide	86	13

incidentally, that they are reference values, perhaps of particular importance, but not exclusive in any way.

● **PQ 6.7** What is likely to happen to the ratio γ of the principal specific heat capacities when pressure rises? And R?

Many reference texts contain tables of specific heat capacity changes with temperature and pressure (hence R can be calculated) and also of the ratio γ,i.e. c_p/c_v. As an example to illustrate, the value of γ for air at room temperature and pressure is 1.40. At atmospheric pressure and about 600°C, this has fallen to around 1.34 and it continues to fall with temperature rise reaching, for instance, about 1.29 at 2000°C. R meanwhile remains constant, as there is no pressure change. At room temperature and about 70 bar, γ for air rises to around 1.6 compared to its atmospheric value at room temperature of about 1.4 and R rises from about 288 to about 417 kJ/kg K.

Whatever the temperature, γ and R for all gases rise with pressure, though the effect on both is less at higher temperatures. Again quoting air as an example, at about 250°C, γ increases by around 7.5% if the pressure rises from atmospheric to around 300 bar. At a temperature of about 2000°C, the increase is reduced to about one half of one per cent. These numbers are presented only for illustration and they have no other special significance. What this sort of variation does tell us, however, is that tabulated values of properties **must** be used to enhance calculation accuracy. There is no global or single value of these properties that is accurate for all circumstances (Figure 6.3).

As before, none of this denies the general applicability of the ideal gas laws when routine calculations are being performed – it serves to improve the accuracy of the outcome, just as the semi-perfect gas concept improved the result. Equally as before, the smaller the changes being handled, the more closely will real gases follow ideal gas laws.

For the next example, check back to earlier chapters if there are points arising on which you are not sure. In the answer, I will also be drawing on earlier information.

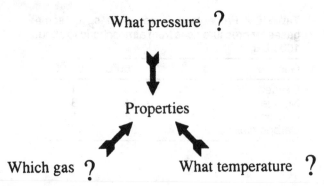

What pressure ?

Properties

Which gas ?   What temperature ?

Figure 6.3 Gas properties are not single values.

Example 6.2 Some nitrogen is used in a process. It enters at 1 bar, 30°C and leaves at 70 bar, 500°C. What would be the change in its specific internal energy under the following circumstances?

(a) Obeying ideal gas laws and thus assuming no influence of either pressure or temperature on properties (perfect gas), so that c_p is constant at 1.040 kJ/kg K and γ is constant at 1.400. These values can be taken as the values applying at 0°C also.

(b) Obeying the semi-perfect gas concept, allowing a change of property with temperature but ignoring pressure effects. In this case the initial gas properties are the same but the final ones are $\gamma = 1.345$ and $c_p = 1.065$ kJ/kg K.

(c) Allowing also for pressure effects, which gives final values of $\gamma = 1.355$.

Where necessary, arithmetic means of properties may be used. The specific internal energy is per unit mass, $\Delta u = c_v\, \Delta T$

(a) Since $\gamma = c_p/c_v$, then

$$c_v = \frac{1.04}{1.4} = 0.743 \text{ kJ/kg K.}$$

Thus

$$\Delta u = 0.743 \times (500 - 30) = 349.21 \text{ kJ/kg.}$$

(b) Now, allowing for the influence of temperature, the initial value of internal energy above the conventional datum of 0°C – call it u_1 – is

$$u_1 = 0.743 \times (30 - 0) = 0.743 \times 30 \text{ kJ/kg.}$$

For the final value of internal energy above the datum, we need the mean value of specific heat capacity at constant volume, which, in turn for the available data, says that we need the mean values of specific heat capacity at constant pressure and of the ratio γ. We are told that we can use the arithmetic averages – this is a reasonable thing to do in the absence of more accurate tables – so

$$\overline{c_p} = (1.04 + 1.065)/2 = 1.053 \text{ kJ/kg K}$$
$$\overline{\gamma} = (1.4 + 1.345)/2 = 1.37,$$

thus

$$\overline{c_v} = 1.053/1.37 = 0.768 \text{ kJ/kg K.}$$

So the final value of specific internal energy above datum, u_2 is

$$u_2 = 0.768 \times (500 - 0) = 0.768 \times 500 \text{ kJ/kg.}$$

So, the change in specific internal energy between 30°C and 500°C is

$$u_2 - u_1 = 0.768 \times 500 - 0.743 \times 30 \text{ kJ/kg}$$
$$= 361.71 \text{ kJ/kg.}$$

● PQ 6.8 A quick bit of revision – should Δu not be kJ rather than kJ/kg? What are the units of γ?

The difference between the ideal gas and semi-perfect gas values (temperature alone) is therefore 12.5 kJ/kg. The more accurate value is 3.58% greater than the ideal value.

(c) Now allowing for pressure. There seems to be a lack of data . . .

● PQ 6.9 So you do it!!

As a reminder, note that the change in internal energy is calculated as (final value − initial value), just as would be done for other property changes. It is a simple and common-sense way of doing it, since it shows immediately whether it is an increase (the answer is positive) or a decrease (the answer is negative) in the property.

There is some difference then between the real gas properties and the ideal gas properties outcome, but much of this is overcome by the semi-perfect approach because of the dominance of temperature effects. Even so, using Example 6.2 as a fairly high-temperature, high-pressure example, the outcome is still within a few per cent difference between perfect and real gas properties. To aid real calculations of a routine nature, there are available tables of mean specific heats over certain temperature ranges, thus simplifying the sums immediately. As you will understand from the repeated mention of reference texts, data tables and so on, available information is an important feature of most real-life science and engineering. It is there to be used (Figure 6.4).

6.3 SOME OTHER PROPERTIES

Very briefly, other gas properties vary with operational conditions also and are similarly tabulated. The **dynamic viscosity** of gases rises with temperature – for example common gases just about double their viscosity over the temperature range 0°C–400°C. Pressure effects are also present – as shown by nitrogen which increases its value by about a quarter, from about 20×10^{-6} to about 25×10^{-6} kg/m s, for a pressure rise from atmospheric to around 250 bar at atmospheric temperature.

● PQ 6.10 Kinematic viscosity is defined as dynamic viscosity divided by density. It is a useful combination of properties in fluid flow calculations. What are the units and dimensions of kinematic viscosity?

Example 6.3 Predict the influence of pressure on the value of the kinematic viscosity of nitrogen in the range 1–250 bar, at atmospheric temperature. Do not use any tabulated actual values of properties.

First, the dynamic viscosity. We have been told just before the last PQ that it rises by about a quarter over this pressure range. As we have to ignore any real values, call the dynamic viscosity at 1 bar '(DV)', so that at 250 bar it will be one quarter higher, 1.25(DV).

Now the density. Boyle's law tells us what happens to density as pressure rises at a fixed temperature. Check back if necessary – never accept statements if you have real doubts. So, if we call the density at 1 bar '(DENS)', then at 250 bar it becomes 250(DENS).

Thus, kinematic viscosity changes from

$$\frac{(DV)}{(DENS)} \text{ at 1 bar, to } \frac{1.25(DV)}{250\,(DENS)} = \frac{(DV)}{200(DENS)} \text{ at 250 bar.}$$

Without recourse to tabulated values, then, we have been able to predict that the kinematic viscosity of nitrogen will fall to about one two-hundredth of its atmospheric value if the pressure rises to 250 bar.

Data Tables = Information

= Better Accuracy

Figure 6.4 Use the data tables.

- PQ 6.11 Do a similar calculation for a common gas at fixed pressure over the temperature range 0°C–400°C.

This example and the PQ show how recorded changes in one property allow calculations for others. While many properties or property changes are quite easy to calculate, much of the work has already been done and tabulated, as noted earlier. When the principles are thoroughly understood, then any such tables or reference material should be sought for routine work. Understanding the principles, though, will allow you to do work which is of a more novel nature.

To round up this section, the ideal gas laws hold good for general use and the properties derived from them are valid. The introduction of the concept of semi-perfect gases, where properties change with temperature, refines the laws but does not invalidate them. Notably, the semi-perfect gas idea dwells on the refinement of specific heat capacity changing with temperature. This is perhaps the most important single improvement but do not restrict yourself to that. Pressure changes will also influence properties and should be included where appropriate.

Finally, once the principles are understood so that you can deal with changes in operation or new applications of the laws, then you can take a lot of the routine out of the work by becoming familiar with and using the tabulated information.

6.4 MORE ACCURATE EQUATIONS OF STATE

In Chapter 2, the *characteristic ideal gas equation* or the *ideal gas equation of state* was met. It is a valuable but relatively simple equation of the form $p V = m R T$. For most everyday routine or industrial plant type calculations, it (and all that follows from it) is an adequate expression, especially when used in conjunction with the semi-perfect or varying properties improvements.

- PQ 6.12 When, in general terms, might the simpler laws be inadequate? When may more precision be needed?

Therefore, while existing expressions may be enough for most purposes, new developments, research, detailed improvements of designs for instance may demand greater precision in the gas laws and that which follows from them. This is not, of course, peculiar to thermodynamics. Most technical disciplines have relatively simple mathematical expressions which are very useful for the majority of applications but fundamental improvements are still sought. In our case, some variation of the term *equation of state* is used for any of the later or more sophisticated equations that have been derived to describe gas behaviour more precisely. There are several and they are usually identified by the names of the researchers involved. Work is continuing as instrumentation and study methods improve but the first improvement of significance was the *van der Waals equation*. The work went back to the concepts of the kinetic theory and refined them in a clear fashion.

The kinetic theory (Chapter 2 and Appendix A) includes the simplifying features:

- that the molecules of the gas take up negligible room, so that the gas can be compressed without the molecules getting in the way;
- that they are inert insofar as there are no significant forces between the molecules, so that pressure is solely due to the molecules bouncing off the container walls.

The van der Waals equation makes an allowance for each of these by saying that the molecules do take up space and that there are intermolecular forces (Figure 6.5).

What this means then is that the volume term V and the pressure term p in the characteristic gas equation have to be modified to accept this observation. The real size of the molecules will reduce the 'free' volume and the molecular forces will change the net forces which give rise to the 'pressure' measurement. The volume term is thus reduced by the actual volume which the molecules would occupy if they were packed together as closely as is theoretically possible. Commonly this molecular

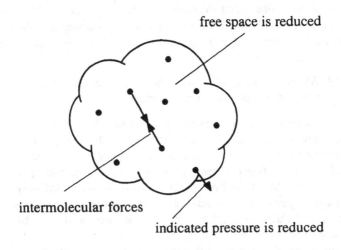

Figure 6.5 Molecules do take up space and there are intermolecular forces.

minimum volume is designated *b*, so that the *van der Waals equation of state* volume term becomes $(V - b)$. Just to put this into context, if I have a cubic metre of air at everyday temperature and pressure, then the molecules themselves occupy about a litre, that is, about one-thousandth of the total volume.

The intermolecular attractive forces are trying to pull the molecules together, thus reducing the gas pressure which is recorded as 'pressure' on a gauge attached to the gas container. Thus the real pressure which affects the gas behaviour must be higher than that read by a gauge, by an amount governed by the value of these intermolecular forces. The letter *a* is often used in this context. The specific volume *v* of the gas also plays a part, since it is related to the number of molecules in a given volume, hence the quantity of molecules producing these intermolecular forces.

Rather like Boyle's and Charles' laws, the van der Waals equation was formulated by observation but this time by very careful and detailed scrutiny of available data. The resulting equation is still easily recognized as an equation of state and clearly related to the original one which was derived from Boyle's and Charles' laws. One common form of the van der Waals equation of state is

$$\left(p + \frac{a}{v^2}\right)(v - b) = RT,$$

where *a* and *b* are the constants derived from the property observations. Note that this form uses the specific volume and the original equation equivalent would be $pv = RT$.

● PQ 6.13 Will the refinements be more or less important at higher pressures? Higher temperatures? Why? Do they both act in the same way, that is, do the *p* term and the *v* term increase together, decrease together or go in opposite directions?

By way of illustration using nitrogen, at room temperature and pressure the corrections have an effect of about 0.2% on the pressure term and about 0.15% on the volume term. What matters, though, is the product of the *p* and the *v* term, whichever equation is used, because we are still concerned with the characteristic equation $pV = mRT$ or the specific volume version of the equation $pv = RT$. We are looking at what happens to the *pv* term when it is improved to $(p + a/v^2)(v - b)$, not just each term on its own.

If we take the straightforward product *pv* from the simple equation and call that 100%, just to give us a handy datum, then Table 6.3, derived from the van der Waals equation, is interesting. The percentage figure is the 'improved' *pv* value.

These improvements or refinements then are fairly modest, but they

Table 6.3 Example of van der Waals improvements

Pressure (bar)	Temperature (K)	Percentage
1	300	100.05
10	300	100.5
100	300	102.0
1	1000	99.98

are the sort of values which could be very important in research work, in space travel and so on. Over a wide range of conditions which may be met in conventional industrial or commercial work, though, the changes are less than 1%, notably because of the contradictory effects of the pressure and the volume refinements.

Thus the prime value of this modification to the equation of state and the later, more detailed and continuing improvements, is for better data acquisition and better predictive abilities. There are several improved versions of the characteristic equation. This one has been used to illustrate the style of improvement and the sort of improved values which are thus introduced.

PROGRESS QUESTION ANSWERS

PQ 6.1

It means that at 30°C I have to supply 743 J to raise the temperature of 1 kg of nitrogen by 1 K, and at 150°C the amount needed to do the same is 747 J.

PQ 6.2

They must be the average values of specific heat capacity between the datum and the required temperature.

PQ 6.3

If in doubt, go back to PQ 5.16.

PQ 6.4

Recall that $R = c_p - c_v$ and that $\gamma = c_p/c_v$. Thus at 300 K $R = 0.287$ kJ/kg K and $\gamma = 1.400$, and at 1000 K $R = 0.287$ kJ/kg K and $\gamma = 1.336$. So the gas constant does not vary over this temperature range whilst the ratio of the principal specific heat capacities falls by about 5%.

PQ 6.5

I have chosen two temperatures to illustrate, 200°C and 600°C. At 200°C, c_p for nitrogen is 1.051 and for oxygen 0.963 kJ/kg K. Per kg air there is 0.767 kg nitrogen and 0.233 kg oxygen, so we can write, where $c_{p\ air}$ is the specific heat capacity of air,

$$0.767 \times 1.051 + 0.233 \times 0.963 = 1 \times c_{p\ air}$$

and by now you should be able to put in the units. This gives then $c_{p\ air}$ = 1.03 kJ/kg K compared to the tabled 1.026 kJ/kg K. At 600°C, c_p for nitrogen is 1.135 and for oxygen 1.068 kJ/kg K. which gives a calculated value for the specific heat capacity of air at 600°C of 1.119 kJ/kg K compared to the tabulated value of 1.114 kJ/kg K.

In either case, the difference is less than one half of one per cent which is well within everyday measurement accuracy. Note carefully that this only applies to unreacting mixtures. If there are chemical reactions involved, then that changes many things, such as we meet in a later chapter.

PQ 6.6

No. We saw in PQ 6.2 that the individual value of R did not change noticeably with temperature because the values of c_p and c_v kept pace with each other. The value of molar mass is constant anyway, so the product of individual gas constant and individual molar mass must stay constant.

PQ 6.7

PQ 6.4 looked at the effect of temperature and Table 6.2 shows some pressure effects on c_p. It has been stated that c_v is almost independent of pressure, so the effect of pressure on the ratio γ will be a direct reflection of the change in c_p.

While R does not change measurably because of temperature, it will change because of pressure, since pressure effects are noticeable on c_p but negligible on c_v. R $(c_p - c_v)$ will therefore rise with pressure.

PQ 6.8

If it was ΔU (large U, total amount) then the answer would be in kJ alone. We were asked about Δu (small u, specific amount, per kg) so the outcome is kJ/kg.

γ is the ratio of the principal specific heat capacities, both of which have the units kJ/kgK, so the units cancel out. γ has no units – it is a number, a ratio. This sort of factor which has no units is often called dimensionless and we will meet dimensionless ratios, as a topic, briefly in the heat transfer chapters.

PQ 6.9

Go back through the points made about pressure effects and you will see that pressure has an almost negligible effect on c_v. It follows then that c_v at 500°C, 1 bar is the same to all intents as c_v at 500°C, 70 bar. The change in internal energy will therefore be the same as in part (b). Note carefully, though, that if we had been discussing enthalpy, there would have been a change because c_p is affected by pressure.

PQ 6.10

From what has just been said of nitrogen, you see that the units of dynamic viscosity are kg/ms. You know already that the units of density are kg/m^3 so

$$\text{units of kinematic viscosity} = \frac{\text{units of dynamic viscosity}}{\text{units of density}}$$
$$= (\text{kg/ms})/(\text{kg/m}^3)$$
$$= \text{m}^2/\text{s}$$

and the dimensions are thus [L^2/T].

PQ 6.11

In this case, we are told that the dynamic viscosity about doubles over the stated temperature range. So (DV) at 0°C becomes 2(DV) at 400°C.

Charles' law applies now, which tells us that volume is proportional to temperature for a given pressure, thus density must be inversely proportional to temperature. Note that this is absolute temperature, of course. Thus the density changes from (DENS) to

$$\frac{273}{273 + 400}(\text{DENS}) = 0.406(\text{DENS}).$$

Thus the kinematic viscosity changes from

$$\frac{(\text{DV})}{(\text{DENS})} \text{ at } 0°C, \text{ to } \frac{2(\text{DV})}{0.406(\text{DENS})} \text{ at } 400°C.$$

That is, the kinematic viscosity of nitrogen increases to around five times its original value.

PQ 6.12

Recall that the laws are quite adequate for routine operations and fairly small changes of conditions. You have seen numbers to support this.

Thus, if there are new processes, precise operations, large changes of conditions, then the convenient simplifications of the ideal gas laws may not give the detailed information needed.

PQ 6.13

As pressure rises, then specific volume falls – the gas is compressed. So the a/v^2 term will rise in value. At low pressures then, this is less significant than at high pressures.

Since b is effectively a fixed amount – the volume occupied by the molecules themselves – then as v falls, the importance of b in $(v - b)$ increases.

When temperature rises, the opposite will be true as specific volume increases with temperature. The actual difference introduced by the van der Waals correction factors will therefore depend upon the temperature and pressure involved.

TUTORIAL QUESTIONS

The ground rules for the following examples are exactly those which have been met for ideal gases. The differences are to do with increased accuracy, not with new fundamental rules.

Note that standard data tables are under frequent review. Tables of different ages may therefore give slightly different property values.

6.1 To a fair approximation over modest temperature ranges, the specific heat capacities of real gases vary progressively according to $c_t = c_0(1 + a\,T + b\,T^2)$ where c_0 and c_t are the values at 0°C and T°C respectively, with a and b being constants.

From this relationship and the air data given below, estimate the specific heat capacity at constant pressure of air at 0°C and at 400°C.

Temp, °C	100	200	300
c_p, kJ/kgK	1.0 1	1.025	1.045

[1.00 kJ/kg K; 1.07 kJ/kg K]

6.2 During the industrial processing of carbon monoxide, estimates are needed of specific enthalpy and specific internal energy changes between 100°C and 600°C. What percentage errors arise, based on the real gas values, if carbon monoxide is taken as a perfect gas using properties at 0°C instead of allowing for temperature effects?

The table gives some carbon monoxide data, and arithmetic means of properties at these temperature levels give good approximations.

$$0°C, \quad c_p = 1.072 \text{ kJ/kg K} \quad \gamma = 1.40$$
$$100°C, \quad c_p = 1.094 \text{ kJ/kg K} \quad \gamma = 1.39$$
$$600°C, \quad c_p = 1.204 \text{ kJ/kg K} \quad \gamma = 1.35$$

[Ideal values are 6.7% and 8.78% low, respectively]

6.3 For 1 kmol of any gas, the van der Waals equation can be written

$$\left(p + \frac{a}{v_m{}^2}\right)(v_m - b) = R_m T$$

where v_m is the molar volume. The constants a and b are often recorded as molar values in tables.

Oxygen at 0°C has a molar volume of 22.4 m³/kmol at a particular pressure. Given $a = 1.37 \times 10^5$ N m⁴/(kmol)² and $b = 0.0318$ m³/kmol for oxygen, what is the percentage difference in pressure calculated by using the van der Waals equation compared to the characteristic ideal gas equation in this case?   [About one eighth of 1%]

6.4 Change the values of constants a and b in the previous example from kmol values to kg values.
$$[a = 1.37 \times 10^5 \text{ kg/m s}^2 \text{ (or N/m}^2\text{)}, b = 9.94 \times 10^{-4} \text{ m}^3\text{/kg}]$$

6.5 The table shows, for hydrogen gas, the specific enthalpy h and specific internal energy u above the usual datum of 0°C. What are the mean values of $\bar{\gamma}$, the ratio of principal specific heat capacities and $\bar{R}$, the gas constant between 0°C and the stated temperatures?

Temp (°C)	100	300	600
h (kJ/kg)	1440	4346	8742
u (kJ/kg)	1031	3122	6353

[The mean values of $\bar{\gamma}$ are 1.397, 1.392 and 1.376 and those of $\bar{R}$ 4.09, 4.08 and 3.98 kJ/kg K]

6.6 This table shows values (actual, not mean) of specific heat capacities at constant pressure, c_p, of hydrogen gas at various temperatures. Use them to prepare a graph and thus estimate the value at 0°C.

Temp (°C)	100	200	300	400
c_p (kJ/kg K)	14.44	14.53	14.57	14.61

[14.24 kJ/kg K]

Liquids, vapours and superheat

7

7.1 SOME DEFINITIONS

Up to now, we have concentrated on gases – mostly ideal gases – as a good starting point to look at some fundamentals of thermodynamics. Even using ideal gases for this, the findings are important for real calculations. This is especially so when real data values are used, the accuracy of the data is respected and instrument reading accuracy is borne in mind. It is no good having very accurate data if the instruments on the plant are used incorrectly!

Gases form only part of the range of materials or working fluids that are met in industry or commerce, of course. There are plenty of solids and plenty of liquids and plenty of mixtures. It is always helpful in any study though, when progressing, to move on to something similar to that which is already understood. **Vapours** are similar to gases and our next area of study is just that.

● PQ 7.1 What is a vapour? What is the main difference between a gas and a vapour?

When talking about liquids or vapours or gases and so on, some everyday words are used in a precise fashion. Because they are everyday, it is easy for them to be used incorrectly. Gas, vapour and fluid can be misunderstood, for instance, as can phase and state. Definitions of these last two are needed before we go much further.

● PQ 7.2 First a brief revision – the words property and system. Define these.

The **state** of a system is strictly the complete description of the system and thus the record of sufficient system properties for the description to be clear and unambiguous. If the state of a system is known, it cannot be confused with any different system. In passing, you may meet the term 'state point', meaning some point on a diagram of properties (a p–V graph, for instance) which identifies those properties at any particular time.

● PQ 7.3 Do you need to know every property to define the state of a system? Are properties connected?

As a relevant incidental, if you wish to offer a technical definition of a **process** – making plastics, boiling water, baking a cake or whatever – then a process is something that causes (usually in a controlled fashion) a **change of state**.

The word **phase**, used thermodynamically, means any identifiable amount of a material or substance that is **homogeneous**. This refers to both physical and chemical properties and is perhaps best illustrated by a simple example which will also include the term 'phase boundary'.

Imagine a steel bucket, filled with water, standing on the floor in a room in a house. The air in the room is in the gaseous phase, the water in the bucket is in the liquid phase and the steel bucket is in the solid phase. Where the water touches the steel of the bucket, there is a phase boundary. Where the water surface meets the air is another phase boundary and where the air of the room touches the bucket is yet another phase boundary (Figure 7.1).

Similar phases can have boundaries – the bottom of the bucket (solid) stands on the floor (solid) and there is a boundary. If some engine oil was poured into the water, the two will not mix and there will be a phase boundary. These boundaries are often identified by the phases involved, such as a solid–liquid phase boundary or a solid–solid phase boundary.

Although the definition of 'phase' is quite precise – physical and chemical homogeneity – it is usually relaxed a little on the grounds of common sense for everyday technical purposes. If I run a bath of water, for instance, using separate hot and cold taps, there is a good chance that the water will be warmer in some parts of the bath than in others. It is not strictly homogeneous therefore, but it would still be described

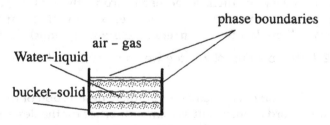

Figure 7.1 Phases and phase boundaries.

normally as being in the liquid phase.

● PQ 7.4 As a vapour is the progressive 'in-between' of a liquid changing to or from a gas, can there be a vapour phase?

Water is an excellent example to use in the study of vapours. It is a widespread material and its thermodynamic properties are very well established. It is important industrially – the most important working fluid. It is heated, boiled and changed into vapour and gas (*steam*) in very large quantities every day. Power stations and process industries depend on it. The most common vapour then, both everyday and industrially, is water vapour or steam. Since it is so common, most of its properties too have been recorded quite accurately and it is thus an ideal material for our present purposes (Figure 7.2). However, change a few words and put the right numbers in the properties, and most of that which follows will apply to any material's liquid–vapour–gas progressive change.

First, a little clarification or definition by use of a conventional name. In common terms, we all know what is meant by ice and by water. Equally, steam is usually taken to mean that whitish cloud of stuff that comes from the spout of a kettle. However, when the word steam is used industrially a little more precision is needed and this spills over into general scientific use. So the term **water substance** is used to cover water in all its forms – solid, liquid, gas or their mixtures. Beyond this, various words which will soon be met are attached to the word steam to identify it more accurately.

Next, take a look at the process of water boiling, which generates steam. The word 'boiling' is often associated, in everyday language, with water boiling because that is the most common example. Other liquids boil when they are heated sufficiently and the boiling point does not have to be at a high temperature. Refrigerants include examples of liquids whose boiling points are below room temperature, for instance the common one R.12 boils at about minus 30°C. The liquids which boil to give oxygen and nitrogen gases have very low boiling points, approaching minus 200°C. Equally there are materials with boiling points higher than water – mercury at 356°C, for example.

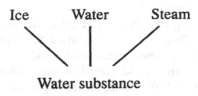

Figure 7.2 Solid, liquid or gas – all water substance.

So the terms 'boiling' or 'boiling point' are to do with some physical phenomenon, not necessarily to do with a high temperature. All the boiling points mentioned refer to a pressure of around 1 bar. Boiling temperatures will be different at different pressures, as will be discussed later.

It is useful to try to describe the process of boiling because that is likely to give some clues to vapour behaviour. It can be done at a simple but quite adequate level by thinking of water in a saucepan being heated on a cooker. Pretend that the heating process produces very smooth changes for even greater simplicity, at least for the present. Heat – called *'sensible heat'* because it is accompanied by a temperature change and can thus be recorded easily – is added to the liquid water.

The water temperature rises to the boiling point but here, even if more heat is added, the temperature will not rise until the next step of the change is completed. The heat still being added is called variously the **heat of change of phase** or the **latent heat of evaporation**. It is the energy required to convert the liquid (at its boiling point) water into gaseous water substance. I will use the term latent heat.

● PQ 7.5 So what is the latent heat supply doing? Note that the water molecules must escape from the liquid mass to become steam in any form.

So the boiling process can be regarded as one of individual molecules having sufficient energy to escape from the liquid phase to become gas. That is what would happen if boiling was a very smooth process. In real life, it is not.

The actual boiling process is quite violent, as can be seen in any kitchen with a pan of water. Liquid is converted to gas at the heated surface – the pan base or the kettle element in a kitchen, for instance – and passes up through the liquid to escape from the liquid surface. Because of the bubbling and the thermals (the recirculating fluid flow currents in the heated liquid) the whole of the escape surface is heavily disrupted. It is easy to see therefore that any gaseous water substance steam leaving the liquid water will inevitably carry with it some amount of the liquid. Other processes are continuing, such as heat transfer between the gas and the carried liquid but the simple model of the steam from this boiling being a mixture of gas and liquid droplets is a reasonable one (Figure 7.3).

The steam escaping like this, in effect a mixture of gas and minute water droplets, is called *'wet steam'*, quite a descriptive term. The steam which comes from a kitchen kettle is a good example. It looks white because the numerous liquid droplets are reflecting any light that happens to be around. If the kettle steam is played onto a cold surface, it condenses back into a liquid.

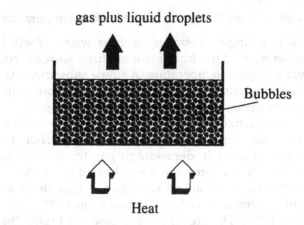

gas plus liquid droplets

Bubbles

Heat

Figure 7.3 A model of boiling water.

Depending upon how much liquid is present in our modelled wet steam, so the vapour – for that is what it is – is given a *dryness fraction*, on a scale of 0 to 1 or 0% to 100% . The higher the number, the drier the steam – that is, the less liquid it carries. If a sample of wet steam can be assessed as (by mass) half liquid, half gas, then its dryness fraction would be 0.5 or 50%.

● PQ 7.6 If the vapour was three-quarters gas, the dryness fraction would be – what?

Any steam below 100% dryness comes into this bracket of wet steam. When steam is just completely changed to gas – no residual liquid content but no extra heat supplied beyond that required for the complete conversion – the steam is known as '*dry steam*' or '*saturated steam*' or even '*dry saturated steam*'. The terms are more or less traditional and are also used interchangeably. Complementary to that, liquid water at its boiling point is often called '*saturated water*'.

This model of wet steam is rather similar to fog or mist which arises in certain weather conditions. The fog is minute water droplets being carried by a gas but this time the gas is air. Apart from that, wet steam and fog or mist are much the same. If the sun comes out, then the water of the fog is heated and evaporated to become gaseous water. The two gases – air and gaseous water substance – mix perfectly well to be a transparent gaseous mixture and the fog disappears. If heat is applied to wet steam, you can imagine the liquid water droplets evaporating to become gaseous water and thus the vapour gets progressively drier until it is all gas.

Thus a simple but effective model of a vapour can be imagined. A liquid is heated to its boiling point – it absorbs sensible heat up to its boiling temperature. Two things about the boiling temperature:

- If the liquid is a single substance, such as water, it will boil at a specific temperature. If the liquid is a mixture, such as crude oil, it will boil over a range of temperature. A single substance has a *boiling point*, a mixture a *boiling range*. For our purposes only single substances are considered, that is boiling points not ranges.
- Boiling point (or range) depends upon the *pressure* at the boiling surface. If I heat an open saucepan of water in the kitchen, it will boil at 100°C or very close to it, depending upon the day's atmospheric pressure. If, though, the same water was heated in a pressure cooker at 1.5 bar, that is half a bar above atmospheric, then the water would not boil until its temperature had reached almost 112°C. At 2 bar, it would be over 120°C. The higher the pressure, the higher the boiling point. Conversely, the lower the pressure, the lower the boiling point. At 0.7 bar for example, the boiling point of water is about 90°C and at 0.2 bar, 60°C.

- PQ 7.7 Why is this? If the molecules have sufficient energy to escape from the liquid, is it only the liquid's ability to retain molecules that is being overcome?

To explore this a little further, the term *vapour pressure* has to be introduced. You know that, for a gas, the gas pressure is caused by the rapidly moving molecules striking a retaining surface. The molecules of a liquid are also moving but the liquid is retained by the walls of its container – the bottle, the pipe, the bucket or whatever. If, however, the liquid has a free surface – like the water surface in a bucket – then the molecules have a chance of escaping. The warmer the liquid, the quicker the molecules move (temperature is a measure of molecular kinetic energy) and thus the greater their chance of escaping. When the liquid is hot enough for the molecules to escape freely and completely, then the liquid is boiling (Figure 7.4).

These molecules trying to escape at a free surface must be, in effect, exerting a pressure at that surface, just as water pressure allows liquid water to escape from the nozzle of a hose pipe. This pressure which the molecules exert is the vapour pressure and, since it is related to the molecules' escape ability, it is related to temperature. The higher the liquid temperature, the higher the vapour pressure.

- PQ 7.8 Think of a couple of everyday examples which illustrate this.

Going back to the free liquid surface, think about some water in a bucket, the bucket standing in a room. The air in the room has pressure, the atmospheric pressure as recorded on a barometer or quoted in the weather forecast. This air pressure – the air molecules striking any

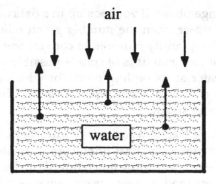

Figure 7.4 Molecules escaping – atmosphere resisting.

available surface – is felt as much at the bucket's water surface as it is at the walls of the room. The air pressure can thus be imagined as a wall holding the water in the bucket. The air pressure is overcoming the liquid's vapour pressure, so the molecules trying to escape are held back. If the air pressure is reduced, then its ability to retain the molecules in the liquid is also reduced. If the applied air pressure is increased, its ability to keep the molecules in the liquid is increased.

● **PQ 7.9** To put PQ 7.7 on a little more technical basis, why is boiling point related to surroundings pressure?

The liquid's vapour pressure, well away from boiling, is still a pressure even though it may be much less than the applied or atmospheric or surroundings pressure. It means that some of the liquid must be able to escape, which is the background to **evaporation**. A puddle on the pavement will dry out without being seen to boil. It will dry far slower than if it was being boiled away but the background process is the same – the vapour pressure is the driving force.

We need to spend a little more time on the heat of change of phase or the latent heat of evaporation. You will recall that these are interchangeable terms and text books of various ages will prefer one or the other.

● **PQ 7.10** Give a definition of latent heat or heat of change of phase that would apply to steam. What are the units?

This definition of latent heat infers the complete conversion of liquid to gas and the definition can be used on all liquids, not just water. So, if some water is heated up by the supply of sensible heat, it will be accompanied by a temperature change up to the boiling point. Addition of sensible heat will only take the liquid up to its boiling point. For the liquid to actually boil, then latent heat is added. The temperature does not change – it stays at the boiling point – because this dose of energy is being used to change phase. If you look up in a data table the latent heat of evaporation of water, then the number given will be some **specific** amount, such as the quantity required to convert one kilogram of water at its boiling point to steam. It is of course possible to supply less than that amount to water at its boiling point, in which case less than the specified amount of liquid water will convert to the gas. Any steam generated will be wet steam.

> Example 7.1 I have 1 kg of water at its atmospheric boiling point. The latent heat of evaporation (heat of change of phase) is 2257 kJ/kg. I supply 1 MJ of latent heat. How much liquid can be changed into gas? If all the fluid escaped as wet steam, what would be its dryness fraction?

First note that the latent heat value was quoted at atmospheric pressure. It will be mentioned later that latent heat values are also pressure dependent, so this particular value would depend upon the atmospheric pressure on the day.

Insufficient heat energy has been supplied (1 MJ) to convert the whole 1 kg of liquid water all the way through the vapour stage into true gas, for which 2.257 MJ is needed. Remember to check units always – is it J, kJ or MJ? So the proportion of the 1 kg of the liquid that can be converted into gas is the ratio of the heat supplied to the heat necessary, i.e. 1/2.257 or just under 45% .

If the water boiled away as wet steam (the gas generated carrying away the rest of the liquid as in our model of wet steam), then this would also represent the dryness fraction, just under 45%. The supplied heat can only convert part of the liquid, so this is the maximum amount of gas that can be generated.

This example leads, then, to a simple rule about dryness fraction. It represents the **proportion** (less than 100% of course) of the latent heat actually supplied in the generation of the vapour. Thus if the dryness fraction was 75% or 0.75, then the vapour would have had 75% of the total conversion latent heat requirement (0.75 × 2257 kJ) supplied to it. The vapour would be modelled as three quarters gas, one quarter liquid.

As the pressure at the boiling surface affects the boiling point, it also affects the latent heat demand.

• PQ 7.11 Why should this be so? If higher pressure has raised the boiling point, think what has happened to the water molecules' energy content when they finally reach boiling point. Will this affect the amount of energy needed to escape from the liquid surface?

Sensible heat demand to reach boiling point rises as pressure rises, and latent heat demand to complete the boiling falls as pressure rises. The two offset one another to some extent. In actual operation, there are some other effects which mean that the total energy demand (sensible plus latent) does vary with pressure. The total rises as pressure rises to a peak at about 40 bar and drops away thereafter. While there is this change in the total, the actual variation is fairly small, as Table 7.1 shows. The pressures quoted cover the vast majority of industrial demand. Note the use of the word 'enthalpy' here in the context of sensible and latent heat. It is a measure of the energy put into the steam during the boiling process and it is the measure of the steam's value as a working fluid – its ability to do work and to transfer heat. So the enthalpy of the table is the specific enthalpy, kJ/kg, of dry steam – the liquid has been converted completely to gas but nothing further has happened.

If energy is now added to the steam beyond that needed to complete the conversion of liquid to gas (dry steam), then the gas, having no more phase changing to do, reacts as any other gas would do. So its temperature will rise or it will expand or any combination of the two, in similar fashion to that which we met when first considering ideal gases. The fact that this gas is water substance is of no consequence in that respect – it is now a gas and it acts as a gas.

Steam is a most important, if not the most important, industrial working fluid. It drives turbines to make electricity; it is a good heat carrier for manufacturing, for instance. So when used industrially, the usual thing that happens when heat is added to steam is that it is done in a way that encourages its temperature to rise and thus increase its value as a hot working fluid. When steam is so heated beyond its vapour state, it becomes a gas at the boiling point and then a gas above

Table 7.1 Some values of specific enthalpy of dry steam

Pressure (bar)	Enthalpy (kJ/kg)	Increase (%)
1	2675	baseline
10	2778	3.85
40	2801	4.71
100	2725	1.87
150	2611	−2.39

the boiling point. As soon as this gas exceeds the boiling point, it is generally called 'superheated steam'.

● PQ 7.12 I have two lots of steam. Both are at 112°C. Are they both superheated?

To identify the steam **condition**, the word usually used for saying whether it is wet or dry or superheated, then it is seen that temperature alone is not enough. We have to know the pressure as well. Since steam is an important commercial and industrial working fluid, let us stay with industrial practice and the way that industrial boilers, more correctly called steam generators, work (Figure 7.5).

What happens in a boiler is that liquid water is pumped at the required pressure in and wet steam is generated, by electricity or by burning fuel, at that pressure. Recall that pressure and temperature are linked and that it will be wet steam because the gas generated at the heated surfaces is bubbling through liquid. If the steam is to be superheated, which is the industrial norm, the superheating also takes place at that pressure. A separate but integral part of the boiler is used for that purpose. There are likely to be some small pressure losses as the fluid flows around the boiler circuit but, to all intents, the steam-generating process in real boilers is a constant-pressure process.

● PQ 7.13 What word can be attached to this heat energy which is added to the water substance to make superheated steam?

Just as a reminder, there are some justifiable academic arguments about heat or energy or enthalpy being added, as if it were money being added to a bank balance. Here, though, I will continue to use 'added' in its common-sense meaning of increasing the measurable amount.

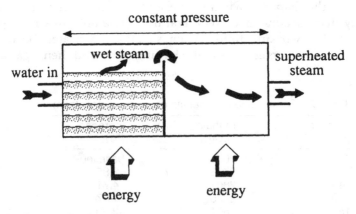

Figure 7.5 Basics of boiler operation.

As enthalpy is added and relevant properties are required for calculations, this emphasizes the need for **both** pressure and temperature being quoted when steam is being discussed.

Example 7.2 Some data tables, which we meet soon, show that the boiling point of water at 2.7 bar is 130°C and at 7 bar it is 165°C. In one set of process plant readings, I have steam at 7 bar, 165°C and in another I have steam at 2.7 bar, 165°C. What can be said of the wetness, dryness or superheat level of the steam in each case?

Wetness is the opposite of dryness, here talking of steam. Dryness is the usual term even though we speak of wet steam. We could thus speak of steam which is 60% dry being 40% wet, therefore.

Since the boiling point at 7 bar is 165°C, then there is no way of knowing from just this information how dry the steam is. The water will certainly be at its 7 bar boiling point and we are told that steam is being generated. The only other thing that we can deduce is that the steam will not be superheated – it has to be above the boiling point for that. So we are dealing here with steam whose condition – a common term, recall – is at best just dry but we can say no more.

For the steam at 2.7 bar, 165°C, then far more definite conclusions can be drawn. The boiling point of water under an applied pressure of 2.7 bar is 130°C, so the steam here is well above that. It is thus gas, not vapour, and it is superheated, not wet nor even dry saturated. In this case, then, the temperature and pressure describe the steam condition completely and, given the proper data tables, many other properties can be deduced immediately. This is not the case for wet steam, where the temperature and pressure do not indicate the degree of dryness in any way.

While it is common to quote temperature and pressure for superheated steam, there is an alternative term which you may meet. In Example 7.2. the superheated steam is at 165°C compared to its boiling point of 130°C at 2.7 bar. That is, it is 35 degrees Celsius above its boiling point at that pressure. In some cases then, this steam would be recorded as having 35 **degrees of superheat**. It is a useful measure industrially, where a plant is operating at a constant pressure. It says something immediately about the steam. For most other purposes though, the simple way to avoid mistakes is to quote the actual temperature and pressure.

Example 7.3 At 7 bar, the latent heat of evaporation of water is 2067 kJ/kg. The dryness of some steam at 7 bar is measured by condensing 2 kg of steam to liquid at its boiling point. The enthalpy recovered is 3390 kJ. What is the steam dryness fraction? Enthalpy losses in the test measurement may be assumed negligible.

If 2 kg of 100% dry steam had been condensed, the enthalpy recovery would have been

$$2 \text{ kg} \times 2067 \text{ kJ/kg} = 4134 \text{ kJ}$$

The actual recovery was 3390 kJ, so the dryness fraction is

$$3390 / 4134 = 0.82 \text{ or } 82$$

In the real test procedure for steam dryness, which is a condensation technique, great care is taken to minimize losses by insulation and to determine any loss correction factors by detailed experiment at the test equipment design stage.

● **PQ 7.14** This PQ and the example which follows it are designed to make you think about a few fundamentals. Define specific heat capacity for steam – wet, dry and superheated.

Example 7.4 Find a specific heat capacity of 80% dry steam at 25°C.

In the data tables, h_f is used conventionally for sensible heat (enthalpy) demand to boiling point and h_{fg} for latent heat (enthalpy) demand to convert to dry steam. Both are specific (small letter h, note) values.

I have chosen this temperature deliberately to show that steam can be generated at temperatures well below the everyday boiling point and that there are data at that level. The generation pressure would be about 0.0317 bar, according to the data tables, with h_f=104.8 kJ/kg and h_{fg}=2441.8 kJ/kg. The specific enthalpy of this wet (80% dry) steam is

$$h = h_f + 0.8 \, h_{fg}$$
$$= 104.8 + 0.8 \times 2441.8 = 2058.24 \text{ kJ/kg}.$$

Raising the steam temperature by 1 K to 26°C, the new values from the complete data tables are $h_f = 108.9$ kJ/kg and $h_{fg} = 2439.5$ kJ/kg. However, note that the pressure must have risen to 0.336 bar to keep the steam wet at this temperature. If the temperature had risen at the original pressure of 0.0317 bar, then the steam could no longer be wet. Make quite sure that you understand that – if not, review it now.

The specific enthalpy of 80% dry steam at 26°C is then

$$h = 108.9 + 0.8 \times 2439.5 = 2060.5 \text{ kJ/kg}.$$

Thus the enthalpy required to raise the temperature of 1 kg of 80% dry steam through 1 K at the stated temperatures, the specific heat capacity, is 2.26 kJ/kg K.

● **PQ 7.15** Having done that, does it mean anything? Is there a flaw to do with constant pressure?

With wet steam, virtually all the industrial calculations are to do with changes of enthalpy or internal energy as the dryness varies. The fact that a definition of specific heat is either vague or messy is little more than unfortunate – it does not matter for most of real-life engineering.

However, as superheated steam is a gas, the gas rules must apply to it. One little problem is that in general, superheated steam is used industrially and commercially quite near to its **condensation** or **liquefaction point** (the opposite of boiling point), the temperature at which the gas condenses back to the liquid. You will recall that when dealing with the ideal gas laws, it was a condition that the gas was well away from its boiling point. In this case then, any ideal gas laws could be invalidated and thus have to be applied with great care. Away from the condensation temperature – the word 'liquefaction', although accurate, tends to be used for low temperatures as in refrigeration and so on – all the usual real or ideal gas laws apply to the same degree as other gases.

7.2 STEAM TABLES

At first sight, this talk of sensible heat varying with pressure, latent heat, the problem of much superheated steam being used too close to the boiling point for the gas laws to apply and so on, may seem to suggest trouble. How can a reasonable attack be made on design needs or safety calculations, for instance? Fortunately, as in much of real-life thermodynamics, the hard work has been done for us and the necessary data accumulated comprehensively. The same goes for other materials which are used in the liquid, vapour and gas form such as ammonia and the range of organic refrigerants. I will stay with water substance for clarity but most of the following is relevant to other materials also.

The tabulation of this necessary information has given rise to the common term of 'steam tables'. More correctly they should be called 'thermodynamic property tables' or a similar title and the commonly available tables include properties for many substances other than water. Observation and measurement provided the basis for these and they are continually being upgraded as instrumentation, data gathering and interpretation improves. Normally you would find them as straightforward tables in booklet form or as part of a reference text but they are now available as a computer program, either free-standing or included in larger programs such as ones developed for energy management. In either case they are simple to use as some examples will soon illustrate.

Table 7.2 is an extract, with permission, from 'Thermodynamic and Transport Properties of Fluids' by G.F.C. Rogers and Y.R. Mayhew (Basil Blackwell, Oxford, 1991 reprint). The significance of the transport properties is that some of the information is relevant to the so-called 'transport processes' of heat, mass and momentum transfer. I recommend that you obtain a full copy of this inexpensive booklet, for the purposes of thermodynamic study and for the remarkable range of generally applicable information which it contains.

For the whole of this chapter, ignore the columns headed s_f, s_{fg} and s_g. They are included at present to illustrate a typical table layout but they are not properties which we shall use or even meet yet. Whatever the source, the tables are usually laid out in similar fashion, with pressure and temperature to the left of the page and other properties, related to those values, next to them. It is worth taking a little time to familiarize, so that the table's use becomes easy.

● **PQ 7.16 Identify each of the symbols, ignoring for the moment the subscripts. You have already met the symbols but perhaps not in this chapter.**

Now take a closer look at the symbols, units and subscripts and what they mean in detail. Taking the first horizontal row of numbers and noting the units below each property letter, the values are:

● p – pressure in bars, 1 bar and this is an absolute value. This pressure value sets the scene for all other properties in the row. They all apply at the stated pressure and only at that stated pressure. In this instance, all the numbers will apply at 1 bar.
● t_s – saturation temperature or boiling point in degrees Celsius, not Kelvin, 99.6°C at pressure p.
● v_g – specific volume of dry saturated steam (just completely gas, no longer wet but not yet superheated) 1.694 m³/kg at pressure p and temperature t_s.
● u_f – specific internal energy of the liquid water, 417 kJ/kg at the stated temperature and pressure. By convention 0°C – more accurately 0.01°C but see later – is used as the datum.
● u_g – specific internal energy of the dry steam (again just gas, not wet but not superheated) 2506 kJ/kg at the stated temperature and pressure. The value of u_g is the total amount from liquid water at datum to dry steam at pressure p and temperature t_s.
● h_f – specific enthalpy of the liquid water, 417 kJ/kg at the stated temperature and pressure and again referred to the datum.
● h_{fg} – specific latent heat or enthalpy of change of phase, 2258 kJ/kg for the water substance at the stated temperature and pressure.
● h_g – specific enthalpy of the dry steam, 2675 kJ/kg at the stated

Table 7.2 Properties of saturated water and steam to 40 bar

p (bar)	t_s (°C)	v_g (m³/kg)	u_f (kJ/kg)	u_g	h_f (kJ/kg)	h_{fg}	h_g	s_f (kJ/kgK)	s_{fg}	s_g
1.0	99.6	1.694	417	2506	417	2258	2675	1.303	6.056	7.359
1.1	102.3	1.549	429	2510	429	2251	2680	1.333	5.994	7.327
1.2	104.8	1.428	439	2512	439	2244	2683	1.361	5.937	7.298
1.3	107.1	1.325	449	2515	449	2238	2687	1.387	5.884	7.271
1.4	109.3	1.236	458	2517	458	2232	2690	1.411	5.835	7.246
1.5	111.4	1.159	467	2519	467	2226	2693	1.434	5.789	7.223
1.6	113.3	1.091	475	2521	475	2221	2696	1.455	5.747	7.202
1.7	115.2	1.031	483	2524	483	2216	2699	1.475	5.707	7.182
1.8	116.9	0.9774	491	2526	491	2211	2702	1.494	5.669	7.163
1.9	118.6	0.9292	498	2528	498	2206	2704	1.513	5.632	7.145
2.0	120.2	0.8856	505	2530	505	2202	2707	1.530	5.597	7.127
2.1	121.8	0.8461	511	2531	511	2198	2709	1.547	5.564	7.111
2.2	123.3	0.8100	518	2533	518	2193	2711	1.563	5.533	7.096
2.3	124.7	0.7770	524	2534	524	2189	2713	1.578	5.503	7.081
2.4	126.1	0.7466	530	2536	530	2185	2715	1.593	5.474	7.067
2.5	127.4	0.7186	535	2537	535	2182	2717	1.607	5.446	7.053
2.6	128.7	0.6927	541	2539	541	178	2719	1.621	5.419	7.040
2.7	130.0	0.6686	546	2540	546	2174	2720	1.634	5.393	7.027
2.8	131.2	0.6462	551	2541	551	2171	2722	1.647	5.368	7.015
2.9	132.4	0.6253	556	2543	556	2168	2724	1.660	5.344	7.004
3.0	133.5	0.6057	561	2544	561	2164	2725	1.672	5.321	6.993
3.5	138.9	0.5241	584	2549	584	2148	2732	1.727	5.214	6.941
4.0	143.6	0.4623	605	2554	605	2134	2739	1.776	5.121	6.897
4.5	147.9	0.4139	623	2558	623	2121	2744	1.820	5.037	6.857
5.0	151.8	0.3748	639	2562	640	2109	2749	1.860	4.962	6.822
5.5	155.5	0.3427	655	2565	656	2097	2753	1.897	4.893	6.790
6	158.8	0.3156	669	2568	670	2087	2757	1.931	4.830	6.761
7	165.0	0.2728	696	2573	697	2067	2764	1.992	4.717	6.709
8	170.4	0.2403	720	2577	721	2048	2769	2.046	4.617	6.663
9	175.4	0.2149	742	2581	743	2031	2774	2.094	4.529	6.623
10	179.9	0.1944	762	2584	763	2015	2778	2.138	4.448	6.586
11	184.1	0.1774	780	2586	781	2000	2781	2.179	4.375	6.554
12	188.0	0.1632	797	2588	798	1986	2784	2.216	4.307	6.523
13	191.6	0.1512	813	2590	815	1972	2787	2.251	4.244	6.495
14	195.0	0.1408	828	2593	830	1960	2790	2.284	4.185	6.469
15	198.3	0.1317	843	2595	845	1947	2792	2.315	4.130	6.445
16	201.4	0.1237	857	2596	859	1935	2794	2.344	4.078	6.422
17	204.3	0.1167	870	2597	872	1923	2795	2.372	4.028	6.400
18	207.1	0.1104	883	2598	885	1912	2797	2.398	3.981	6.379
19	209.8	0.1047	895	2599	897	1901	2798	2.423	3.936	6.359

Table 7.2 continued

p (bar)	t_s (°C)	v_g (m³/kg)	u_f (kJ/kg)	u_g	h_f	h_{fg} (kJ/kg)	h_g	s_f	s_{fg} (kJ/kgK)	s_g
20	212.4	0.09957	907	2600	909	1890	2799	2.447	3.893	6.340
22	217.2	0.09069	928	2601	931	1870	2801	2.492	3.813	6.305
24	221.8	0.08323	949	2602	952	1850	2802	2.534	3.738	6.272
26	226.0	0.07689	969	2603	972	1831	2803	2.574	3.668	6.242
28	230.0	0.07142	988	2603	991	1812	2803	2.611	3.602	6.213
30	233.8	0.06665	1004	2603	1008	1795	2803	2.645	3.541	6.186
32	237.4	0.06246	1021	2603	1025	1778	2803	2.679	3.482	6.161
34	240.9	0.05875	1038	2603	1042	1761	2803	2.710	3.426	6.136
36	244.2	0.05544	1054	2602	1058	1744	2802	2.740	3.373	6.113
38	247.3	0.05246	1068	2602	1073	1729	2802	2.769	3.322	6.091
40	250.3	0.04977	1082	2602	1087	1714	2801	2.797	3.273	6.070

temperature and pressure. As with u_g, this is the total from liquid water at datum to dry steam at p and t_s.

The properties are **specific** – they refer to unit mass of water substance. u_f and h_f are values for the liquid water at the boiling point, taking a datum of 0°C. Strictly, the *datum temperature* is 0.01°C which is a temperature of interest and which will be explored later. For our immediate purposes and with little practical error, the datum can be taken as 0°C. The values u_f and h_f thus refer to energy inputs to the liquid between 0°C and the boiling point at whatever pressure is being exerted on the working fluid. Notice that the values vary depending upon the pressure, as you see if you look up and down the columns of the table.

v_g, u_g and h_g refer to the dry steam, (also called saturated steam, or dry saturated steam, recall) at the boiling temperature and pressure in question. The latter two are also based on a datum of 0°C, (0.01°C) so that the specific enthalpy and the specific internal energy of the dry steam include that required to raise the water to the boiling point. That is, u_g includes u_f and h_g includes h_f.

h_{fg}, the enthalpy of change of phase or latent heat as it is called perhaps a little casually, is the enthalpy required to convert the saturated liquid – liquid water substance at its boiling point – into saturated gas – dry steam at the boiling point for the pressure in question.

In industrial steam usage, these enthalpy values are very significant because, as noted earlier, boiler plant runs at substantially constant pressure for any given operation and, of course, enthalpy is a measure of a working fluid's heat transfer and work transfer capabilities. Boiler pressures can be varied readily within design limits as necessary but they usually run at a preset pressure level.

● PQ 7.17 Write down the relationship between h_f, h_{fg} and h_g, using the numbers to justify your statement. What would be the meaning of u_{fg} and what is its value at 1 bar?

The enthalpy or the internal energy of the dry steam is therefore the sum of values for the saturated water plus that for the change of phase. There are one or two other points to notice as you scan the tables. First, the values of u_f and h_f are very similar over a wide range of pressures and boiling points. The difference between enthalpy and internal energy is the effect of the pV term as in

$$H = U + pV \text{ and } \triangle H = \triangle U + \triangle(pV)$$

You have met these or similar terms but review them now if necessary. As liquid water substance is heated, it will expand but it is a modest expansion. For instance, from 0°C to 100°C the volume change is about 4%. Compared to the increase of internal energy over the same temperature range then, the change of the pV term is quite small. Hence for liquid water, the internal energy and enthalpy changes are very similar over a wide range of operating conditions. Even at 40 bar, the difference is still less than half of one per cent.

Second, the value of h_f is also almost totally dependent upon the temperature and not the pressure.

● PQ 7.18 Why? You have already met the reason.

You can see that also by the way in which the enthalpy and the internal energy match each other. Internal energy is a direct reflection of temperature influence ($U = mc_vT$ and so on) so pressure is only making the contribution of half a per cent at 40 bar. Here, h_f is 1087 kJ/kg and u_f is 1082 kJ/kg. The pressure influence has contributed 5 kJ/kg. As a further illustration, if liquid water at 90°C 1 bar is compressed isothermally to 90°C 100 bar, the change of specific enthalpy is about 1.6%, about 6 kJ/kg. This is roughly the same as a change of a couple of degrees or so in temperature, so the pressure and temperature effects are two or three orders of magnitude different. While these numbers are given for illustration only and have no other significance, they do indicate the sort of information which can be drawn from the routine property tables.

Third, both the internal energy and the enthalpy associated with the change of phase, u_{fg} and h_{fg}, fall progressively with rising pressure but these changes are accompanied by increases in the values of u_f and h_f as boiling temperature rises. The amount by which one change offsets the other varies as boiling point rises – refer back to PQ 7.11.

Notice that the total values u_g and h_g increase up to about 30 bar and then start to decrease. Although the sample table only goes up to 40 bar, you will see from a complete copy that this decrease continues. This is a reflection of the progressive change in water substance properties as temperatures and pressures rise. Since the changes are so progressive, they can be represented in a simple and characteristic fashion on a temperature–enthalpy (or temperature–internal energy) graph, such as that in Figure 7.6

Example 7.5 What are the specific internal energy and enthalpy values of steam at 8 bar, 94% dry?

From the tables, 8 bar corresponds to 170.4°C boiling point and the important information, all in kJ/kg, is:

$$u_f = 720; \ u_g = 2577; \ h_f = 721; \ h_{fg} = 2048 \ \text{and} \ h_g = 2769.$$

In fact, as we will see, h_g and the boiling temperature are spare pieces of information.

First the enthalpy, as that is the easier. When the steam is at boiling point, all the sensible heat h_f, 721 kJ/kg, will have been supplied, so all this contributes to the enthalpy of the wet steam. The dryness fraction is 94%, so only 94% of the change of phase enthalpy h_{fg}, i.e. $0.94 \times h_{fg}$, will have been absorbed by the steam. So the specific enthalpy of the wet steam, usually designated h kJ/kg with no subscripts, is

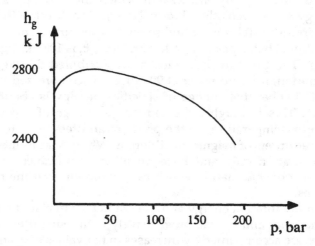

Figure 7.6 Progressive change of dry steam enthalpy h_g with pressure p.

$$h = 721 + (0.94 \times 2048) \text{ kJ/kg} = 2646.1 \text{ kJ/kg}.$$

Take care to avoid numerous decimal places, since the tables are normally to round numbers.

The internal energy calculation is identical in style but, again commonly, tables tend not to give values of u_{fg} as separate columns since those values are of limited use in industrial steam usage. Their omission is of little consequence, of course, as $u_{fg} = u_g - u_f$. Thus

$$u_{fg} = (2577 - 720) \text{ kJ/kg} = 1857 \text{ kJ/kg},$$

and the calculation as before will give

$$u = 720 + (0.94 \times 1857) \text{ kJ/kg} = 2465.6 \text{ kJ/kg}.$$

u for wet steam, in parallel to h, is presented without subscript.

Example 7.6 For the same steam, what is its specific volume?

The table gives v_g, the specific volume of dry steam as 0.2403 m^3/kg. Wet steam is modelled successfully, you recall, as a gas carrying some liquid. In this case, the wet steam is 94% dry gas and 6% liquid. The letter x (little x, lower case) is used commonly to identify this dryness factor, the fractional amount of dry gas in the total amount of wet steam.

Taking 1 kg of wet steam, x kg will be dry gas and thus $(1 - x)$ kg will be liquid. If the volume of 1 kg of dry gas is v_g (the specific volume), then the volume of x kg will be xv_g. Similarly, if the volume of 1 kg of liquid is v_f (the liquid specific volume), the volume of $(1 - x)$ kg will be $(1 - x)v_f$. All these are, of course, in m^3. Thus the volume v (no subscript) of 1 kg of wet steam is given by

$$v = xv_g + (1 - x)v_f.$$

This expression can be simplified in many cases. At modest pressures, say up to around 40 bar, the specific volume of liquid water is so much smaller than that of dry steam at boiling point that it can be ignored for all practical purposes. For instance, at 40 bar v_g is greater than v_f coincidentally by a factor of about 40, which may mean about a 1% error by this simplification. At other pressures and their relevant boiling points, the error will be different. For instance at 1 bar the ratio v_g/v_f is about 1600, at 10 bar it is about 170 and even at 100 bar it is still over 12. The information for these ratios is all contained in typical thermodynamic property tables, such as the ones already quoted.

It can, therefore, be argued that the liquid volume under these conditions is of little consequence. This conclusion is certainly acceptable under everyday industrial conditions, where steam dryness is usually in the upper part of its range, say 0.7 upwards. Thus, to a good approximation

$$v = xv_g.$$

Then, the specific volume of steam at 8 bar, 94% dry, is 0.94 × 0.2403 m³/kg, i.e. 0.2259 m³/kg. Since this is an approximation, it would be sensible to round the value to 0.226 at best and not claim any greater accuracy.

Take care to note that this is a good approximation only so long as the specific volume of the liquid is insignificant for the purposes of the calculation. If in doubt, then make an allowance.

This is easily done from a knowledge of the specific volume of water under a stated condition and its coefficient of cubical expansion or by reference to the standard property tables. The volume change of the liquid is dominated by temperature. For example, the specific volume of water at 1 bar is 0.0010 m³/kg near its freezing point. At 100°C, it is about 0.00105 m³/kg and at 300°C it is 0.00145 m³/kg or so.

The same water taken to around 1000 bar will be compressed by less than half of one per cent at room temperature and by about 13% at 300°C. That sort of pressure is a very unusual one in everyday industry, where levels may go up to but seldom exceed several tens of bars. So the specific volume of water for most applications is very much controlled by temperature rather than pressure.

By way of completeness, if the liquid water volume had been included in the answer to Example 7.6, v_f under those conditions is about 0.0011 m³/kg, so $(1 - x)v_f$ would be 0.000066 m³, to give an error of less than 0.03%.

Whilst the differences between liquid enthalpy and internal energy were small, when we look particularly at steam enthalpy and internal energy, then quite measurable differences arise. Compared to the liquid, the pV term for changes in the vapour is large. For instance, at 1 bar the difference between h_g and u_g is 169 kJ/kg, whereas the difference between h_f and u_f is negligible. At 6 bar the specific differences are 189 and 1 kJ/kg, and at 30 bar they are 200 and 4 kJ/kg.

Example 7.7 Look at the steam table values at 1 bar and deduce the pV contribution to dry steam specific enthalpy. Now prove it by looking at the p and v_g values. Why is there a difference? Do the same for dry steam at 10 bar.

Using the nomenclature with which you are now familiar,

$$h = u + pv.$$

Thus

$$pv = h - u$$

and directly from the data tables, $h = 2675$ kJ/kg and $u = 2506$ kJ/kg. These are, of course, for dry steam the values h_g and u_g at 1 bar. Thus

$$pv = 2675 - 2506 = 169 \text{ kJ/kg}.$$

Now look at the p and v values from the data table. p is 1 bar but we must convert this to fundamental units, 10^5 N/m². v_g is 1.694 m³. Thus

$$pv_g = 10^5 \times 1.694 = 169.4 \text{ kNm or } 169.4 \text{ kJ}.$$

The difference is essentially one of rounding up or rounding down for the table values. Additionally though, the 'starting point' of water at 0°C is the datum level of nil enthalpy. Real water at that temperature will have a real volume but it is negligible in this context.

You can check the data table for the values for 10 bar, from which you will see

$$pV = h_g - u_g = 2778 - 2584 = 194 \text{ kJ/kg,}$$

and

$$pv_g = (10 \times 10^5) \times 0.1944 = 194.4 \text{ kJ/kg.}$$

The difference is similar for similar reasons.

- **PQ 7.19** Do the same for a change of dry steam from 5 to 15 bar.

Just one last point about liquid and vapour properties from the tables – if you need any values intermediate between tabulated values, it is sufficiently accurate for most purposes to use *arithmetic or linear interpolation*. The next example shows how.

Example 7.8 What is the specific enthalpy of steam, 77% dry at 15.4 bar?

That pressure is not in typical tables – which cannot cover every possibility or they would be infinitely large – but there are values at 15 and 16 bar. For this wet steam, both h_f and h_{fg} are needed. Let h_{15} be the enthalpy of wet steam at 15 bar and h_{16} be it at 16 bar. Since

$$h = h_f + xh_{fg},$$

we have

$$h_{15} = 845 + 0.77 \times 1947 = 2344.19 \text{ kJ/kg}$$

$$h_{16} = 857 + 0.77 \times 1935 = 2346.95 \text{ kJ/kg.}$$

Note that the decimal places are retained, to round off at the end. The difference between the two values is 2346.95–2344.19, i.e. 2.76 kJ/kg. So the step between 15 and 16 bar adds 2.76 kJ/kg to the 15 bar amount, thus stepping up part way from 15 to 15.4 bar will add 0.4 times this to the 15 bar amount. (If the pressure had been 15.6 bar, then it would have added 0.6 times.) Thus,

$$h = 2344.19 + 0.4 \times 2.76 = 2345.29 \text{ kJ/kg.}$$

Bearing in mind that the table values are rounded a little, this needs rounding off, say to 2345.3 kJ/kg.

7.3 SUPERHEAT TABLES

Superheated steam is any steam which has exceeded the 'just dry' condition. As soon as the steam temperature exceeds the saturation temperature – the temperature at which water boils at a given pressure – the steam cannot be wet. At a given pressure, the temperature of steam cannot increase if the steam is wet, so any energy input goes towards increasing the steam dryness. Only when that has been completed can the steam temperature rise and this is the onset of the superheated condition. Superheated steam is a *gas*.

The steam tables or tables of thermodynamic properties will normally include values for superheated steam – steam that is heated *beyond* the point of dryness. Table 7.3 is a sample of superheat information from *Thermodynamic and Transport Properties of Fluids*, acknowledged earlier. Formats do vary from one publisher or source to another but the contents are similar. As before, for the present ignore the property identified as s_g.

Looking at the sample table, at the very top left is p(bar), the pressure, and t_s (°C), the saturation or boiling temperature at that pressure. Thus the vertical column of figures below these two give pressure and boiling point – for instance 1 bar (99.6°C) or 3 bar (133.5°C). Then across the top is t °C and a series of numbers – 50, 100, 150, 200 and so on. These numbers represent the actual operational temperature (°C) irrespective of pressure. They have nothing to do with the boiling points. They are read in conjunction with the columns of figures directly below them, as will be illustrated.

As an example, look at the horizontal row for 2 bar and start at the far left of that row.

- Pressure and temperature are noted as p bar and t_s °C. These are the corresponding boiling point and pressure values. At 2 bar, water boils at 120.2°C.
- The next column gives the specific properties v_g, u_g and h_g for dry

Table 7.3 Properties of superheated steam from 1 to 4 bar

p(bar) [t$_s$(°C)]	Dry steam properties		50	100	150	200	250	300	400	500
0	u=h−RT at p=0	u	2446	2517	2589	2662	2737	2812	2969	3132
		k	2595	2689	2784	2880	2978	3077	3280	3489
1 [99.6]	v$_g$ 1.694	v		1.696	1.937	2.173	2.406	2.639	3.103	3.565
	u$_g$ 2506	v		2506	2583	2659	2734	2811	2968	3131
	h$_g$ 2675	h		2676	2777	2876	2975	3075	3278	3488
	s$_g$ 7.359	s		7.360	7.614	8.033	8.215	8.543	8.834	
1.01325 [1000.0]	v$_g$ 1.673	v			1.912	2.145	2375	2.604	3.062	3.519
	u$_g$ 2506	u			2583	2659	734	811	968	3131
	h$_g$ 2676	h			2777	2876	2975	3075	3278	3488
	s$_g$ 7.355	s			7.608	7.828	8.027	8.209	8.209	8.828
1.5 [111.4]	v$_g$ 1.159	v			1.286	1.445	1.601	1.757	2.067	2.376
	u$_g$ 2519	u			2580	2656	2733	2809	2967	3131
	h$_g$ 2693	h			2773	2873	973	3073	3277	3488
	s$_g$ 7.223	s			7.420	7.643	7.843	8.027	8.355	8.646
2 [120.2]	v$_g$ 0.8856	v			0.9602	1.081	1.199	1.316	1.549	1.781
	u$_g$ 2530	u			2578	2655	2731	2809	2967	3131
	h$_g$ 2707	h			2770	2871	2971	3072	3277	3487
	s$_g$ 7.127	s			7.280	7.507	7.708	7.892	8.221	8.513
3 [133.5]	v$_g$ 0.6057	v			0.6342	0.7166	0.7965	0.8754	1.031	1.187
	u$_g$ 2544	u			2572	2651	2729	807	966	3130
	h$_g$ 2725	h			2762	2866	2968	3070	3275	3486
	s$_g$ 6.993	s			7.078	7.312	7.517	7.702	8.032	8.324
4 [143.6]	v$_g$ 0.4623	v			0.4710	0.5345	0.5953	0.6549	0.7725	0.8893
	u$_g$ 2554	u			2565	2648	2727	805	965	3129
	h$_g$ 739	h			2753	2862	2965	3067	3274	3485
	s$_g$ 6.897	s			6.929	7.172	7.379	7.566	7.898	8.191

v in m^3/kg; u in kJ/kg; h in kJ/kg; s in kJ/kg K.

steam at that pressure and temperature. You have already met this sort of information so they need no elaboration. Again, ignore the property identified as s_g.

- The remaining information tells how the dry gas values v_g, u_g and h_g change with temperature after boiling and whilst the pressure remains at the stated value.

Some numerical examples will show the way of using the superheat tables.

Example 7.9 For superheated steam at 2 bar, 250°C, what is its specific volume and specific enthalpy?

The left-hand column lists the pressure together with the boiling temperature at that pressure. Since the steam in this case is superheated, its temperature will be above the 2 bar tabulated boiling temperature of 120.2° C. Looking across the row of figures, the next column gives the specific values at the boiling point, the so-called saturation values as they refer to dry steam. The subsequent sets of figures refer to the superheated steam at whatever temperature is read down from the top. The blank spaces are simply explained – they are in the temperature columns which are below the boiling point for the pressure in question. Thus at 2 bar there is no superheat information under the 50°C and 100°C columns because the boiling point is 120.2°C.

Looking down from the 250°C heading, the fourth temperature column at the top of the table, gives the required information across from the 2 bar identifier. The letters have their standard meaning, so v is 1.199 m^3/kg and h is 2971 kJ/kg. As with wet steam, there are no subscripts to the identifying letters for superheated steam. The reason is the same – h_g and so on identifies a very particular dry saturated value but wet steam and superheated steam can cover quite a range of values.

Example 7.10 What is the specific internal energy of steam at 2 bar, 300°C?

First, check whether or not the steam is superheated. If the quoted temperature is higher than saturation (boiling point at that pressure) temperature, then it is superheated.

Looking at the 2 bar horizontal rows of values, identify the specific internal energy one. It is marked u_g for dry steam (2530 kJ/kg) and plain u for the superheated values. Look across this u row and look down from the 300°C at the top of the table. This gives u for steam at 2 bar, 300°C as 2809 kJ/kg.

Example 7.11 What would be the specific enthalpy at 2 bar, 270°C?

You will see that there is no 270°C temperature column in the table so there is no direct reading of specific enthalpy. The temperature of 270°C lies between two tabulated values, 250°C and 300°C, so there has to be interpolation of the properties – the estimation of a value which lies between two others – just as was done in Example 7.8 for wet steam. The values change progressively so, strictly speaking, a graph should be drawn carefully using the established points and thus allowing the selection of intermediate ones.

However, for most practical or industrial purposes it is adequate to do a simple *linear interpolation* – assume regular step changes between one value and the next just as in Example 7.8 – especially as the commonly available tables are rounded numbers. In this case, there is a 50-degC step between 250°C and 300°C and 270°C is 20 deg C along that difference step so the values can be assumed to change by 20/50 of the total change.

The specific enthalpy at 2 bar, 300°C is 3072 kJ/kg and at 250°C, 2 bar is 2971 kJ/kg. Note that both temperature and pressure are quoted to identify the steam completely. The change in specific enthalpy from 250°C to 300°C is therefore

$$3072 - 2971 = 101 \text{ kJ/kg,}$$

so the change from 250°C to 270°C is 20/50 (0.4) of this i.e.

$$0.4 \times 101 = 40.4 \text{ kJ/kg.}$$

The specific enthalpy of superheated steam at 2 bar, 270°C is thus the value at 250°C plus the increase due to the extra 20 deg C, that is

$$2971 + 40.4 = 3011.4 \text{ kJ/kg.}$$

• PQ 7.20 Go through the similar calculation for specific volume.

Two points arise from this last PQ. First, take care to be quite sure and to record whether the values are rising or falling wherever you interpolate. In the answer to PQ 7.20, the changes of specific volume were written as + 0.117 and +0.0468 m³/kg. Depending upon the change of conditions, the change of property value may not always be positive.

• PQ 7.21 What is the specific volume of saturated steam at 250°C, 2.6 bar? Use linear interpolation.

Second, a reminder to note the use of the term '*deg C*'. By convention, the symbol °C is generally taken to refer to a **particular** temperature, such as 270°C, and the term 'deg C' is taken to refer to a **difference** of temperature – the difference between 250°C and 300°C is 50 deg C. The same distinction does not arise in the use of Kelvin, since the simple, plain abbreviation K is used for actual temperatures and for temperature differences.

This linear interpolation can be continued for values which lie between tabulated temperatures and tabulated pressures, such as 320°C and 8.2 bar. The principle is exactly the same but the process has to be repeated, for instance by interpolating values at 320°C for 8 bar and 9 bar, then re-interpolating for 8.2 bar. Should you try that one, the values are $v = 0.3210$ m³/kg, $u = 2829.7$ kJ/kg and $h = 3098.6$ kJ/kg.

Thus the thermodynamic property tables – the steam tables – are straightforward to use and save much calculation. Although they have been illustrated here as applicable to steam, there are parallel tables for many other materials, particularly but not exclusively those materials used in the refrigeration and air-conditioning world. As noted, the tables proper title is *Thermodynamic and Transport Properties of Fluids* and a good set of tables will contain a very wide range of useful information. Their successful use is simply a matter of familiarity.

7.4 TWO SPECIAL PROPERTY POINTS

The conversion of a liquid to a gas via the vapour phase is a progressive process. It is part of the interrelationship between solids, liquids and gases, and its progress is dependent upon both temperature and pressure. We have already seen some of this when looking at the example of the influence of pressure on the boiling point and the influence of temperature on various properties of both liquid and gaseous water substance.

Still using steam as a good illustrative example – but only for that purpose – most of what has been said relates to industrial levels of operation in an everyday fashion. That, of course, does not limit the range of study of solids and fluids, it is simply a convenient focus. If, though, we start to take the materials to pressures and temperatures or combinations of these which are seldom met in everyday industry, then there are other phenomena to consider and two of them are particularly appropriate here. They are called the 'triple point' and the 'critical point'

We have seen that whether water is a liquid or a vapour or a gas depends upon both temperature and pressure. Figure 7.7 and 7.8 illustrate the sort of pattern that this temperature and pressure dependence makes, the location of the equilibrium lines depending upon both properties.

When looking at the pressure, temperature, enthalpy and enthalpy of change of phase (latent heat) for water substance, it was noted that the latent heat demand fell as the pressure rose. Whilst our sample table did not go that far, a full set would show that this trend continued until the enthalpy of change of phase from liquid to gas fell to *zero*. This actually occurs at 221.2 bar, 374.15°C and this point is called the **critical point**. It is marked as C on Figure 7.8. At this critical condition, there would be no wet steam part of the liquid-to-gas change. At temperatures above the critical value there is no effective difference between the gaseous and the liquid phase, and gases cannot be liquefied, however high the pressure. Look at the so-called envelope which the extremes of the vapour plateaux make on Figure 7.8.

To the right the line shows where wet steam becomes gas at various pressures. To the left are the points where liquid becomes wet steam. At

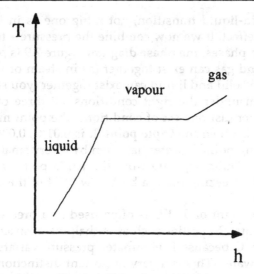

Figure 7.7 Temperature *T* and enthalpy *h* are related.

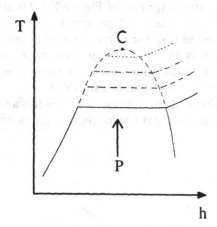

Figure 7.8 Pressure moves the line.

the critical point therefore, when wet steam does not exist, the liquid and the gas appear the same. Although this may sound strange at first, remember that they are both fluids and that the normal recognition of the fluid states is at everyday temperature and pressure. The critical point is certainly not an everyday condition.

Just as an *equilibrium line* can be drawn between liquid and gas or liquid and vapour, so a similar one can be drawn between liquid and solid. Continuing to use water as the best common example, it is easy to think of the transition from solid to liquid as being at 0°C, the temperature when we usually meet ice. There is however a pressure effect on this solid-liquid transition, not a big one as in liquids and gases but still an effect. If we now combine the pressure – temperature effects on all three phases, the **phase diagram** Figure 7.9 is produced.

Just as liquid and gas can exist together (as in steam or water at the boiling point) and solid and liquid can exist together (you see this on a winter's day) then under the right conditions, all three can co-exist. This happens under just one set of conditions, the point marked T on Figure 7.9, which is called the **triple point**. It is 0.01°C, 0.0061 bar. You see that the freezing point of water has risen by a very small amount at this low pressure. Quite separate from the triple point, a very high pressure moves the freezing point a bit below 0°C but the changes are very small indeed.

This fixed triple point of 0.01°C is often used as a **precise** *datum* for many measurements. Properties such as enthalpy are, strictly, referred to it rather than 0°C because it eliminates pressure variations on the freezing point of water. This is a very important distinction in precise measurements but it is realistic to use 0°C generally, as already noted.

As a final comment, the diagrams of Figure 7.7–7.9 are examples of the ways in which phase relationships can be presented. They are common but they are in no way exclusive. So long as conditions can be identified completely, it is only a matter of convenience which properties are used for this purpose. Most reference texts or more advanced books on thermodynamics will include variations.

For vapours generally and their associated liquids and gases, there are plenty of thermodynamic data recorded in tables and, latterly, in

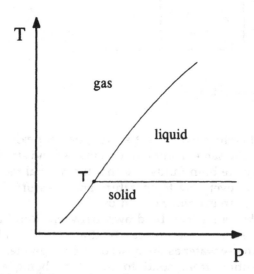

Figure 7.9 Water substance phase diagram.

computer program form. For most applications therefore, calculations to do with vapours are more of the data processing type than of the first principles type and the data have direct industrial value.

PROGRESS QUESTION ANSWERS

PQ 7.1

You may have to go back to a school physics book or similar for this. A vapour is produced as a liquid boils and changes into a gas. It is an intermediate stage as it is neither liquid nor gas – it is really a mixture of the two. Energy, usually as heat, is required to change a liquid into a gas – that is the complete change. If the change is only partly complete, that is, not all the necessary heat has been added, then a vapour results. Obviously, a gas condensing to a liquid also provides vapour.

PQ 7.2

Check back if necessary – they have already been defined and used.

PQ 7.3

Yes, many properties are connected – think of $pV = mRT$ – so no, you do not need to know every separate property to define a state.

PQ 7.4

On the basis of technical common sense, a vapour phase can be defined, even though it is effectively part liquid, part gas. If vapour is the progressively changing intermediate stage between liquid and gas, it can be identified and it will have properties which define it. We will meet some soon. Although there may be some lack of physical homogeneity if it is changing progressively, there can thus be a common sense use of the term 'vapour phase'.

PQ 7.5

For anything to separate from anything else, some energy has to be used; the latent heat is that energy. It is increasing the molecular kinetic energy so that the molecules can escape from the liquid into the gaseous phase.

PQ 7.6

The figure would be 0.75 or 75%. Both decimal number and per cent are used interchangeably.

PQ 7.7

The molecules also have to overcome the resistance of the surrounding atmosphere to anything else entering it. So the higher the surrounding atmospheric pressure, the more energy the molecules need to enter the atmosphere. Higher temperature means higher kinetic energy, so a higher atmospheric pressure means that the boiling will take place at a higher temperature. The converse also applies.

PQ 7.8

Here are my two examples. A puddle dries quicker on a hot day than on a cool day. You can usually smell a volatile liquid – petrol for instance – more noticeably on a warm day than on a cool one.

PQ 7.9

If the surroundings pressure exceeds the vapour pressure, the liquid will not boil. If the liquid temperature rises, then a point is reached where vapour pressure and surroundings pressure are equal and the liquid is at its boiling point. Any further energy addition – the latent heat – will allow boiling to proceed, with the liquid molecules escaping into the gaseous phase. Change the surroundings pressure and the vapour pressure demand changes, hence boiling temperature and surroundings pressure are inseparable.

PQ 7.10

The latent heat of evaporation or heat of change of phase is the amount of energy, usually expressed as kJ/kg, required to change liquid water at its boiling point completely into gas at the same temperature.

PQ 7.11

Sensible heat raises the water to its boiling point and the latent heat converts liquid to gas. If the pressure rises, so does the boiling temperature and thus the sensible heat demand. This sensible heat is used to raise the molecular kinetic energy, parallel to a gas internal energy. Thus when boiling point is reached, the extra energy (latent heat) required for the molecules to escape the liquid is reduced because some has already been added as sensible heat. If pressure is reduced, the opposite happens.

PQ 7.12

I don't know. If the steam was generated at atmospheric pressure, boiling point 100°C or thereabouts, then the steam at 112°C must be superheated – it is well above the boiling point. If the steam was generated at 1.5 bar, boiling point 112°C (look back!) then it is either wet steam or, at best, just dry (saturated, dry saturated). It cannot be superheated.

PQ 7.13

You have already met it and the clue is in the constant pressure – it is enthalpy.

PQ 7.14

First the easy ones. If steam is 100% dry or superheated, then it is a gas and the same definition as for any other gas will apply.

Now for wet steam. If the definition is the amount of heat required to raise the temperature by 1K, then it does not work. If I add heat to wet steam, it makes it more dry – it does not raise its temperature. So to make sense of raising the temperature whilst keeping the dryness fraction the same, it follows that there must be an accompanying rise of pressure. Look at the next example.

PQ 7.15

It is a number which could have value in very special circumstances but as most steam operations, particularly with wet steam, are at constant pressure, then the only reason to add energy to wet steam is to increase dryness. Apart from that, there is the obvious conflict if you try to draw a parallel with gases – enthalpy for gases is mc_pT and we have not derived a specific heat capacity at constant pressure, have we?

PQ 7.16

p – pressure, t – temperature, v – specific volume, u – specific internal energy, h – specific enthalpy. Check back if any of these terms are unfamiliar.

PQ 7.17

Since h_g is the specific enthalpy of dry steam, h_f is the specific enthalpy of the liquid water substance up to the boiling point, both measured from the datum, and h_{fg} is the specific latent heat; then

$$h_g = h_f + h_{fg}.$$

That is, the dry steam enthalpy is equal to the liquid water enthalpy plus the phase change enthalpy.

The parallel for internal energy would be

$$u_g = u_f + u_{fg}$$

so at 1 bar

$$u_{fg} = 2506 - 417 = 2089 \text{ kJ/kg}.$$

Since enthalpy rather than internal energy dominates the industrial use of common tables, it is unusual to have a separate u_{fg} column.

PQ 7.18

The pV term effect, which is the pressure influence part of enthalpy, is so small that it leads to that second point.

PQ 7.19

From the data tables, the specific enthalpy of dry steam at 5 bar is 2749 kJ/kg and at 15 bar is 2792 kJ/kg. The increase is 43 kJ/kg. The specific internal energies are 2562 and 2595 kJ/kg respectively. The increase is 33 kJ/kg.

Therefore, the increase of the $p V$ contribution to the specific enthalpy of dry steam between these two pressures is $43 - 33 = 10$ kJ/kg.

PQ 7.20

Using v with no subscript for the specific volume

v at 250°C, 2 bar	=	1.199 m³/kg
v at 300°C, 2 bar	=	1.316 m³/kg,

so that change for the 50 deg C step = + 0.117 m³/kg.

Thus,

change for the 20 deg C step = + 0.0468 m³/kg
and v at 270°C, 2 bar = 1.199 + 0.0468
  = 1.246 m³/kg.

PQ 7.21

Without going through all the detail procedure, with which you are now familiar,

$$v \text{ at 2 bar, } 250°C = 1.199 \text{ m}^3/\text{kg}$$
$$v \text{ at 3 bar, } 250°C = 0.7965 \text{ m}^3/\text{kg,}$$
so that change of v from 2 to 3 bar $= -0.4025 \text{ m}^3/\text{kg}$.

Thus

$$\text{change of } v \text{ from 2 to 2.6 bar} = -0.2415 \text{ m}^3/\text{kg}$$
$$\text{and } v \text{ at 2.6 bar, } 250°C = 1.199 - 0.2415$$
$$= 0.9575 \text{ m}^3/\text{kg.}$$

TUTORIAL QUESTIONS

You will need access to a full set of tables, rather than the present text extracts, for some of these examples. Use the tables wherever appropriate.

7.1 Determine (a) from the tables and (b) from the ideal gas laws the specific volume of steam at 3.4 bar, 400°C. [0.9276 m³/kg; 0.9143 m³/kg]

7.2 What is (a) the specific enthalpy of liquid water substance at 30°C and (b) the increase in specific enthalpy when this liquid is used to make 70% dry steam at 80°C? [125.7 kJ/kg; 1825 kJ/kg]

7.3 What is the mean specific heat capacity of liquid water substance between (a) 0°C and 100°C (b) 0°C and 50°C (c) 50°C and 100°C? [4.191 kJ/kg K; 4.186 kJ/kg K; 4.196 kJ/kg K]

7.4 Superheated steam at 7 bar is heated from 400°C to 600°C. Show from table values that $\Delta h = \Delta u + \Delta(pV)$ in the usual nomenclature. [Look up relevant table values, noting that p is constant.]

7.5 For the same change, determine the value of R, the gas constant, for superheated steam in that range and compare it with the ideal gas value. [Value from tabular data 470 J/kg K; ideal gas, 462 J/kg k]

7.6 Ammonia liquid enters a commercial chiller cabinet at −18°C and leaves with 20K superheat at 3 bar. What is the specific enthalpy of the ammonia at exit and what has been the increase of specific enthalpy during the process? [1481.5 kJ/kg; 1382.7 kJ/kg]

Steady flow energy equation applied to liquids and vapours

8

8.1 SFEE – A REMINDER

This is a brief chapter but it makes use of some of the points arising in Chapters 5 and 7. Recall that the **steady flow energy equation**, in the usual notation, says

$$\Delta Q = \Delta W + \Delta E$$

or something similar. It is easy to write down, but there is detail needed to cover all the energy terms arising in ΔE:

- changes in kinetic energy, enthalpy and potential energy, plus losses;
- all in self-consistent units.

- PQ 8.1 Remind yourself of the units for both amounts of energy and amounts per unit time.

The steady flow energy equation (SFEE) was introduced in Chapter 5 by referring to working fluids at large, so it applies equally as well to liquids and vapours as it does to gases and their systems of use. It is worth taking a brief separate look at liquids and vapours, since there are one or two particular points to make. For much of that which follows, water substance will again be used to illustrate but, as before, this is in no way an exclusive treatment. Exactly the same techniques of calculation would be used for any other liquids or vapours but using, naturally, the properties of those other materials.

The thermodynamic tables, such as those mentioned in Chapter 7, and *Engineering Tables and Data* by Howatson, Lund and Todd, 2nd edn, Chapman and Hall, London, 1991, contain most of the necessary data for many common applications. Access to such tables will be needed in this chapter, as some data have to be used which are not in the sample tables of Chapter 7.

As this chapter is not primarily new material, a good way to deal with the topic is by worked example. Where appropriate, the worked examples will be developed to cover various relevant points in a little detail.

8.2 SFEE AND LIQUIDS

Over much of their working range, especially that of most industrial and commercial significance – a few tens of bars – many of the properties of liquids are not affected to any great extent by pressure changes alone. This more or less removes one variable. For example, the specific heat capacity of liquid water substance is about 0.13% higher at 10 bar than at 1 bar and about 0.74% higher at 100 bar than at 1 bar. While such differences should not be ignored, their importance depends entirely upon the accuracy of accompanying data and measurements.

Example 8.1 What is the specific enthalpy of liquid water h at 130°C?

Refer to the table of saturated water and steam properties in Table 7.2 and look down the t column until you find 130.0°C. You will find it against the pressure of 2.7 bar and you will see that h is 546 kJ/kg. This, strictly, is the specific enthalpy of liquid water above the datum, at its 2.7 bar boiling point. That is what the table records.

However – and this is significant – pressure has little effect on the specific heat capacity of liquid water over a range of a few tens of bars so it can have little effect on specific enthalpy over the same range. So whatever the applied pressure in this sort of range, the specific enthalpy of water at 130°C will be much the same, 546 kJ/kg. It makes no difference whether or not the water is at 2.7 bar or that 130°C is not the boiling point at other pressures – although it will have boiled already if the pressure is lower!!

So, if the pressure was 6 bar or 14 bar or 25 bar, h would be 546 kJ/kg within the limits of routine accuracy. For the sake of numbers, the pressure effect at 25 bar would add about one-tenth of one per cent, maybe half a kilojoule, which is the rounding-off limit to the whole number value of the table.

Since *pressure effects* on liquid properties are so small in the everyday industrial measurement context for most of the typical working range, this means that we only need to consider *one* specific heat capacity for liquids, rather than having two principal ones as for gases. As enthalpy is an important property and as the letter c is sometimes used to indicate velocity, the abbreviation c_p will be used here to indicate liquid specific heat capacity. That is an individual value at a given temperature, so it will still be necessary to find the mean value $\overline{c_p}$. The effect of temperature on specific heat capacity of liquids certainly **cannot** be ignored. While the variation for water is less than 1% up to about 100°C, thereafter, it becomes more important. At 200°C, for example, it has risen by over 6%.

Remember – pressure effects on properties do matter and should always be considered. However, the difference that pressure makes is very often far smaller than many other influences. Note in this example, however, that this is only the effect of pressure on the liquid property. The actual liquid pressure may be very important when doing an energy balance for the SFEE. A good everyday illustration of the value of liquid pressure in energy transformations is the garden hosepipe – water pressure is transformed into kinetic energy at the hose nozzle.

Some examples of the SFEE and liquids have already been met, for instance the waterfall in Chapter 5. There are many similar sorts of examples – water flowing through pipes for instance – but they are very close to isothermal and are more appropriate to the study of fluid mechanics than thermodynamics. Beyond this, the most common application of the SFEE to liquids alone is to fairly routine heating or cooling processes. In such cases, the SFEE here is usually dominated by the energy demands associated with an enthalpy transfer.

Example 8.2 Water flows down a short run of well insulated horizontal pipe of constant diameter at 0.15 kg/s, entering at 5 m/s. Electric elements in the pipe are used to raise the water temperature to 60°C from its inlet value of 20°C (Figure 8.1) What is the necessary energy input from the elements?

The information allows us to remove some parts of the SFEE. The pipe is horizontal, so there is no change of gravitational potential energy – the inlet and outlet are at the same height. The pipe is well insulated, so heat losses can be ignored. There is no external work being done, so this too is deleted.

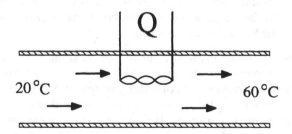

Figure 8.1 Pipe and heating element.

Strictly speaking, there will be some energy used in making the water flow through the pipe – that is, to overcome fluid flow resistance (Figure 8.2).

- **PQ 8.2 So, where does this energy reappear?**

Experience suggests that this resistance will be negligible over short lengths – think of rainwater running along a gutter. Always be prepared to draw on your experience, whatever the technical discipline! The pressure loss (difference in fluid pressure between inlet and outlet) required to maintain the flow is an indication of the friction involved, when such calculations are necessary.

So, the SFEE reduces to the energy going in, that is the energy supplied by the heating elements plus the enthalpy and the kinetic energy of the incoming water, equalling the energy going out, the enthalpy and kinetic energy of the outgoing water.

With the usual letters and remembering about $\dot{m}$ (m dot) and so on, the complete equation could be written as

$$Q + \dot{m}gz_1 + \dot{m}c_1^2/2 + \dot{m}c_pT_1$$
$$= \dot{m}gz_2 + \dot{m}c_2^2/2 + \dot{m}c_pT_2 + \text{losses} + \dot{W}.$$

However, this now reduces to

$$\dot{Q} + \dot{m}c_1^2/2 + \dot{m}c_pT_1 = \dot{m}c_2^2/2 + \dot{m}c_pT_2.$$

The question gives little information, other than the flow conditions, so some water properties are to be found from the tables. Values for specific heat capacities can be read for the temperatures in question, or even simpler, the specific enthalpies h_f can be taken.

- **PQ 8.3 What are the values of the specific enthalpy of the water before and after the heating element?**

As the water is warming, it will expand and, since the pipe is of constant diameter, the water must therefore accelerate. The coefficient of volume or cubical expansion is thus needed and again

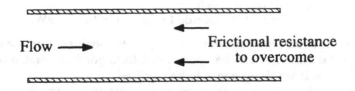

Flow ⟶   Frictional resistance
                              to overcome

Figure 8.2 Fluid resistance uses energy.

various data tables give either the direct value of the coefficient or specific volumes at various reference temperatures.

The coefficient is 0.44×10^{-6} m^3/m^3 K and two reference values are, at 0.01°C (the triple point or standard reference point), the specific volume is 1.000×10^{-3} m^3/kg and, at 100°C, 1.044×10^{-3} m^3/kg. Note that this is routine data table information.

• PQ 8.4 Calculate the specific volume of water at 20°C and 60°C.

The incoming velocity is given as 5 m/s so the exit velocity must be

$$5 \times (1.0264/1.0088) \text{ m/s} = 5.087 \text{ m/s}.$$

Remember to keep the rounding-off to the end. If those specific volumes had been rounded to 1.026 and 1.009, the new velocity would have been 5.084 m/s – not a large difference in this case but it could be important in others.

The SFEE now becomes, where Q is the energy input rate from the heating elements for the water flow rate of 0.15 kg/s and everything has been brought to consistent units,

$$Q + 0.15 \times 5^2/2 + 0.15 \times 83.9 \times 10^3 =$$
$$0.15 \times 5.087^2/2 + 0.15 \times 251.1 \times 10^3$$

$$Q + 1.875 + 12585 = 1.941 + 37665.$$

The enthalpy is in joules, not kilojoules, for consistency. Missing that conversion is an easy mistake so take care. **Always** use the basic units if you have doubt.

Notice the big difference in the components. The enthalpy increase consumes 25 080 J (37 665–12 585) for each 0.15 kg or 167.2 kJ/kg. The kinetic energy increase absorbs 0.066 J (1.941–1.875) for each 0.15 kg or 0.44 J/kg.

Obviously in this example the kinetic energy demand is negligible. It isn't even in the same range as the number of decimal places in the enthalpy, 167.2 kJ/kg compared to 0.44 J/kg. It is important though that all terms are considered in a reasonable manner if there is any doubt whatsoever about their likely contribution to the whole.

Returning to this case, the energy demand is

$$Q = 25\ 080 + 0.066$$
$$= 25.08 \text{ kW, rounded reasonably to 25.1 kW.}$$

This has been a long answer but the points considered on the way are important and need to be appreciated for a good understanding. Go back now if anything needs to be clarified.

As a relevant aside, an alternative route to the specific enthalpy values would be by finding specific heat capacities, which many tables give. A word of caution here. Your tables may give the specific heats to greater accuracy, perhaps as 4.183 and 4.185 kJ/kg K respectively. These are however the specific heat capacities at those temperatures, not the mean specific heat capacities between datum and the stated temperature. A simple multiplication of the specific heat capacity and the temperature rise above zero datum will **not** therefore give a true value of the water enthalpy at that temperature.

For instance, the specific enthalpies calculated by multiplying the specific heat capacity by temperature above datum are 83.7 kJ/kg and 251.2 kJ/Kg at 20°C and 60°C respectively. Table values are 83.9 and 251.1 kJ/kg respectively. In this case, the difference is small but under other circumstances it could be very important.

If your tables give a mean specific heat capacity from zero to a stated temperature then that can be used. In either case, reading h or the mean value of $\overline{c_p}$, someone has already done the hard work for us in the tables.

● PQ 8.5 Explain why the values of h read at 20°C and 60°C are applicable. 20°C is the boiling point of water at 0.0234 bar and 60°C at 0.1992 bar. Neither value is taken from a column of values relating to the actual pressure of water in the pipe – which is not given anyway.

● PQ 8.6 From the given values h at 20°C and 60°C, estimate a mean value of the specific heat capacity of liquid water substance between 0°C and 60°C and between 20°C and 60°C.

8.3 SFEE AND VAPOURS

Perhaps the most important individual feature of the use of a vapour for any process is that it may yield or consume energy without changing its temperature or pressure. The reason is, of course, that the **dryness fraction** plays such a vital part. If energy is lost or gained, the first thing to alter is the dryness fraction. Only when that has reached 0% (the vapour has returned to its liquid state) or 100% (the vapour is dry, it is a true gas) may other changes occur. This shows particularly in SFEE problems and much of the industrial use of vapours, especially steam, is in the SFEE mode.

Example 8.3 A steam turbine is a machine which uses steam as the working fluid. In action and design, it is similar to the turbine part of the gas turbine which we met in Chapter 5. The flowing steam turns bladed wheels so that some of the steam's energy (*enthalpy*) is used to do work. The size and complexity of the steam turbine ranges from the simplest possible configuration (Figure 8.3) to the major units used in electricity generation.

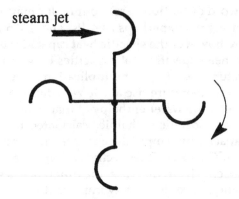

Figure 8.3 The simplest steam turbine.

Steam at 20 bar, 400°C is travelling at 40 m/s as it enters an experimental steam turbine. It leaves the turbine at 2 bar, 16 m/s and 0.3 dryness fraction, the exit port being 1.5 m above the inlet.

Although well insulated, heat losses do occur and they account for 1% of the inlet enthalpy. What is the specific work rate of the turbine, assuming steady flow conditions?

No overall flow rate is given for the steam but the question asks for the specific work rate – work output rate per kg of working fluid. So, on the basis of 1 kg/s of steam, the amount of work per second is required. Without recording every detail, with which you are now familiar, and taking information as required from the data tables:

Ingoing energy per second:
      steam enthalpy   from the superheat tables = 3248 kJ
      kinetic energy $mc_1^2/2 = 1 \times 40^2/2$ = 800 J.

Outgoing energy per second:
      steam enthalpy   from $h_f + x h_{fg}$ = 1165.6 kJ
      kinetic energy $mc_2^2/2 = 1 \times 16^2/2$ = 128 J
      potential energy change $mgh = 1 \times 9.81 \times 1.5$ = 14.7 J
      losses   1% of ingoing enthalpy = 32.48 kJ

Notice carefully that some values are in kJ and others are in J !!

$$W = 3248 + 0.8 - 1165.6 - 0.128 - 0.0147 - 32.5$$
$$= 2050.6 \text{ kNm/s} = 2.05 \text{ MW}.$$

As noted earlier, the principles illustrated by reference to water substance vapour, i.e. steam, apply equally well to other materials. Only the property values have to be changed. The final example involves ammonia, used for many years as a refrigerant because of its low boiling point of around −33°C at 1 bar. As an item of interest, its *critical point* is just over 130°C at 113 bar, compared to water of nearly 375°C at 221 bar.

There is an important detail when dealing with low-temperature materials, that of the datum for properties such as enthalpy and internal energy. The common datum for many materials is 0°C (0.01°C) because it is a common temperature, apart from being the freezing point of water. Where materials are used habitually below 0°C, the use of that as the datum would mean that there would be a lot of negative values to determine and apply. A lower datum temperature is therefore used for refrigerants, −40°C being common. That is a very handy choice because −40°C is also −40° F.

Recall, though, that datum levels are only levels of convenience for routine data application and calculation. They are **arbitrary temperatures** where some properties are assumed to be zero, anything below the datum being negative. The only true basis is still absolute zero and that should be used if you have any doubts.

● PQ 8.7 A quick review. My steam tables say that at a saturation temperature of 0.01°C the saturation pressure is 0.0061 bar. h_f is zero, h_{fg} is 2501 kJ/kg and h_g is 2501 kJ/kg. What does that all mean?

Example 8.4 A simple water chiller works by ammonia being drawn into a pump and then being pumped through pipes through a well insulated water container. The ammonia flow rate is 0.22 kg/s and the water 2.0 kg/s.

The water, c_p 4.17 kJ/kg K, is reduced from 15°C to 5°C. System losses are negligible, as are changes in ammonia kinetic and gravitational potential energies, compared to the enthalpy changes.

If the ammonia is drawn steadily into the pump at −10°C, 0.7 dry and leaves the chiller at 0°C dry saturated, what is the work input rate from the pump to the ammonia? The following ammonia data apply. At −10°C, $h_f = 135.4$, $h_g = 1433$; at 0°C, $h_f = 181.2$, $h_g = 1444.4$ kJ/kg.

Regard the flowing ammonia as a *system*. The inputs to the ammonia system are the ingoing enthalpy of the ammonia, the work done on it by the pump and the enthalpy supplied by the water which the ammonia is cooling. The output is the single one of the enthalpy of the outgoing ammonia, since all other items are negligible.

Note carefully that the work input is only that work actually going into the ammonia. The pump itself will be consuming more energy than that but the pump efficiency means that not all will appear in the ammonia.

- **PQ 8.8** So what happens to the pump energy requirement that does not appear in the ammonia?

In real life, pump efficiencies are well established and the manufacturers normally supply any required performance data.

Now apply the SFEE, as we are told that the ammonia flows steadily. First the ingoing energies:

water enthalpy change = 2 kg/s × 4.17 kJ/kg K × (15 − 5)K
= 83.4 kW;

ingoing ammonia enthalpy
= 0.22 kg/s × [135.4 + 0.7 (1433 − 135.4)] kJ/kg
= 229.62 kW.

The sole outgoing energy is that of ammonia enthalpy:

outgoing ammonia enthalpy = 0.22 kg/s × 1444.4 kJ/kg
= 317.77 kW.

Thus pump work – to be found

$$W = 317.77 - (83.4 + 229.62) \text{ kW}$$
$$= 4.75 \text{ kW}$$

In some abbreviated tables, it is common to quote only h_f and h_g, leaving out h_{fg}, for some materials. That is the case here for ammonia so that the enthalpy of the wet vapour had to be found from

$$h_{fg} = h_g - h_f$$

thus

$$h = h_f + x (h_g - h_f)$$

For the dry vapour, of course, h_g was exactly the figure required.

As indicated at the beginning, this chapter has been brief because the principles had already been established in Chapters 5 and 7. The steady flow energy equation is a fundamental one, so it can be applied widely with confidence.

PROGRESS QUESTION ANSWERS

PQ 8.1

While the individual energy contributions may have differing initial units, they must come down to a common base and the usual ones are J (= Nm) for the amounts and W (= J/s = Nm/s) for flow rates or amounts per unit time. Depending upon the numbers involved, these may be kJ and kW.

PQ 8.2

As heat. The resistance may be in the body of the liquid, like the energy needed to stir paint or as friction at the wall. All friction ultimately reappears as heat, whether it is water in pipes or brake shoes slowing a car. Note the words on flow work in Appendix B.

PQ 8.3

Refer back to Example 8.1. You will need to use the complete property tables, not just the sample in Chapter 7. In the complete tables, there is direct tabulation of h_f against temperature for temperatures up to 100°C.

At 20°C and 60°C respectively the values are 83.9 and 251.1 kJ/kg.

PQ 8.4

You will recall the way of doing this from school physics. The specific volume rises with temperature according to the normal sort of expansion relationship

$$V_T = V_o (1 + AT),$$

where A is the coefficient and T the temperature rise from volume V_o to volume V_T.

Thus the specific volume at 20°C is 1.0088×10^{-3} m³/kg, and at 60°C it is 1.0264×10^{-3} m³/kg.

For completeness, arithmetic interpolation between the two quoted reference values gives exactly the same answer.

PQ 8.5

The values can be used because the pressure effect on specific enthalpy is very small and can be neglected for most common industrial applications. Thus the specific enthalpy of liquid water at 20°C will be virtually the same at 0.01 bar, 1 bar and 10 bar. So it does not matter in everyday practice that the water is not at its boiling point nor that the tabled pressure is not the operational pressure. Note that this is only a statement that this property is not affected substantially by pressure over the quoted range. It does not dismiss pressure from anything else.

PQ 8.6

Enthalpy and specific heat capacity are related in the form

$$H = m \times c_p \times T \text{ or } \Delta H = m \times c_p \times \Delta T \text{ or similar.}$$

Thus between 0°C and 60°C

$$251.1 = 1 \times \overline{c_p} \times (60 - 0),$$
$$\text{so that } \overline{c_p} = 4.185 \text{ kJ/kg K.}$$

Similarly between 20°C and 60°C

$$251.1 - 83.9 = 1 \times \overline{c_p} \times (60 - 20),$$
$$\text{so that } \overline{c_p} = 4.180 \text{ kJ/kg K.}$$

PQ 8.7

Saturation in this case means boiling point, so the boiling temperature of water at 0.0061 bar is 0.01°C. The specific enthalpy of liquid water h_f at that point is zero. If I want to change 1 kg of liquid water into dry (dry saturated) steam at that temperature and, of course, at that pressure, then the latent heat of evaporation or enthalpy of change of phase h_{fg} is 2501 kJ/kg.

Since $h_g = h_f + h_{fg}$, then at this point h_g and h_{fg} are equal values.

PQ 8.8

Any energy which does not reappear in the working fluid must have been dissipated to the surroundings. It will include, for instance, friction in bearings which ultimately is given out from the pump casing or body as heat.

TUTORIAL QUESTIONS

The principles of the SFEE to apply are those already met and used in Chapter 5 and the data are processed as in Chapter 7. Any differences are mainly in sourcing the information.

8.1

(a) What is the specific volume of water at 100 bar, 100°C?
$$[1.038 \times 10^{-3} \text{ m}^3/\text{kg}]$$

(b) Water is admitted slowly at 15°C to a heating and pressurizing device (such as an industrial pressure washer) and leaves at 100 bar, 100°C and 100 m/s. What is the specific enthalpy increase, the exit specific kinetic energy and the ratio of these two values? [363.2 kJ/kg; 5 kJ/kg; 72.6:1]

8.2 In older steam engines, most component and working fluid velocities were low and the inlet and outlet steam ports were in or near the
same horizontal plane. In one such engine, steam was admitted at
3 kg/s, 4 bar, 150°C and exhausted at 1.2 bar, 70% dry.

If casual heat losses were equivalent to 5% of the ingoing enthalpy,
what was the thermal efficiency (conversion of heat energy supply to
useful work energy) of the engine and the work output rate?

[22%; 1.817 MW]

8.3 Steam is admitted at 10 bar, 300°C through a 30 cm diameter inlet to
an experimental turbine. It leaves at 2 bar, 130°C through a duct 80 cm
in diameter. The flow rate is 15 kg/s and the outlet duct is 3 m above the
inlet. General losses account for 80 kW.

This turbine powers an electricity generator of 95% (work supply to
electricity output) conversion efficiency. What is the rate of electricity
generation?   [4.56 MW]

8.4 Water is pumped from a static reservoir at 3 kg/s upwards through
a 4 m long vertical pipe, the pump delivering 250 W. An electric heating
element rated at 1 kW is fitted into the pipe. If the water temperature
rises by 0.05 K and the total losses from this pump + pipe + heating
element system are 0.5 kW, what is the outlet velocity of the water?
Take the specific heat capacity of water as 4.18 kJ/kg K.

[About 1.88 m/s]

A first look at entropy and the Second Law of Thermodynamics

9

9.1 ENTROPY

Most thermodynamic properties can be measured fairly readily. Pressure with a gauge and temperature with a thermometer, for instance, are straightforward and the interpretation of the gauge or the thermometer is quite clear. Some properties may need a controlled laboratory test, like specific heat capacity. Others need deriving from combinations of properties – a heat flow meter is one example, combining thermal properties and temperatures to give another property. Even where a direct measure is not always practicable, the typical thermodynamic property can be sensed or appreciated. The assessment of work transfer in some cases may need derivation rather than simple measurement but the idea of work transfer is still understood without difficulty (Figure 9.1).

We are coming to one property however – **entropy** – which, while being a true thermodynamic property, cannot be measured by a simple

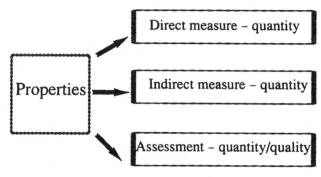

Figure 9.1 Not all properties can be measured simply.

device. There is no 'entropy meter'. One of the reasons is that most properties are to do with quantity, or quantity change, almost exclusively. Entropy can have numbers attached to it but it is also to do with quality – the **quality of energy**, how usable it is.

Once quality enters an assessment – as in saying 'What a beautiful view' or 'What a good car' – then the assessment is not so simple. These few words are intended to set the scene, for we have yet to deal with entropy and yet to try to define it. They are a little warning – we are going to meet something that is quite useful but which is rather abstract and which requires very careful understanding. A good start can be made by looking at something which **can** be measured and appreciated simply – **energy**.

Energy can be changed from one form into another – heat can be used to do work, potential energy can be converted into kinetic, electricity can generate heat and so on. Since, generally, energy is **quantifiable**, that is it is easy to measure given the right equipment, then it is not too difficult to estimate how much energy is needed for a particular purpose.

● PQ 9.1 Why estimate? Why not an accurate answer? Think perhaps of what portion of the fuel in a car is actually used to drive the car forward.

Such a calculation to estimate the amount of energy needed for conversion into the form needed for a given task will be a mixture of the analytical and the empirical. The **analytical** will be the accurate calculation of energy demand for a carefully specified task – what work is done if a mass of 150 kg is raised through 2 m, for instance. The **empirical** will include the conversion efficiency – what is the required electrical supply to the hoist that will lift the mass? However, even that calculation makes an automatic assumption that the energy we are hoping to use will be *available* in the necessary form.

Taking the heat-into-work change as an example, then the process of using heat to do work is simple and well established. It arose in Chapter 3, the non-flow energy equation. Heat up some gas held by a piston in a cylinder, so that the gas expands and forces the piston down the cylinder. It is quite easy to connect something to this piston so that its movement does some useful external work. That's basically how car and lorry engines work (Figure 9.2).

Figure 9.2 Usable heat becoming external work.

This is easy to say and, like the calculation, it makes the tacit assumption that the heat is not just simply there but that it is available for the job. Those last dozen words may seem contradictory – if the heat is there, surely it is available? To answer this, we must look at the **quality** – for want of a better expression – of the heat.

As an example, take two very well insulated pieces of mild steel, one of 5 kg and the other of 100 kg. Let's say that they are both at 0°C to start. Add to each, by some heat transfer process, 960 kJ and assume that an even temperature is achieved throughout each piece.

● PQ 9.2 Taking the mean specific heat capacity of mild steel for this example as 0.48 kJ/kg K, what are the temperatures achieved in each piece? Could I use either of these now-hotter pieces of steel to boil some water at 1 bar?

This use of the word quality is a bit loose in the scientific sense but its use here is obvious. The term '*low grade heat*' is often used to describe heat which is available at relatively low temperatures and which is of little use industrially or commercially.

So, we have the same amount of heat being used – 960 kJ – but the resultant temperatures are quite different. In one case, the 960 kJ is representative of so-called *low grade heat* – meaning for our purposes low temperature – and this would not boil any water at 1 bar. In the other case it is '*high grade heat*', again for our purposes meaning high temperature and very usable. Thus a heat source has to be specified as adequate in both quantity and quality to do a job (Figure 9.3).

As an aside, one of the most important problems in energy conservation is what to do with low grade heat, if anything of value can ever be done with it in large quantities. There is a lot of it about, notably because it represents the ultimate fate of much energy usage. Friction in bearings, heat loss from houses, exhaust from lorry engines, all are examples of the dissipation of low grade heat.

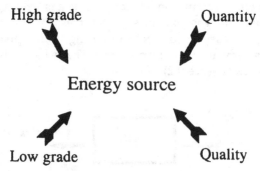

Figure 9.3 A heat source has to be usable.

Thus not only the quantity but also the quality of energy must be specified in some way which indicates its availability for use. In some cases this is simple, such as in a furnace where something is being heated. Fuel is burnt in the furnace, making flame and hot gases which are at a design temperature higher than the furnace contents, so that the contents will be heated. Heat transfer takes place and the furnace temperature is measured and then maintained by continued addition of fuel. The calculations which tell how much fuel is needed and at what temperature for this furnace process are fairly simple, so the energy quality is covered. The fuel energy is released as heat and is used as heat.

In other circumstances, especially where heat energy is transformed into work, it may not be quite so easy to cover this question of energy quality in a simple way, such as by measuring and maintaining temperature. Fortunately, a **thermodynamic property** called **entropy** has been introduced so that energy quality and availability can be studied from first principles. Take a careful note of the spelling, so that it does not become confused with enthalpy. They are separate properties and are used separately but they may also be used together!

● PQ 9.3 Now that we have met entropy, go back to the beginning of this chapter and review what was said about it.

For engineering needs, however, it is generally insufficient to describe a property in qualitative terms alone. Engineering suggests calculated values, quantified designs and so on. Thus if entropy is going to be a truly useful thermodynamic property, there has to be some way of using it with numbers. Remember though that there is no entropy meter – the numbers have to be derived.

9.2 ENERGY AND EFFICIENCY

If we are going to talk about energy usage, often called energy conversion, then the question of **efficiency** comes up.

● PQ 9.4 Give a simple definition of efficiency that would fit any energy conversion process.

Take as an example for looking at efficiency an engine, changing heat (maybe by fuel combustion) into work. There are two bases on which an operational efficiency can be assessed – relative and absolute.

Relative here means compared to similar engines, to see whether a redesign or a new component affects the efficiency. This comparison is good for immediate practical purposes, such as testing a new carburettor, but it is quite limiting. For instance, if the original engine was poor and the modified one less poor, the relative change may be encouraging but you still have a poor engine.

So there really needs to be some measure of efficiency compared to the ideal engine, that is an **absolute** rather than a relative measure (Figure 9.4). In this example, perhaps an idealized engine would be specified – not a real working one but one whose real specification has been given some ideal features. It's the engine equivalent of an ideal gas.

For instance, some conditions are laid down such as: all the fuel (the heat provider for the heat-to-work conversion) burns perfectly; there are no frictional losses; the maximum amount of heat for the operational conditions is used and so on. The ideal can never be achieved but it helps to find the real shortcomings and to avoid any misunderstanding of relative improvements. We will meet this again when looking at engine cycles.

As engineering thermodynamics is about the thermodynamic background to real operations, then some **idealised model** of a conversion process or a machine or an engine for efficiency studies is the one to look at. Start with a pretend engine (no need to specify what) which converts heat into work. An ideal gas is the working fluid. This gas is, for our present purposes, quite imaginary in that it stays as a gas whatever the temperature and its properties are constant over the whole of the temperature range.

Remember that this is just a starting point and it is totally imaginary. The gas is not completely ridiculous though – at atmospheric pressure, nitrogen is still gaseous at 78K and hydrogen is still a gas at 15K.

Whatever the engine does, let the working fluid – the ideal gas – be introduced at a starting temperature of T_1K. As it flows through the heat-to-work engine, the gas uses some of its enthalpy to do work and cools down to T_2K by the end of the process. The gas then flows out of the system or is **rejected**.

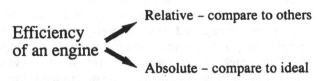

Figure 9.4 Improvements may be relative or absolute.

● PQ 9.5 For this perfect system with constant gas properties and no losses other than the enthalpy of the reject gas, what is the heat energy to work energy conversion efficiency? Call the gas specific heat capacity, which is constant for this ideal gas, c_p and the gas flow rate $\dot{m}$.

Note carefully the use of the absolute zero datum when previously we have often used 0°C as an enthalpy datum. This is because we are looking for an absolute measure of efficiency, not one based on an arbitrary scale such as Celsius. The 0°C datum is used because it is convenient and properties are measured easily at or above that temperature. It is only an arbitrary datum and materials do have properties below 0°C!

Now this is quite a simple treatment but the outcome is the same as the more sophisticated approach which will be met later when we deal with engine cycles. It is a valid conclusion, that the maximum efficiency for a process of this single-phase nature is

$$\eta = \frac{T_1 - T_2}{T_1}.$$

This maximum efficiency for heat-to-work conversion processes, however achieved, is the **Carnot efficiency**, named after the investigator who did some of the early work on ideal engines.

● PQ 9.6 Why single-phase? Would the same relationship apply if the heat source was dry steam (gas – one phase) and the exhaust wet steam (vapour – another phase) at the same pressure?

This efficiency equation is a good example of how simply-derived relationships can give real practical guidance. In this case, it may well indicate an absolute or unachievable efficiency but it does highlight an important point for heat energy (enthalpy) conversion. The bigger the difference between the energy supply temperature T_1 and the energy reject temperature T_2, the higher the efficiency. If this applies to the absolute or theoretical case, it must apply to real cases. This increase of $T_1 - T_2$ can be achieved in two ways.

● PQ 9.7 What are they?

If T_1 is raised, the numerator ($T_1 - T_2$) rises but so does the denominator T_1. If T_2 is lowered, only the numerator changes.

Example 9.1 A heat-to-work engine operates between an energy supply temperature T_1 of 800K and an energy reject temperature T_2 of 500K. I have the choice of raising T_1 by 50K and then 100K or reducing T_2 by 50K and then 100K. What are the results?

Calculating from the equation just derived, $\eta = (T_1 - T_2)/T_1$ gives

T_1 (K)	T_2 (K)	η (%)
800	500	37.5
850	500	41.2
900	500	44.4
800	450	43.8
800	400	50.0

So, if possible, a reduction of reject temperature looks preferable.

A practical example is the use of steam in a steam turbine such as may be found in an electricity generating station. Turbines are very practical machines and, however well made, are not ideal. However, much research and development work has been done to raise the steam supply temperature T_1 and to reduce the steam reject temperature T_2.

● **PQ 9.8** Without knowing anything of steam turbines, what could be done to reduce T_2 without returning to the liquid phase?

We are now in a position to consider the effect of **quality** on the usability of heat energy and to put numbers to it. Staying with the idealized engine, the heat energy to work energy conversion device using an ideal gas as the working fluid, the amount of work W yielded from a supply of heat energy Q is

$$W = \text{supply energy} \times \text{conversion efficiency}$$
$$= Q \times \eta$$
$$= Q \times (T_1 - T_2)/T_1$$

The casual use in this section of the word 'heat', which more accurately may be enthalpy or even heat energy, is deliberate. The word is used casually in real-life engineering, so it will be good practice to learn to translate heat into a more technically accurate term. Having said that, the discipline of heat transfer, which we meet soon, is very highly regarded and that is its accepted title.

Just as $(T_1 - T_2)/T_1$ can give a guide to improvements in real processes, so this second equation can give a guide to the maximum amount of conversion of one energy form (heat) into another (work). Going right back to the beginning, we now have an example of an absolute measure against which designs may be judged.

This is an important step, moving towards a general expression involving heat, work and temperature which shows the importance of quality or availability as well as quantity. Care is needed, though, for it is easy to say that everything now needs converting to practical efficiencies, including all the losses and all the imperfections of real processes. While that is ultimately vital for individual designs, it does limit any such expression to a particular case where the real efficiencies have been assessed.

What is needed first, therefore, is a wide-ranging factor which is applicable at the concept or design stage, even of new processes. Such a factor or combination of properties must recognize:

- that heat supplied with a working fluid can yield work;
- that operational temperatures control the quality (high or low grade) of that heat energy and thus its convertibility into work;
- for a given heat supply, the higher the temperature demand, the less the proportion of that heat which can be used.

By way of illustration, if a central heating boiler supplies water at 85°C to warm a room and the room thermostat is set initially at 65°C and later at 55°C, then there is more heat available from the water supplied to satisfy the latter temperature than the former.

- **PQ 9.9** Leading up to this factor or combination, how will Q and T feature? Is it likely to be something to do with Q/T or $Q - T$, or what?

The factor – really a thermodynamic property – that has been exploited to deal with this need for a wide-ranging design aid is called **entropy**. It is useful to continue with the ideas of changing or transforming heat into work to develop the concept of entropy. While the use of calculus has been avoided as far as possible or given separate treatment in this text, here is one area where it is most beneficial to use it from the outset.

The property entropy is usually given the identifying letter S for total amounts and s for specific amounts – the upper and lower case as in other properties such as enthalpy. If a small amount of heat dQ (the change of heat in the work-producing process) is transformed at a temperature T, then this may be recorded as a small change of this property entropy dS in the form

$$dS = \frac{dQ}{T}.$$

Note that both the **quantity** dQ and some indication of **quality** T, are included. It implies that some heat usage or transfer dQ has been possible, so the energy supply must have both the necessary quantity and the necessary quality; also that the temperature is important and that there must be some **degradation** of the heat. If high quality heat dQ has been used, then that left must be of lower quality. Since this concept of entropy or entropy change has been introduced without reference to any particular working fluid in any particular operation, then it is likely to be widely applicable (Figure 9.5).

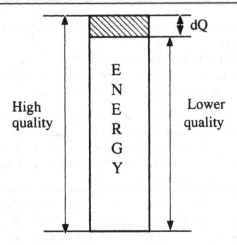

Figure 9.5 Change of entropy – change of quality.

Note that the equation refers to a **change** of entropy dS. As with some other properties, 0°C is commonly used as a datum for entropy. That is nothing more than a practical convenience, rather like enthalpy in the steam tables. Clearly, heat transfer and work usage can take place at far lower temperatures, so the entropy concept is not limited to temperatures above 0°C. There can be negative values.

Since the entropy change dS in the equation is identified by reference to two other thermodynamic properties, energy dQ and temperature T, then entropy S itself is related to those properties. In turn, any combination of properties must also be a property, so entropy is thus shown to be a thermodynamic property.

● **PQ 9.10** For a given amount of heat usage dQ, is entropy change dS larger at a higher temperature, say 500°C, than at a lower temperature, say 100°C?

Therefore, for a given amount of energy usage, the lower the operational temperature, the greater the change of entropy. If this is regarded as a change in quality of energy, then taking a little high quality energy away at high temperature leaves the remainder a bit lower in quality but still useful for other purposes. Taking the same from lower quality energy makes the remainder even lower and there aren't many industrial uses for low quality energy. Thus the change in entropy can be used as an indicator of change of **usefulness** – the bigger the change, the greater the drop in usefulness.

9.3   ENTROPY CHANGE IN SOME IDEAL GAS NON-FLOW PROCESSES

As ideal gases and non-flow processes are quite simple, they are useful here in illustrating a few entropy relationships and as a start in using this abstract property.

Since entropy changes are associated with heat energy usage or heat transfer, any calculations or equations must involve finding or measuring that heat usage. The easy way for doing the calculations is to use the calculus, so it is included quite deliberately here rather than putting it with others in Appendix B.

Take first a constant volume change with an ideal gas and recall the relationship $\Delta Q = mc_v \Delta T$ for a constant volume non-flow process. The calculus style of equation is

$$dQ = mc_v dT.$$

Thus

$$dS = \frac{dQ}{T} = \frac{mc_v dT}{T}$$

which, when integrated for a change from temperature T_1 to temperature T_2, gives

$$S_2 - S_1 = mc_v \ln (T_2/T_1)$$

● PQ 9.11 So what would be the change of entropy for a constant pressure non-flow process?

● PQ 9.12 A constant volume non-flow process changes heat transfer ΔQ into internal energy ΔU. There is no external work, so what has this to do with entropy?

Some numbers will show the difference in entropy change for the two processes. Notice that entropy change rather than entropy alone (rather like enthalpy change) is the important feature.

Example 9.2 Nitrogen is heated from 27°C to 400°C in a closed container in two separate experiments. The two conditions are constant volume and constant pressure. Assuming ideal gas properties and given $c_p = 1.05$ kJ/kg K, what are the two specific entropy changes? What are the units of entropy, specific entropy and entropy change?

There appears to be a minimum of information here! However, as the gas is nitrogen, we can find the value of R from a knowledge of its molar mass or molecular weight (28 from standard data tables) and

R_m, also from standard tables, 8.314 kJ/kmol. This gives R for nitrogen as 297 J/kg K and hence c_v ($R - c_p$) as 0.753 kJ/kg K. If you have any doubts about these derivations, please go back to Chapter 2 and revise them.

The temperatures must be converted to absolute temperatures since the fundamental entropy equation is to do with absolute values. Thus the temperatures are T_1 = (273 + 27)K = 300K, and T_2 = (273 + 400)K = 673K. For the constant volume change then, using s to denote specific entropy,

$$s_2 - s_1 = c_v \ln (T_2/T_1)$$
$$= 0.753 \times \ln (673/300)$$
$$= 0.753 \times 0.808 = 0.608 \text{ kJ/kg K.}$$

Similarly for the constant pressure change,

$$s_2 - s_1 = 1.050 \times \ln (673/300)$$
$$= 1.050 \times 0.808 = 0.848 \text{ kJ/kg K.}$$

The units fall out of the calculation. Let us review them. For total entropy change, the equation would be written

$$S_2 - S_1 = m \text{ kg} \times c_v \text{ kJ/kg K} \times \ln (T_2/T_1)$$

So the units of total entropy change are kJ/K. For specific entropy change – i.e. per kg – the units are thus kJ/kg K. The units of entropy and of entropy change are the same, just as the units are the same for enthalpy and enthalpy change – but do not get entropy and enthalpy mixed up!

Staying with non-flow ideal gas processes, isothermal and reversible adiabatic changes are next. Review these terms in Chapter 4 if you have any doubts because it is sometimes very easy to confuse them. Adiabatic is simple – there is no heat transfer, as that is what adiabatic means, so $dQ = 0$. Thus for an adiabatic process, dS is also zero and this leads to an alternative name for reversible adiabatic changes of *isentropic*, meaning at constant entropy. As a reminder, recall that the word reversible appears, to show that the process for ideal treatment has no losses and that the process – the machine or the engine if one is involved – could be driven backwards, as it were, to return to the starting condition.

For an isothermal (meaning constant temperature – Boyle's law) change, there is of course no change of internal energy, since dU and dT are directly related and in this case dT is zero. Thus all the heat supply dQ for our non-flow isothermal process goes into external work production, dW.

So we have

$$dS = \frac{dQ}{T} = \frac{dW}{T} = \frac{pdV}{T}.$$

Also, $pV = mRT$, or

$$T = \frac{pV}{mR}.$$

Therefore

$$dS = \frac{p\,dV\,m\,R}{pV} = \frac{m\,R\,dV}{V}$$

and

$$S_2 - S_1 = mR \ln (V_2/V_1).$$

Since $p_1 V_1 = p_2 V_2$ for an isothermal change, then we can write

$$S_2 - S_1 = mR \ln (p_1/p_2).$$

Both are equally applicable for the stated conditions. Yet again, please refer back to Chapters 2 and 3 now if any of the above is unclear.

Example 9.3 What is the entropy change associated with 2 kg of nitrogen expanding isothermally from 2.5 to 4 m^3?

Two equations were derived for isothermal change, one based on volumes, one on pressure and both equally valid. Volumes are quoted here, so use that equation

$$S_2 - S_1 = mR \ln (V_2/V_1).$$

The value of R for nitrogen was found earlier, 297 J/kg K, so

$$\Delta S = 2 \times 297 \times \ln (4/2.5) \text{ J/K}$$
$$= 279.2 \text{ J/K}.$$

Finally in this section, look at the general change $pV^n = $ constant which was met in Chapter 4. The external work transfer was shown to be

$$W = \frac{p_1 V_1 - p_2 V_2}{n - 1} = \frac{mR (T_1 - T_2)}{n - 1}$$

and we know $\Delta U = mc_v (T_2 - T_1)$. So

$$Q = \frac{mR (T_1 - T_2)}{n - 1} + mc_v (T_2 - T_1),$$

which rearranges to

$$Q = \frac{m\,(nc_v - c_p)\,(T_2 - T_1)}{n - 1}.$$

• **PQ 9.13** If you are familiar with the calculus, change this expression to a calculus form and thus find $S_2 - S_1$. If you are not, then please accept the conclusion.

The change of entropy for a general change of the form $pV^n =$ constant is therefore

$$S_2 - S_1 = \frac{m(nc_v - c_p)}{n - 1}\ln(T_2/T_1).$$

The limitations of the general change still apply. It is a general equation which is worth a try if there are no alternatives. That it is a general equation does not mean that it will always work, however. For instance, putting $n = 1$ and/or $T_1 = T_2$ for an isothermal change gives a useless answer. Putting $n = \gamma$ for an adiabatic change does however give $\Delta S = 0$, which was deduced earlier.

It is often convenient to represent gas relationships on a graph of pressure against volume, the so-called p–V graphs. We will see in a later chapter that, when the gases are used in a cycle of operations, as happens in engines, these p–V graphs help to determine outputs and efficiencies. These graphs are often called 'diagrams' (as in p–V diagram) but that word is more descriptive when several successive curves are put together to show a cycle of operations (Figure 9.6).

Since the pressure–volume equations lead to entropy–temperature

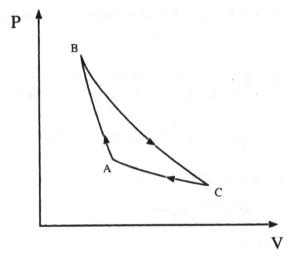

Figure 9.6 P–V curves join to show a cycle of operations.

equations of the types just derived, then these p–V changes or relationships can also be represented on temperature–entropy graphs, again called T–S diagrams, especially if they form a cycle. Whereas p–V graphs are almost always curves, two of the T–S graphs are certainly straight lines (Figure 9.7).

- PQ 9.14 Which? You have already met them.

This can be very handy, as we will see when looking at cycles of change or operation.

9.4 ENTROPY CHANGE IN LIQUIDS AND VAPOURS

While entropy was introduced through ideal gases, it is a thermodynamic property so it must apply to other phases, just as it applies to real as well as ideal gases. Since we have met steam as a very handy vehicle to look at phase changes and vapours – but only as a convenient example, not exclusive in any way – we will stay with water substance for the present.

Looking first at the liquid phase for entropy changes, the same fundamental equation $dS = dQ/T$ applies. Remember that we are often looking at entropy changes and that entropy is commonly, for convenience, reckoned from a 0°C datum. The temperatures involved are, recall, absolute temperatures because that comes from our derivation of the entropy idea. The rule always applies in any case – if ever in any doubt, use absolute values. The only other property that is needed for studying entropy here is the specific heat capacity of the liquid, since the value of Q or dQ for the liquid is the product of mass, specific heat and temperature or temperature change. An example is far better than words for this purpose.

Example 9.4 Taking the specific heat capacity of water c as 4.17 kJ/

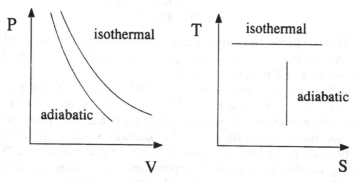

Figure 9.7 P–V and T–S graphs for isothermal and reversible adiabatic changes.

kg K, find the specific entropy change for liquid water substance between 0°C and 90°C. Using 0°C as the conventional datum, what is the entropy of saturated water at 1 bar?

Note that we are dealing with just one definition of specific heat capacity for liquids (the same would apply to solids) rather than the range of specific heat capacities available to gases. Go back now to remind yourself why there are many possible definitions of gas specific heat capacities, if necessary. For liquids and solids, the simple definition of the amount of heat energy required to raise the temperature of unit mass through unit temperature applies.

The heat input here (little q for specific or 1 kg amount) is

$$q = 4.17 \times (90 - 0) \text{ kJ}$$
$$= 357.3 \text{ kJ.}$$

The change of specific entropy is given, as for gases, by

$$ds = \frac{dq}{T}$$

and $dq = c\,dT$, thus

$$ds = \frac{c\,dT}{T}.$$

Integrating, $s_2 - s_1 = c\ln(T_2/T_1)$, giving

$$s_2 - s_1 = 4.17 \times \ln(363/273)$$
$$= 1.188 \text{ kJ/kg K.}$$

From the steam tables at 1 bar, the temperature of saturated water (water that is on the point of boiling but has just not started boiling) is 99.6°C. The specific entropy change from 0°C to 99.6°C, by the same process as above, is

$$s_2 - s_1 = 4.17 \times \ln(372.6/273)$$
$$= 1.297 \text{ kJ/kg K.}$$

While this is strictly the entropy change of the liquid from 0 – 99.6°C it is also, on the 0°C datum, the specific entropy of liquid water substance at its 1 bar boiling point.

If you look up the value in the steam tables, you will see a minor difference, 1.303 compared to 1.297 kJ/kg K. This is due to the tables being constructed from the most accurate available data, whereas this calculation has been on the basis of a constant value of specific heat capacity and, of course, the use of the rounded value of 273 for changing Celsius to absolute temperature. The tables are used in exactly the same way for entropy as they are enthalpy.

● PQ 9.15 What is the table value for the specific entropy change for liquid water substance between 0°C and 90°C?

For most industrial purposes and to a great extent theoretically, the influence of pressure on many liquid properties is not too important. That is not to say that it should always be ignored but, as in Chapter 7, the effect of temperature generally overwhelms any pressure influence under most circumstances. Thus, if the entropy or entropy change of a liquid (or, for the same reason, of a solid) is required under conditions of varying pressure, it is likely that the temperature effect will dominate any property changes. Fortunately for most common materials – solids, liquids, vapours and gases – values have been found either by experiment or by use of accurate data and then tabulated, just as in the steam tables for enthalpy. Indeed, bearing in mind the proper title of 'thermodynamic and transport property tables', entropy values are usually included along with many other properties such as specific heat capacities, viscosities and thermal conductivities for many materials.

When the water starts to boil, it releases steam as a wet vapour which, with further heat transfer, dries progressively. As this drying process enhances the energy quality – there is more per kg the drier the steam, of course – then there must be an entropy change. There are two important unchanging features of this stage of the normal steam-raising process – both the pressure and the temperature stay constant. Any energy supplied is therefore used for the sole purpose of changing phase.

Example 9.5 During the drying of steam at 1 bar, 2000 kJ/kg of latent heat are added. What is the change in specific entropy? What would the change be at 1.5 bar and at 2 bar?

The equation $dS = dQ/T$ still holds and the boiling points at 1, 1.5 and 2 bar are 99.6°C, 111.4°C and 120.2°C respectively. As the temperature at any particular boiling pressure stays steady during the vapour phase, this simplifies the whole. Thus the specific entropy changes are:

$$\text{at 1 bar } \Delta s = \frac{2000}{(99.6 + 273.2)} = 5.365 \text{ kJ/kg K}$$

$$\text{at 1.5 bar } \Delta s = \frac{2000}{(111.4 + 273.2)} = 5.200 \text{ kJ/kg K}$$

$$\text{at 2 bar } \Delta s = \frac{2000}{(120.2 + 273.2)} = 5.084 \text{ kJ/kg K}$$

Note the use of 273.2 rather than 273 for the temperature conversion, since the Celsius temperatures are given to one decimal place.

Be quite clear of the true significance of the lower entropy change value at the higher temperatures, not just the arithmetic reason of dividing by a bigger number. It comes down to the energy quality concept which was a starting point for the introduction of entropy.

● PQ 9.16 In terms of the heat of change of phase or latent heat of evaporation and the boiling temperature, what is the specific change of entropy from 0% to 100% dryness for a steam raising operation at 1 bar? At 1.1 bar?

At 1.1 bar the increase of specific entropy is less than the same operation at 1 bar. There are two clear arithmetic reasons – h_{fg} is less and T is higher for the higher pressure. The basic technical reason about higher quality is thus reinforced.

So the entropy change for vapour, in this case wet steam, gives a good example of this importance of usability or quality of energy, as illustrated by the entropy property. The latent heat of the steam can be used in some process but the higher the temperature at which it is used, the smaller the proportion that can be used before the rest has to be rejected or passed on to some other, less demanding, duty. Add to this the fact that the latent heat reduces as pressures and thus boiling temperatures rise (2258 kJ/kg at 1 bar, 2015 at 10 bar, 1317 at 100 bar for example), and you begin to see why so much effort has to be put into the complex design of power stations using very high pressure and temperature steam!

With wet steam, the entropy or entropy change is related to the degree of dryness since that dictates the proportion of latent heat supplied. Values can be calculated or read from tables, so here is a quick review.

Example 9.6 What is the specific entropy change for steam at 3.5 bar, 0.4 dry, being heated to 0.7 dry?

The steam tables give the necessary enthalpy information and at 3.5 bar the heat of change of phase h_{fg} is 2148 kJ/kg. The heat input value required is enthalpy, as the process is at constant pressure. This is typical of most industrial processes using a vapour. The temperature of the wet steam at 3.5 bar, also from the tables, is 138.9°C, 412.1 K.

As the steam changes from dryness 0.4 to dryness 0.7, the amount of enthalpy used to do this must be $(0.7 - 0.4) \times 2148$ kJ/kg. Thus,

$$s_2 - s_1 = \frac{(0.7 - 0.4) \times 2148}{412.1}$$
$$= 1.564 \text{ kJ/kg}$$

• PQ 9.17 Compare this with the value from the steam tables. That is, use the entropy columns in exactly the same fashion as you would use the enthalpy columns. Any minor differences are again due to exactness of data.

When the steam starts to become superheated (when it passes the 100% dryness level) and more heat or enthalpy is added, then it can be treated as the gas which it is and subjected to the calculus-based relationships which have been derived already. Remember, though, that the just-superheated steam is a gas very close to its condensation or liquefaction point and will therefore not be obeying the ideal gas laws as closely as when well away from that condition.

Exactly as with steam enthalpy and steam internal energy, there is plenty of tabulated information to cover this shortcoming of not always matching the ideal gas laws. Interpolation can be used between tabulated values, just as with enthalpy. The same goes for most other materials used as liquids, vapours and gases, such as refrigerants – there is a lot of tabulated information, ready to use. This does not reduce the value of calculations, for they show what the topic is all about and thus lead to a proper understanding.

Example 9.7 Use the transport property tables (steam tables) to find the specific entropy of superheated steam at 5 bar, 325°C.

The tables which I have record properties at 300°C and 350°C, so I will have to interpolate. Strictly, a graph should be plotted from known values and intermediate values read from it but a straight line interpolation is reasonably accurate between adjacent values. This is the same argument as used for enthalpy interpolation.

The two tabulated values are 7.460 kJ/kg K at 300°C and 7.633 kJ/kg K at 350°C. So at 325°C, the interpolated value is 7.547 kJ/kg K.

Example 9.8 Use the tables to compare enthalpy rise against entropy rise from 300°C to 350°C and from 450°C to 500°C for steam at 5 bar.

Between 300°C and 350°C

$$h \text{ rise is } (3168 - 3065) = 103 \text{ kJ/kg}$$
$$s \text{ rise is } (7.633 - 7.460) = 0.173 \text{ kJ/kg K.}$$

Thus the ratio is $\triangle h / \triangle s = 595.4$ K.
Between 450°C and 500°C

$$h \text{ rise is } (3484 - 3377) = 107 \text{ kJ/kg}$$
$$s \text{ rise is } (8.087 - 7.944) = 0.143 \text{ kJ/kg K}$$

and this ratio is $\triangle h / \triangle s = 748.3$ K.

Interpret this in terms of energy quality – as temperatures rise (higher grade heat), so more enthalpy is required per unit change of entropy.

9.5 THE SECOND LAW OF THERMODYNAMICS

The *Second Law of Thermodynamics* is usually associated with entropy but they are not totally inseparable. It is just that much of the analytical study associated with the second law is made easier by the idea of entropy.

Broadly speaking, if you read four books on thermodynamics, there is a chance that you will get four apparently different expressions of the second law. This is perhaps because the law is not the outcome of a single piece of isolated work, but arises from a coming together of many. The names of Clausius, Kelvin and Planck are met, for instance. A simple statement of the law – some would say a consequence rather than a statement but the end result is the same – is that from the work of Clausius. It is this. Heat flows from a hot substance to a cold one unaided but it cannot flow from a cold to a hot one without outside assistance (Figure 9.8).

That is a fairly simple statement, indeed an obvious one to anybody with any experience but it is what follows from it that makes it important, whether in this or any other form. It has to be taken in conjunction with either analytical or practical application since thermodynamics, like any other topic, may be interesting but it is what you do with it that matters.

For instance, first it means that heat will be lost from any system at a higher temperature than its surroundings. The *perfect thermal insulator* is thus impossible because if it is cooler than the body it insulates, then heat will flow from the body into it. If the insulator is at the same

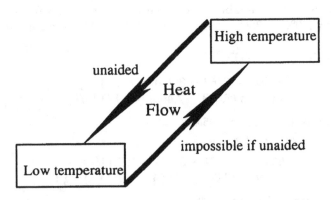

Figure 9.8 The Second Law in action.

temperature as the body and thus warmer than the surroundings, heat must flow from it to the surroundings. In practice, thermal insulation reduces the rate of heat loss – dramatically so in some cases – but cannot stop it. The hot body will ultimately cool into equilibrium with its surroundings, insulated or not.

Second, it means that there can be no practical, 100% efficient heat-driven engine or heat-to-work converter and this includes those where the heat comes from a fuel. At the very least, some of the heat supplied to the engine must be lost to the cooler surroundings. More important though, reject heat from the engine cycle (we meet these soon) at exhaust temperature, being lower than supply or main operational temperature, cannot flow directly against the temperature gradient to be used again in the engine.

More formal statements of the Second Law are given shortly but even this simple version is seen to be important because of its consequences. Notice too the implication for entropy – heat energy is being degraded as it flows down the temperature gradient. The original energy, whether as high temperature heat or as a fuel or whatever, becomes low grade heat and is dissipated to the general surroundings.

● **PQ 9.18** Provide three other examples of different forms of energy ultimately being dissipated as low grade heat.

This lowering of the quality of the energy from whatever source gives rise to an alternative concept of entropy, that of *disorder*. If various energy forms are regarded as being in a state of disorder, then they are very energetic indeed. They can thus be marshalled and used. The only way in which all forms can be brought into order is to use them, therefore. Their use inevitably means reducing their ultimate availability for other purposes.

It is rather like having a heap of bricks and a heap of sand and some cement. They could be transformed into a house, a garage, a wall. They were in disorderly heaps but their use orders them. Unfortunately, unless a lot more outside energy is expended, those bricks, sand and cement are no longer available for any other use.

Entropy is seen as representing disorder and the more orderly the energy form, the less entropy it has. Unfortunately, with machines, central heating, car brakes, lights, electricity generation and so on, the *ultimate* orderly form is that of *low grade heat*. As was noted earlier, there's a lot of that about but nobody has yet worked out what to do with it!

Here are the more formal Second Law statements, with the names usually associated with or attached to them.

● Clausius – No system or device may operate so that the only process is heat transfer from a cooler to a hotter body.

- Planck – It is impossible to make an engine cycle which will use only one heat reservoir and thence convert all the heat into work.
- Kelvin – If a body is at a temperature below that of its surroundings, none of its heat may be transferred into work unaided.

Some random practical conclusions from these statements are:

(a) As a real engine cannot work with one reservoir only (heat source only, say) then there has to be a second one at least – a heat sink. These words 'source' and 'sink' are descriptive and in common engineering use.

Thus there has to be a heat transfer to that sink and that heat is therefore not available to do work. The sink has to be at a lower temperature than the source, or heat will not flow unaided. So it does not matter how good the engine is, how much friction is eliminated, how sophisticated the design, the heat-to-work converter can never be 100% efficient.

- PQ 9.19 Although heat cannot be converted completely into work in a real engine, can work ever be converted completely into heat?

(b) For any real system to work, it has to obey both the First and the Second Laws. Heat and work are mutually interchangeable (1st Law) but not necessarily completely (2nd Law).

(c) If we wish to design a refrigerator, there is no way that we can cool anything below ambient temperature without another energy source. No heat will transfer from the colder-than-surroundings fridge contents unaided.

The Second Law also leads to an important analytical point called the **Clausius inequality**, but first a further look at entropy transfer, as occurs in energy conversion, is needed.

Example 9.9 One kg of gaseous carbon monoxide cools at constant pressure from 120°C to 90°C. The enthalpy from this cooling is transferred entirely to another 1 kg of the gas, initially at 30°C, to heat it also at constant pressure. What is the entropy change in each case? Take c_p for carbon monoxide to be constant at 1.044 kJ/kg K for this example.

Enthalpy change is given by $m\,c_p\,\Delta T$ and the loss from the warmer gas is the gain by the cooler gas. Mass m is 1 kg in either case. Call the final temperature of the cooler gas T. Therefore

$$1 \times 1.044 \times (120 - 90) = 1 \times 1.044 \times (T - 30)$$

thus $T = 60°C$.

The entropy change in a constant pressure operation is

$$S_2 - S_1 = mc_p\ln (T_2/T_1).$$

So, for the warmer gas,

$$\Delta S = 1 \times 1.044 \times \ln(363/393) \text{ kJ/K}.$$
$$= -0.0829 \text{ kJ/K} = -82.9 \text{ J/K}.$$

Absolute temperatures and the answer is negative because the gas is cooling.

For the cooler gas

$$\Delta S = 1 \times 1.044 \times \ln(333/303) \text{ kJ/K}$$
$$= +0.0986 \text{ kJ/K} = +98.6 \text{ J/K}.$$

Exactly the same amount of energy (enthalpy) is involved but the entropy differs. The difference may be argued as availability. If I am supplying an amount of energy, say X kJ provided as hot gas at 750°C to a furnace working at 500°C, then I can only use that bit of the energy which will bring the hot supplied gas down to 500°C – one-third if we refer to 0°C datum.

Now if I supplied exactly the same hot gas to a furnace running at 250°C, I can use two-thirds. The quality of the source and the quality of the demand say how much I can use directly. The entropy values show this. The entropy loss of the hotter source is less than the entropy gain of the cooler sink.

● PQ 9.20 If the enthalpy was transferred instead to 10 kg of the gas initially at 30°C which is then heated at constant pressure, what would be the entropy increase of this 10 kg?

Therefore, for a given amount of energy transfer, the lower the recipient working temperature, the higher the entropy change. The entropy is more ordered and the energy is lower grade.

Note, however, that this is an **entropy change**. It is not saying total entropy. It is not saying that the entropy of 1 kg of gas at 10°C is therefore greater than at 100°C, for instance. It is saying that yielding entropy at a higher temperature increases the entropy of a lower temperature recipient by a larger numerical amount than that lost by the higher temperature source. Check this by reference to the property tables and work through the calculation again if necessary. Remember energy quality. There is an entropy inequality.

In financial terms it is the equivalent of one person with £1000 being able to buy something that 200 individuals cannot buy with £10 each.

The 200 individuals may have more money in total but it is spread very thinly.

● **PQ 9.21 Could the entropy inequality conclusion lead to any measure of an energy conversion system's efficiency?**

So, we have now met the idea that the *best* theoretical performance of an energy converter would be if the entropy change was zero. That is, entropy in = entropy out, an **isentropic process**. If the entropy change of a real system is measured then, this gives a point of comparison called the **isentropic efficiency** which we meet again later.

Now think of a simple engine – not its construction or operational details but just the fact that it is a device for converting heat into work. The second law says that no such engine can be perfect, so the heat supplied is converted partly into work and partly into heat losses. For a small amount of heat being supplied, dq_1 say, dq_2 becomes work and dq_3 is lost. The steady flow energy equation applies also

$$dq_1 = dq_2 + dq_3$$

and the system has an energy balance, (energy in) − (energy out) = 0.

Now the entropy. If it was possible for the engine to work at one temperature T throughout the whole of its operation, then

$$\frac{dq_1}{T} = \frac{dq_2}{T} + \frac{dq_3}{T}$$

and

$$ds_1 = ds_2 + ds_3.$$

So for this perfect operation there is an entropy balance also:

$$\text{entropy in} = \text{entropy out,}$$
$$ds_1 - (ds_2 + ds_3) = 0.$$

This is the **best** that could be attained, even theoretically. If we added up all the entropy changes for this perfect engine's cycles of operation, all the little bits of ds, we could write in the usual calculus fashion

$$\int \frac{dq}{T} = 0.$$

However, the Second Law says that no engine can operate at one reservoir temperature. The sink for losses must be at a lower temperature. This being so, take the least-imperfect possibility, that only the lost or reject heat is related to an infinitesimally lower sink temperature, $T - dT$, while the work is related to the original temperature T. Now the entropy equation becomes

$$\frac{dq_1}{T} = \frac{dq_2}{T} + \frac{dq_3}{T - dT}$$

and ds_3 has risen because $(T - dT)$ is smaller than T. The total entropy leaving the system is *greater* than the total entropy entering and this is a statement of the **Clausius inequality**. Since this has been demonstrated by reference to the least-imperfect engine, it most certainly applies to all real engines.

So, entropy in − entropy out <0, which can be expressed mathematically as

$$\int \frac{dq}{T} < 0.$$

Now, if we interpret this more widely, it means that whenever energy is used, it is degraded. It applies to the engine we have just used but if we had used a furnace or a central heating example, the conclusion would have been the same. On the larger scale, it applies to the earth's fuel resources also − as a fuel (coal, oil, gas) is used, it undergoes energy conversion processes and the ultimate resultant heat is dissipated and lost to future use.

So we have met and used entropy, an abstract thermodynamic property. The fact that it cannot be measured on an entropy meter or that it has to be assessed from other thermodynamic properties does not reduce its usefulness. We will meet it again and it will prove to be a valuable tool.

The Second Law which often accompanies it looks very much like a set of common-sense statements − but then, common sense really is a good scientific and engineering property!

PROGRESS QUESTION ANSWERS

PQ 9.1

It comes down to being able to measure accurately the efficiency of conversion. 100% of any energy form into any other is most unlikely. There are heat losses, friction to overcome, chemical equilibrium to satisfy and so on. If the losses are not known completely, then there has to be some estimation. For the car, there will not be quoted a fuel consumption of, say, 37.268 mpg for all occasions. You may read something like 41 mpg on journeys and 33 mpg around town. These are good estimates which can be put to practical use.

PQ 9.2

Remember what has been said about 'heat addition'. It is a simplification which we are continuing to use. The heat addition or enthalpy increase Q, the temperature rise ΔT, the mass m and the specific heat capacity c are connected by

$$Q = m \, c \Delta T$$

For the 5 kg piece, in consistent units,

$$960 = 5 \times 0.48 \times \Delta T$$
$$\Delta T = 400K.$$

For the 100 kg piece, in consistent units,

$$960 = 100 \times 0.48 \times \Delta T$$
$$\Delta T = 20K.$$

The 5 kg piece is certainly hot enough ($0°C + 400K = 400°C$) to boil some water but the 100 kg piece ($20°C$) is not. There is exactly the same amount of heat but its quality is very different.

PQ 9.3

Some particular points were made. It is a thermodynamic property; it is related to quality but we can attach numbers to it; it is abstract in that it cannot be measured simply, as can say mass or velocity or specific heat capacity. Entropy is rather like clothing – very useful, size can be measured but the quality cannot.

PQ 9.4

Of the energy supplied, what proportion is ultimately used for the intended purpose?

PQ 9.5

Using the usual abbreviations the enthalpy (heat energy) going into the engine is

$$Q_1 = \dot{m} \, c_p T_1$$

and being rejected is

$$Q_2 = \dot{m} \, c_p T_2.$$

So the enthalpy converted to work, $Q_1 - Q_2$, gives the work done W:

$$W = \dot{m} \, c_p \, (T_1 - T_2).$$

Thus the efficiency, work done compared to energy originally supplied,

$$\frac{W}{\dot{Q}_1} = \frac{\dot{m}\,c_p\,(T_1 - T_2)}{\dot{m}\,c_p\,T_1}$$

Therefore, ideal efficiency, usually designated by the Greek letter η is

$$\eta = \frac{T_1 - T_2.}{T_1}$$

PQ 9.6

For dry steam yielding enthalpy to give wet steam at the same pressure, the temperature change is zero. Thus $T_1 - T_2$ is zero and so the equation cannot be applied directly.

PQ 9.7

Raise T_1 or lower T_2.

PQ 9.8

Reduce the pressure. In general, T_1 is limited by material properties in the steam generator (the boiler), so emphasis is placed on reducing T_2.

PQ 9.9

The quantity of heat required for any particular change must feature directly. If there is twice as much heat of given quality, then there is likely to be twice as much work yielded. The higher the operational temperature, the higher must be the quality of the heat.

So the 'availability' of the heat is less at the higher temperature or, as a first try, it seems reasonable to write availability as $1/T$. Thus this new factor, combining Q and T, will be something like Q/T.

PQ 9.10

Use absolute temperatures, not the arbitrary 0°C!! At the higher,

$$dS = \frac{dQ}{(500 + 273)} = 0.0013\ dQ.$$

At the lower,

$$dS = \frac{dQ}{(100 + 273)} = 0.0027\ dQ.$$

So, the entropy change at the lower temperature is about twice that of the higher. The actual values will depend upon the actual numbers. These ones are simply examples.

PQ 9.11

The calculation is exactly the same except that for a constant pressure non-flow process $\Delta Q = m\, c_p \Delta T$ applies. Refer back now to Chapters 3 and 4 if necessary. Thus

$$S_2 - S_1 = m\, c_p \ln (T_2/T_1)$$

applies for a constant pressure non-flow process.

PQ 9.12

Remember that heat-to-work was used as a way of introducing the idea of heat quality and the concept of entropy. Since entropy is a thermodynamic property, then it is always there. The constant volume change is to do with heat and temperature change, so it is to do with entropy. It is asking and answering the question – will a constant volume change affect the energy quality?

PQ 9.13

Changing to the calculus form will involve dQ instead of Q and dT instead of $T_2 - T_1$ to give

$$dQ = \frac{m(n\, c_v - c_p)}{n - 1}\, dT.$$

Therefore,

$$dS = \frac{m(n\, c_v - c_p)}{n - 1}\, \frac{dT}{T}$$

and

$$S_2 - S_1 = \frac{m(n\, c_v - c_p)}{n - 1}\, \ln(T_2/T_1).$$

PQ 9.14

The adiabatic (isentropic, constant entropy) and the isothermal (Boyle's law, constant temperature).

PQ 9.15

1.192 kJ/kg K.

PQ 9.16

From the steam tables, h_{fg} at 1 bar is 2258 kJ/kg and saturation temperature is 99.6°C, 372.8 K. Thus

$$\Delta s = \frac{2258}{372.8} = 6.057 \text{ kJ/kg K.}$$

Similarly, at 1.1 bar

$$\Delta s = \frac{2251}{375.5} = 5.995 \text{ kJ/kg K}$$

PQ 9.17

The specific entropy of change of phase from saturated liquid to saturated vapour at 3.5 bar is 5.214 kJ/kg K. Thus the specific change of entropy from 0.4 to 07 dryness fraction is

$$s_2 - s_1 = (0.7 - 0.4) \times 5.214 = 1.564 \text{ kJ/kg K.}$$

The numbers may not always agree so well – it depends upon data accuracy.

PQ 9.18

Electrical energy fed to a light bulb; some of a car's fuel being used to overcome friction; heat escaping from a room; the sound from a radio speaker, are some examples.

PQ 9.19

Yes. Think of friction, such as a car brake. The work done in bringing the car to a halt is dissipated as heat.

PQ 9.20

Using the same sort of enthalpy balance gives the final temperature of this 10 kg of gas as 33°C, 306K. Thus the entropy increase of this 10 kg of cooler gas is

$$\Delta S = 10 \times 1.044 \ln (306/303) \text{ kJ/K}$$
$$= + 0.1029 \text{ kJ/K} = 102.9 \text{ J/K.}$$

PQ 9.21

Yes. Since the best theoretical possibility is that there is no entropy change – no low grade sink increasing its entropy – then the system which stays at the same entropy is the most efficient possible. How close any conversion system is to isentropic looks to be a good measure.

TUTORIAL QUESTIONS

You will need access to data tables for some of these questions. Recall that if the state of a substance is defined (its pressure, temperature, volume or whatever is necessary) then its properties are fixed, however that state is reached. Make sure that calculation and data accuracies are compatible.

9.1 Determine from property tables the specific entropy of steam at 4 bar, 200°C. If this steam expands isentropically to 1 bar, what will be its temperature? If it has become vapour, what is its dryness fraction?   [7.172 kJ/kg K; 99.6°C; 0.969]

9.2 Estimate from the specific enthalpy tabular values the change of specific entropy when saturated water at 2 bar is converted to steam 70% dry. Compare the value derived to the specific entropy tabular value.   [Both 3.918 kJ/kg K to 3 decimal places]

9.3 Air is heated and compressed from 10°C, 1 bar to 150°C, 2.5 bar. What are the initial and final values of specific entropy compared to the conventional 0°C, 1 bar datum? Take the mean values of the specific heat capacity for air as 1.005 kJ/kg K between 0°C and 10°C and 1.011 between 0°C and 150°C. Take the gas constant for air as 0.287 kJ/kg K.   [0.036 kJ/kg K; 0.18 kJ/kg K]

9.4 1 kg of nitrogen is initially at 15°C, 1.05 bar. What is its change of entropy if it is (a) compressed isothermally at 15°C to 3 bar, (b) heated isobarically at 1.05 bar to 125°C, (c) heated and compressed to 3 bar at 125°C, (d) compressed reversibly and adiabatically to 3 bar?
   [(a) − 0.3118 kJ/kg K; (b) +0.3372 kJ/kg K; (c) +0.0254 kJ/kg K; (d) 0]

9.5 Air is heated according to pV^n = constant from 50°C to 100°C. Given that the mean specific heat capacity at constant pressure for air between these temperatures is 1.01 kJ/kg K, the gas constant for air is 287 J/kg K and the specific entropy change is 63 J/kg K, what is the value of the index n?   [2.01]

9.6 A copper pipe of mass 1 kg is being heat-treated by raising its temperature to 400°C and then plunging it into a well-insulated tub containing 5 kg of water, which was initially at 10°C. The copper pipe is left in the water until thermal equilibrium is reached – that is, the temperature of the copper and the water are the same.

If the specific heat capacity of the copper is 0.38 kJ/kg K and of the water is 4.187 kJ/kg K, estimate the change in specific entropy of the water and of the copper and thus estimate the change in entropy of the insulated copper/water system.

[Water +0.504 kJ/kg K; copper −0.320 kJ/kg K; system +0.18 kJ/K]

Heat transfer I – conduction and convection

10

10.1 BACKGROUND

Up to now, whenever there has been heat usage for any reason, non-flow or steady flow, it has been an acceptable simplification to say that the heat is supplied. That has been adequate for our purposes and there have been no conditions attached other than the implied ones that the heat was available in the necessary quantities and with the necessary quality.

In reality there has to be some mechanism by which the heat is supplied for the intended purpose – some way of performing the transfer of heat from its source to the point of use. The word 'sink' is often taken as the point or place of use, so that there are heat sources and heat sinks. The process of supplying heat is called **heat transfer**.

There are three ways in which heat can be transferred, as heat that is, rather than as a fuel or as microwave energy for instance which are converted to and then transferred as heat. These three ways or modes of heat transfer are **conduction**, **convection** and **radiation** (Figure 10.1). Strictly, the word 'thermal' should appear with each mode, as in thermal conduction, so that they are clearly identified as heat transfer modes and not, for instance, electrical conduction. Commonly, the word thermal is omitted.

Figure 10.1 There are three modes of heat transfer.

Conduction is typical of heat flow through solids, such as resting your hand on a hot central heating radiator and your hand becoming warmed. The heat flows through the solid wall of the radiator and the solid of your hand. Conduction does not disturb the solid through which it flows significantly, so that there is no material movement associated with the conduction heat transfer process.

Convection is typical of fluids, such as a fan heater warming a room's atmosphere and thence its contents. Here there is a very definite movement – the movement of the fluid like the air flow in the room being warmed. The moving fluid transfers the heat – it convects it from one place to another – and most commonly it is convecting it from one solid surface – the fan heater elements – to another, perhaps your body or your furniture.

Radiation is typical of heat transfer from a remote source – the sun warming you. There does not have to be anything between the source and the sink – no solid through which to conduct, no moving fluid for convection currents. In fact, the best thermal radiation takes place when there is absolutely nothing between source and sink, exactly as the most sunshine falls on you when there are no clouds.

● PQ 10.1 Identify from your everyday experience an industrial or commercial example of each mode of heat transfer.

For simplicity, these modes are usually studied separately but in fact they often occur together. As examples of modes occurring together, a domestic central heating radiator is both convecting and radiating as it warms a room. The flames and hot gases in a furnace radiate and convect at the same time. Having said that, each mode has a definite role in any given situation, frequently with one being dominant, which makes it useful to look at them individually.

Just as with any process or operation in real-life engineering, especially energy usage, heat transfer can be steady, unsteady, transient or intermittent. The analytical or practical treatment may vary or become more sophisticated accordingly. For our purposes in this text – obtaining a foundation and appreciation of thermodynamic knowledge – we will concentrate on **steady state heat transfer**. That means that heat flows are continuous and do not vary with time.

Similarly, for simplicity we will deal mostly with transfer in one dimension – that is, dimension as in one, two or three dimensions rather than mass, length and time! Although this steady state is the simplest, it lays the foundation for any other study of heat transfer.

10.2 CONDUCTION

Conduction is the simplest form of heat transfer to understand, to predict and to measure. It is primarily the passing of energy (heat) through a material from one particle of the body of the material to the next. The position of each particle in relation to the rest of the material is not affected by the conduction process and there is no large-scale movement associated with this mode of heat transfer. Assuming that the heat transfer does not damage the material or the body involved, every part of it would be in the same relative position at the end of the heat transfer as at the beginning. It is because nothing moves that conduction is the easiest mode to investigate (Figure 10.2).

It is the characteristic mode of heat transfer through a solid and if radiation cannot pass through the solid body (like sunshine through glass) or material in question, then it is the only mode of significance.

While conduction is characteristic of solids, it is not exclusive to them and it will occur in gases and liquids. In fluids, though, convection dominates as the heat transfer mode by a very large margin, so although conduction may be present, its contribution is somewhere between small and insignificant. There are some special circumstances, such as heat transfer in very slow moving fluids or fluids adjacent to a wall, where it may be important. However, this introduction will not include those special conditions and conduction through solids will do all that we need.

When starting to study conduction, it is easiest to look first at a plane surface – a simple flat wall as in PQ 10.2 below – where heat is flowing from one side to the other, perpendicular to the flat faces. It is also usual and valuable to talk of **heat flow rates**, since this reflects industrial practice of supplying fuel (hence heat) at so many litres or cubic metres or kilograms per second or per hour to real-life processes. The running of a central heating system (gas bills per quarter) or a car engine (mpg) are examples of the rate of energy supply being important. It follows that total flow can be calculated from the rate and the time. It is also common to use the letter q for this purpose so, unless stated otherwise, $\dot{q}$ (q dot) in this chapter means rate of flow of heat, normally J/s which is watts, W.

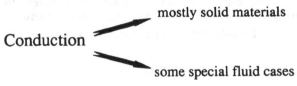

Figure 10.2 Conduction is typical of heat flow through solid materials.

As an aside, very strictly the letter used for heat flow should be capital letter Q as it normally refers to a total amount not small letter $\dot{q}$ referring to a specific (per unit something) amount. However, many heat transfer textbooks do use the small letter, which is why it is used here. This is an exception because the capital letter and small letter convention is followed elsewhere.

● PQ 10.2 If heat flows through a solid wall, from one side to the other, what factors may influence the rate of conductive heat transfer? They are common-sense ones! (Figure 10.3.) What will be the units of the rate of heat transfer?

Look a little further at those three factors. The temperature difference, often designated $\triangle T$, is the **driving force** of heat flow, just as voltage is the driving force of electrical current or water pressure is the driving force of flow from a tap. Both common sense (useful in thermodynamics, remember) and the Second Law say that a temperature difference must be present for heat to flow. The higher the difference, the greater the flow rate.

● PQ 10.3 Where precisely will these temperature be measured?

Just as temperature difference and heat flow are parallel to voltage and electrical current flow, so the wall thickness is parallel to electrical resistance. The thicker the wall (like loft insulation in a house) the less the heat flow rate. There is, though, another contributor to **thermal resistance**.

Different materials have different thermodynamic properties. If the wall was of copper, you would expect the heat to flow through it easier than if it was made of brick. This thermodynamic property which

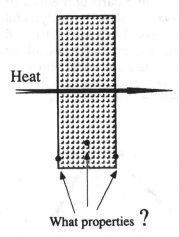

Figure 10.3 Heat flow through a flat wall.

indicates how well or otherwise a material will allow heat to be conducted through it is called the **thermal conductivity**.

So, for a plane wall, the rate of flow of heat through the wall depends upon the thickness of the wall, the temperature difference across the wall and a thermal property of the material of the wall, its thermal conductivity. While there are tables of thermal conductivity for many materials, just like tables of other properties, there is a small complicating factor. Its value is influenced by both the material and the type of material construction being used (Figure 10.4).

Think of two woollen garments, both using the same type of wool and both of the same thickness. One is knitted quite tightly so that it fits closely. It will be warm but not particularly so in cold weather. The other is knitted loosely and is not a tight fit so that it does keep you warm in cold weather. It is not just the wool then, it is the way in which the wool is used.

In the same way, an industrial insulant or a ceramic block will have a range of effective thermal conductivities depending upon the material itself and how it is being used. An outstanding example is the use of glass. As a pane of glass, its thermal conductivity is high – it will conduct a lot of heat for given conditions. As the insulating material glass fibre, its thermal conductivity is low and it will retard the flow of heat. The difference can be as high 1000:1 so **great** care has to be taken in choosing the correct value when using property tables for calculations. Like so many other thermodynamic properties, it is also temperature dependent but that variation is well tabulated. For most real needs, the proper value of thermal conductivity is normally freely available. It is important though to make sure that the proper value is chosen.

With heat transfer as in other parts of thermodynamics, much of the early work was by observation, with the analysis following. For each of the modes, conduction, convection and radiation, there is attached the names of one or two investigators. In the case of conduction, one name is Fourier, the same man as in the mathematical work, the Fourier series. In fact, that work was pursued to aid his solving of the complex heat transfer problems which he investigated.

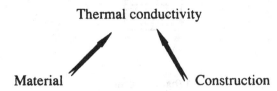

Figure 10.4 Both the basic material and its way of use affect thermal conduction.

In heat transfer, Fourier formulated the fundamental conduction equation. At its simplest, the **Fourier equation** for the rate of conductive heat transfer through a plane wall is written

$$\dot{q} = - k A \, \frac{\Delta T}{x}$$

where k is the thermal conductivity (k is the usual letter for that property), A is the face area of the plane wall, ΔT is the temperature difference through the wall, measured at the inlet and outlet faces and x is the wall thickness.

● PQ 10.4 Draw a simple sketch, based on Figure 10.3, which identifies the factors of the Fourier equation. Put SI-consistent units to this equation and thus state the units and dimensions of thermal conductivity.

So why the minus sign? Really, it is a matter of consistency when doing various sorts of conduction calculations, its origin and prime use being in calculus-related work. Having mentioned the calculus, it is worth putting the simple equation in that form for future reference:

$$\dot{q} = - k A \, \frac{dt}{dx}$$

To maintain consistency, it is therefore common to retain the minus sign and to put temperature differences as (final condition) − (initial condition) in all equations. Thus, if heat was flowing through a wall, from a warm side at T_1 to a cool side at T_2, the equation would be

$$\dot{q} = - k A \frac{(T_2 - T_1)}{x}$$

It is similar to equations for enthalpy and internal energy changes which include temperature differences in the same way. As long as the Second Law is remembered – heat flows down a **temperature gradient** – then the minus sign will not cause confusion. However the equation is written, q must flow from the higher temperature to the lower.

This equation is quantifying the 'driving force' and 'resistance' aspects of conductive heat transfer and there will be similar expressions met soon for the other modes. The driving force is ΔT or dt and the resistance depends on the remaining factors – the thermal conductivity, the thickness and the area.

As an aside, all the basic **transport laws** (the laws which govern the flow of anything) have common elements of flow rate, resistance and driving force. The laws are analogous. In practice, this is useful for solving complex heat transfer problems that get rather messy when relying solely on mathematics. The use of such analogies is generally called **modelling**. One popular model involves setting up an electrical network to simulate a heat transfer problem. The currents and voltages are measured easily and are then translated into heat flows and temperatures.

Example 10.1 A wall 0.2 m thick is made of housebrick whose thermal conductivity is 1.25 W/mK. One face of the wall is at a temperature of 20°C and the other at 5°C. What is the rate of loss of heat through the wall?

There is no wall area given so the only sensible way to answer is to calculate a heat flow per unit area of wall face. This is quite common, just as enthalpy changes may be worked out per kg if a total mass is not known. So here, the area A will be 1 m^2.

$$\dot{q} = -kA\frac{\Delta T}{x} = -kA\frac{(T_2 - T_1)}{x}$$

$$= -1.25 \times 1 \times \frac{(5 - 20)}{0.2} \text{ W}$$

$$= 93.75 \text{ W}.$$

You will see that the common $(T_2 - T_1)$ has been used but the minus sign makes sure that the answer is sensible – there is a positive heat flow from the warm to the cool side! Note clearly that this is not a total heat loss rate from any brick building. It is the rate of loss per square metre of wall. If the term A had been left out of the equation, this would have the same effect as specifying an area of 1 m^2 except that the answer would come out as 93.75 W/m^2. Check it.

Example 10.2 A roof area of 100 m^2 is built of insulating sheets. The inside surface of the roof is at 25°C and the outside surface is at 0°C. If the insulating sheet material has a thermal conductivity of 0.06 W/mK how thick must the sheet be if the roof heat loss rate is not to exceed 15 kW?

Heat transfer conditions, especially temperature, often vary in practice so it is common to set some sort of design limit for expected common extreme temperatures. In this case it is 'losses not to exceed 15 kW' for the given temperatures. Change the loss into watts rather than kilowatts so the units are self-consistent and call the material thickness x.

$$15\,000 = -0.06 \times 100 \times \frac{(0 - 25)}{x} \text{ W}$$

thus, thickness = 0.01 m or 10 cm.

The use of walls made up of two or three layers of different materials is quite common industrially. A furnace may have an inner wall of a refractory brick that can withstand hard knocks. This is likely to be backed up with a more fragile insulating brick. It is similar in fact to a house wall, which may have an inner skin of breeze-block, a space filled with cavity wall insulation and then an outer wall of conventional brick. It is important, therefore, to be able to assess heat loss or flow through these composite walls.

This is done by looking at each layer in turn and remembering that, since the heat flow rate is steady (steady state conditions), it must be exactly the same through each layer. From the house, the heat loss or heat flow rate through the breeze-block, the cavity wall insulation and the outer brick layers is the same loss at the same rate (Figure 10.5).

So if the Fourier equation was applied to each of these layers, using the proper values of k, T and x for each, the answer $\dot{q}$ would be the same. There is, however, a simple practical problem. It is not always easy to get the intermediate face temperatures. Those are the temperatures where the layers meet, such as where the cavity wall insulation meets the brick. This means then that an equation is needed which relies on the easy-to-measure temperatures, the inner and outer faces for which there is free thermometer access.

Example 10.3 A plane (meaning flat) wall is made up of two materials, thicknesses x_1 and x_2 and thermal conductivities k_1 and k_2. The warm face temperature is T_1 and the cool T_2. What is the steady state heat loss per m^2 of wall face?

Call the intermediate temperature where the two materials meet T and then write down equations for the two materials in terms of $(T - T_1)$ and $(T_2 - T)$ as a start. Recall that the formal Fourier equation requires the temperature difference as (final temperature) − (initial temperature). If the wall area is A, then since steady state means that the heat flow rate through each material is the same,

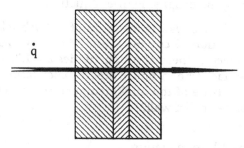

Figure 10.5 Composite wall, same heat flow through each layer.

$$\dot q = - k_1 A \frac{(T - T_1)}{x_1} = - k_2 A \frac{(T_2 - T)}{x_2}$$

thus,

$$T = T_1 - \frac{\dot q\, x_1}{A\, k_1} \quad \text{and} \quad T = T_2 + \frac{\dot q\, x_2}{A\, k_2}.$$

so that

$$T_1 - \frac{\dot q\, x_1}{A\, k_1} = T_2 + \frac{\dot q\, x_2}{A\, k_2}.$$

Therefore,

$$T_1 - T_2 = \frac{\dot q}{A} \left(\frac{x_1}{k_1} + \frac{x_2}{k_2} \right)$$

and

$$\dot q = A \frac{(T_1 - T_2)}{x_1/k_1 + x_2/k_2}$$

Notice that this has come out as $(T_1 - T_2)$, the warmer temperature minus the cooler temperature. Now you see another reason for the minus sign in the starting equation. The more complex cases come out as a positive heat flow being driven down a temperature gradient.

From this two-layer wall example it is a simple step to say that for plane walls of several layers, going up to n layers,

$$\dot q = \frac{A\,(T_1 - T_n)}{x_1/k_1 + x_2/k_2 + \ldots + x_n/k_n}.$$

Remember that the heat flow will be in the direction of the temperature gradient. So if T_1 is larger than T_n, the heat flow will be in the direction from T_1 towards T_n. If T_n is the larger, then the heat flow is in the other direction. Common sense is the guide, but if any calculus derivations are used, then you have to use the formality of the minus sign in the original Fourier equation.

Example 10.4 A house wall is made up of outer brick, 10 cm thick, cavity wall insulation, 5 cm thick and inner breeze-block, 12 cm thick. Their respective thermal conductivities are 1.2, 0.03 and 0.8 W/mK. If the inner face (the in-house breeze-block face) is 20°C and the outer face (the outside of the brick) temperature is –3°C, what is the rate of heat loss per m^2 of wall?

We have the multilayer equation

$$\dot{q} = \frac{A\,(T_1 - T_n)}{x_1/k_1 + x_2/k_2 + \ldots + x_n/k_n}$$

so for the three layers of the house wall this becomes

$$\dot{q} = \frac{A\,(T_1 - T_n)}{x_1/k_1 + x_2/k_2 + x_3/k_3}$$

First the denominator of that equation, the x/k. Look at just the numbers to see the contribution which each makes. Change thicknesses into m rather than cm to be consistent with the other units.

outer brick	+	cavity wall	+	breeze-block	
$\dfrac{0.1\ \text{m}}{1.2\ \text{W/m K}}$	+	$\dfrac{0.05\ \text{m}}{0.03\ \text{W/m K}}$	+	$\dfrac{0.12\ \text{m}}{0.8\ \text{W/m K}}$	
0.083	+	1.667	+	0.15	= 1.90.

These x/k terms are part of the *thermal resistance* – the resistance to heat flow which is driven by the temperature difference. The higher the resistance, the lower the heat flow.

Look at the contribution of each layer – the x/k term – to the total resistance. The breeze-block and the brick make up about 13% and the cavity wall insulation provides about 87%. That is, about 87% of the resistance to heat flow is provided by one feature of the composite wall, the insulation. Although this is a very specific example, a house wall, it is quite common in industry to find one heat transfer feature dominating the others. Where this happens, it is often called the *rate controlling factor*.

● PQ 10.5 How thick would the breeze-block wall have to be to have the same insulating effect as the 5 cm cavity wall?

Therefore, the heat loss per m² ($A = 1$) of wall is

$$\dot{q} = \frac{1 \times (20 - (-3))}{1.9} = \frac{23}{1.9} = 12.1\ \text{W}.$$

Note carefully that the temperature difference is 20 minus (–3) deg C.

● PQ 10.6 If the breeze-block was the outer skin and the brick the inner but with the same inside and outside temperatures, what would be the heat loss rate?

There are plenty of plane walls around through which heat is flowing – houses, furnaces, for instance – but they are not the only shapes. Perhaps the most common shape, in terms of amount used, is in fact a tube – an insulated water pipe as found in most houses, the steam or hot water supply of many industrial and commercial premises, for example. It is certainly important to insulate many pipes, such as to prevent frost damage to cold water ones or to retain heat in steam ones. A straightforward calculation will show how much heat or cold can be kept in or out and the simplest way is to use calculus. As before, either work through the calculus or accept the findings, as appropriate.

Figure 10.6 is a cross-section of a pipe, length L, internal radius r_1, external radius r_2, thermal conductivity k. The inner face temperature is T_1 and the outer T_2. For the calculation, take some radius r and an element of the pipe thickness dr. Look at the steady state heat flow rate through this element. As it is steady, it is the same as that flowing through the inner skin and the outer skin of the pipe, radially from the centre. The element has a circumference $2 \pi r$ and thus an area for the length L of $2 \pi rL$.

If the temperature drop across the element is dT, then the Fourier equation gives

$$\dot{q} = - k \, 2\pi \, r \, L \, \frac{dT}{dr}$$

$$\dot{q} \, \frac{dr}{r} = - k \, 2\pi \, L \, dT,$$

which integrates to

$$\dot{q} \ln (r_2/r_1) = - k \, 2\pi \, L \, (T_2 - T_1).$$

Thus

$$\dot{q} = \frac{- k \, 2\pi \, L \, (T_2 - T_1)}{\ln(r_2/r_1)}$$

$$= \frac{+ k \, 2\pi \, L \, (T_1 - T_2)}{\ln(r_2/r_1)}.$$

In exactly the same way that a simple steady state conduction equation can be derived for a plane wall of several layers, so one can be derived for a pipe of several layers.

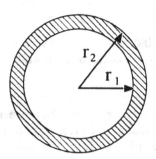

Figure 10.6 Cross-section of a pipe.

Say the pipe was a steel pipe internal radius r_1, external radius r_2 and thermal conductivity k_1. It has insulation of thermal conductivity k_2 added to a final radius r_3, as in Figure 10.7. Call the innermost face temperature T_1, the common temperature where steel and insulant meet T_2 and the outermost face temperature T_3.

As with the plane wall, the heat flow rates through each layer are identical, so

$$\dot{q} = -k_1 \frac{2\pi L\,(T_2 - T_1)}{\ln(r_2/r_1)} = -k_2 \frac{2\pi L\,(T_3 - T_2)}{\ln(r_3/r_2)}.$$

T_2 could be measured but perhaps with difficulty, so the easy way is to eliminate it.

● PQ 10.7 Do the next step.

If there are several layers to the pipe construction, the equation just stretches to include all the terms. If the final outer skin temperature was T_n, then for several layers

$$\dot{q} = \frac{2\pi L\,(T_1 - T_n)}{[\ln(r_2/r_1)]/k_1 + [\ln(r_3/r_2)]/k_2 + [\ln(r_4/r_3)]/k_3 + \ldots}.$$

Example 10.5 A steel pipe of thermal conductivity 71 W/m K carries steam. The inside surface of the pipe is at 140°C, its inner diameter is 10 cm, and outer diameter 12 cm. It is covered by a 5 cm-thick layer of insulation of thermal conductivity 0.04 W/m K and the outer face of this insulation is at 20°C. What is the heat loss per metre run of pipe?

Just as with enthalpy per kg of substance or plane wall heat transfer per m₂, so heat loss from pipes is often recorded as 'per unit length' or

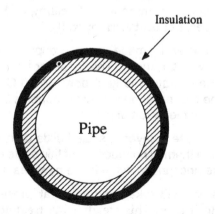

Insulation

Pipe

Figure 10.7 An insulated pipe.

'per metre run'. There are tables or graphs freely available which give heat losses per metre run for various pipes, insulants and insulation thicknesses. Here, though, we are calculating the loss, not reading it from a graph.

The equation just derived is the one to use, so the ratios of radii r_2/r_1 and r_3/r_2 are needed. For the steel pipe, the ratio is of course exactly the same as the diameter ratio, (12 cm)/(10 cm) or 1.2. As it is a ratio, the answer is a plain number with no units, so it does not matter that we have used cm rather than m to get the answer. If ever in doubt, use the prime units.

For the insulation, the external diameter is (12 + 5 + 5) cm, i.e. 22 cm. There are two layer thicknesses making up the diameter – it is easy to say the outside diameter is 12 + 5 cm, so take care. The internal diameter is the same as the outside diameter of the steel pipe, 12 cm so the ratio for the insulation is (22 cm)/(12 cm) or 1.833.

For one metre pipe run, the length L in the equation is 1 m, so with $\dot{q}$ in this case being the rate of heat loss per metre run of pipe,

$$\dot{q} = \frac{2\pi\,(140 - 20)}{(\ln 1.2)/71 + (\ln 1.833)/0.04}$$

$$= \frac{753.98}{0.0257 + 15.15} = 49.68 \text{ W/m.}$$

Two points to note. First, the thermal resistance of the steel is just about negligible compared to that of the insulating layer. If the thermal resistance of the steel had been ignored, it would have made less than one quarter of one per cent difference to the answer. The insulation is the rate controlling factor. Secondly, the length L for the pipe has been left out of the equation quite deliberately as no total length was specified. The answer comes out as a heat loss per unit length W/m, which is a common way of dealing with heat losses – per unit length for pipes, per unit area for walls.

● PQ 10.8 An oil cooler is made up of copper pipe 2 cm internal diameter and 2 mm wall thickness. The inner face temperature is 90°C and the outer face temperature is kept down to 40°C by a fan blowing across the pipe. If the thermal conductivity of copper is 380 W/mK, what is the heat dissipation per metre run of pipe?

● PQ 10.9 After long use, a layer of scale builds up on the inside of the pipe, 1 mm thick. If the thermal conductivity of this scale is 1.25 W/mK and the temperatures are unchanged, what is the new heat flow?

That is as far as we will take conduction at present but there are further references to it later. This steady state treatment has plenty of real engineering use and it covers the important features for an introduction to unsteady or transient conduction whenever necessary, later in your studies. The same sort of treatment will introduce convection.

10.3 CONVECTION

Convective heat transfer is usually the most important mode associated with fluids. Remember that a fluid is anything that is not solid, so convection is to do with gases, liquids, vapours and slurries. It involves movement of the fluid, often on quite a large scale such as the heat flow associated with major ocean currents. Normally, however, convection is studied on a far smaller scale – maybe the flow through a *heat exchanger* such as a car radiator or hot gases in a furnace. The fundamental rules are the same, however. A fluid moves and is the agent of heat transfer.

As fluid movement is the important feature, the style of that movement is significant. It can be **natural convection**, driven by the density differences which may arise in a fluid due to temperature differences. Many weather effects are due to this. It may be **forced convection**, such as by the use of an electric fan. It may be convection to or from a solid – a fan heater warming your body. It may be convection between fluids – hot and cold water mixing in a bath. Wherever heat is transferred by the movement of a fluid, then convection is involved (Figure 10.8).

The fundamental equation for the calculation of heat transfer by convection has, like so many other thermodynamic laws, an investigator's name attached to it. In this case it is **Newton's law of convection**. As before, we will only look at steady state heat transfer and the simplest steady state convection to study is where there is heat transfer between a fluid and a solid surface. Not only is it the simplest but it is also the most common and, in engineering generally, the most important.

Since this convection, as the Second Law says and just like conduction, is driven by a temperature difference, then the surface tempera-

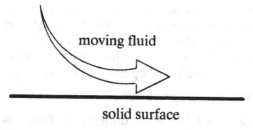

moving fluid

solid surface

Figure 10.8 Fluid movement – convection.

ture and the fluid temperature will be important. The amount of surface available for the heat transfer will count also – more surface, more heat transfer.

● PQ 10.10 If there is a wall of area A at temperature T_1 and this is being washed by a hot gas at temperature T_2, how are these factors likely to influence the convective heat transfer rate? Do not bother about the format of an equation, just say how they might be linked.

A note about that statement 'a hot gas at T_2'. It is somewhere between unlikely and impossible that any large amount of fluid is all at the same temperature. Where a hot fluid is transferring heat to a cool surface, then certainly the fluid temperature will be lower near the surface than a long way from it. In any case, for heat to be transferred at all, there has to be a temperature gradient and thus temperature differences. What is normally used for convective heat transfer calculations is the temperature of the general mass of fluid, the so-called **bulk fluid temperature**. It follows that any necessary thermodynamic properties must be evaluated at the bulk temperature and not some local or random temperature. Throughout this chapter, fluid temperature means the bulk temperature.

The Newton equation for the rate of convective heat transfer $\dot{q}$ is a very simple one and its form applies to most of the convection problems normally met. It involves the area and temperature factors, as in PQ 10.10, and also a factor called the **convective heat transfer coefficient**. This in turn depends on the properties of the fluid concerned – is it air, water or steam? – and the system in which the transfer is taking place – meaning is it a flat wall, some round tubes, a bath or what? The coefficient is commonly identified by the letter h or the letter U and the Newton equation at its simplest is

$$\dot{q} = h \, A \, (T_2 - T_1).$$

Just like thermal conduction, it is obvious which way the heat is flowing – down the temperature gradient from the high temperature to the low. The heat flow rate $\dot{q}$ is from T_2 to T_1, both of which can be measured. The available area A can be measured also but, rather like thermal conductivity k, the value of h is the awkward one. k depends on both the material and the way in which it is used. h depends on the fluid and on the system. There is no rule which says. 'The fluid is air, so the heat transfer coefficient is this value.' It has to be derived from the properties of the fluid and the properties of the system.

● PQ 10.11 What are the units of h?

As there has been a lot of work done on all forms of heat transfer, there is for most cases either a reasonably representative value of the convective heat transfer coefficient tabulated or it can be assessed without too much trouble. At worst, such as studying a new problem, a relatively simple mixture of analysis and experiment yields a lot of information and the techniques for doing this are well established.

● **PQ 10.12** Thinking generally about fluids – warm air, warm water, hot cooking oil maybe – that have either warmed you or splashed on you, what fluid properties might influence the rate or amount of heat transfer from them to you?

As can be imagined, there is likely to be quite a difference between the rates of heat transfer under natural or forced conditions. With natural convection, the heat transfer rate is more or less governed by the relatively gentle movement of a fluid as it responds to density differences, in turn due to temperature differences. The way that heat is convected around a room from a central heating radiator is an example (Figure 10.9).

The air by the radiator gets warm and rises because its density falls as its temperature increases. It moves up and away from the radiator, to be replaced by cooler air. The process is repeated and the air circulates around the room – natural convection is the agent.

If instead a fan was blowing the air across the radiator surface, then the rate of heat removal from the radiator would increase. This is an example of forced convection – the fluid flow does not depend upon natural or simple temperature-driven movement. Look at the compactness of a fan heater (forced convection) compared to a normal (natural convection) radiator for the same room-heating ability. The difference between forced and natural heat transfer coefficients is quite significant.

For example, with air the natural convection coefficients are roughly in the range 5 to 25 W/m²K and the forced, equally roughly, in the range

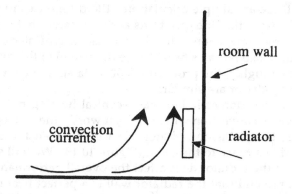

Figure 10.9 Natural convection from a radiator.

10 to 500 W/m²K. They do overlap because both the fluid and the system are important but, in general, turbulent convection coefficients are an order of magnitude or more higher than the natural ones. Just to emphasize the importance of the fluid, turbulent coefficients for water under the same conditions would be an order of magnitude higher again.

So where fluid temperature difference is the only driving force, this is natural convection and coefficient values are relatively low. Where a fluid is being driven mechanically or electrically, such as by a fan or a pump, this is forced convection and the coefficients are higher.

For the moment, that is enough on the background to convective heat transfer coefficients. Values are tabulated for many common applications so the examples which follow have taken values from the various data tables. Later, we take a closer look into the determination of coefficients.

Example 10.6 Air at 10°C flows under natural convection across a central heating radiator of surface temperature 50°C and surface area 0.4 m². Data tables give a value of heat transfer coefficient of 11 W/m²K. What is the rate of heat transfer by convection?

The temperature gradient is from the hot surface to the cool air, thus

$$\dot{q} = h A (T_2 - T_1) = 11 \times 0.4 \times (50 - 10) = 176 \text{ W}.$$

Example 10.7 Now an electric fan is put in front of the radiator to blow the air across. The convection is forced instead of natural and the data tables give a coefficient value of 150 W/m²K. What is the new rate of convective heat transfer?

The only item that has changed is the coefficient value, so

$$\dot{q} = h A (T_2 - T_1) = 150 \times 0.4 \times (50 - 10) = 2400 \text{ W}.$$

Although these are simple calculations, the differences in values are quite realistic. Note that the questions ask very carefully for the rate of heat transfer by convection. This is because a typical room central heating radiator also radiates heat (hence the name) to the room and its contents. Very roughly, the proportions of radiated and convected heat for a domestic radiator are similar.

Staying with the domestic hot water central heating radiator, there are two aspects of convection which make it work – the convection from the hot water to the inside radiator surface and then the convection from the outside surface to the room (Figure 10.10). We will see shortly that the wall of the radiator does affect the overall performance but, for the moment, pretend that the radiator wall is a perfect heat conductor. Any heat convected to the wall is transferred immediately through it, to be picked up by convection on the other side. The wall would be said to have *negligible thermal resistance*.

Suppose that the water temperature in the radiator is T_1, the radiator wall temperature is T_2 and the air temperature is T_3. If the convective heat transfer coefficient from the water to the wall is h_1 and from the wall to the air is h_2, with the radiator wall having area A, then for steady state heat transfer

$$\dot{q} = h_1 A (T_1 - T_2) = h_2 A (T_2 - T_3).$$

Make sure that those temperatures are in the right order.

● **PQ 10.13** Now eliminate the wall temperature T_2 to give $\dot{q}$ in terms of A, h_1, h_2, T_1 and T_3. It's just like the conduction, for Example 10.3.

This gives rise to the idea of an overall coefficient based on the values of the two individual coefficients. If h (no subscripts) is the overall convective heat transfer coefficient from the water to the air, then the equation derived in PQ 10.13 can also be written

$$\dot{q} = h A (T_1 - T_3),$$

where the overall coefficient h is

● **PQ 10.14** What? Write h in terms of h_1 and h_2.

This overall coefficient can also be expressed as

$$\frac{1}{h} = \frac{1}{h_1} + \frac{1}{h_2}$$

which helps to illustrate the possibility of rate controlling factors in convection just as met in conduction – the wall insulation of Example 10.4.

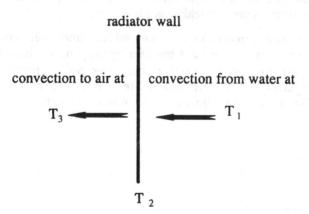

radiator wall

convection to air at   convection from water at

T_3 ⟵   ⟵ T_1

T_2

Figure 10.10 Convection from water to air, through the radiator wall.

Example 10.8 Two fluids are separated by a thin wall of negligible thermal resistance. Their individual convective heat transfer coefficients are 2 W/m² K and 10 W/m²K. What is the overall coefficient?

The overall coefficient is given by

$$\frac{1}{h} = \frac{1}{h_1} + \frac{1}{h_2} = \frac{1}{2} + \frac{1}{10} = 0.6 \text{ m}^2 \text{ K/W.}$$

Thus $h = 1.67$ W/m² K.

The overall coefficient is lower than either of the contributing ones but notice that it is closer to the smaller than the larger.

Example 10.9 What happens if the larger coefficient is raised to 100 W/m² K and then to 1000 W/m² K?

In the first case, the equation becomes

$$\frac{1}{h} = \frac{1}{2} + \frac{1}{100},$$

giving $h = 1.96$ W/m²K, and in the second

$$\frac{1}{h} = \frac{1}{2} + \frac{1}{1000},$$

which gives $h = 1.996$ W/m²K.

Even if h_2 became infinitely large, the overall coefficient would not exceed the value of the lower coefficient, 2 W/m²K. The lower one is the rate controlling factor. Whatever is done to the larger, the absolute limit is imposed by the smaller. It is rather like a football ground having gates and turnstiles. It does not matter how wide the gates are, it is the rate of flow of people through the smaller, slower turnstiles that really controls how quickly people get in.

Conduction and convection are met widely and their controlling laws, however complex the heat transfer operation, are based on the steady state ones met here. Even in transient or multidimensional heat transfer calculations, the techniques are extensions of these ones. Thermal radiation and heat transfer modes in combination are next.

PROGRESS QUESTION ANSWERS

PQ 10.1

There are plenty and my three are:

- conduction – heat loss through the wall of an oven or furnace;
- convection – a gas fired space heater in a warehouse;
- radiation – infra-red paint driers.

PQ 10.2

There are three important factors:

- the temperatures on either side of the wall;
- the thickness of the wall;
- the material of the wall.

Rate of means per unit time, per second in SI units. Heat energy is measured in joules J, so the rate of heat flow is J/s or watts W.

PQ 10.3

They must be the temperatures at the inlet and outlet face of the wall. They cannot be any intermediate temperature nor can they be some random outside temperature.

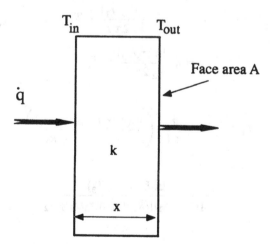

Figure 10.11 Fourier equation.

PQ 10.4

Consult Figure 10.11 $\dot{q}$ is a rate of flow, W, so

$$W = (k) \times m^2 \times \frac{K}{m}$$

from which the units of k are

$$\frac{W \times m}{m^2 \times K} = W/mK .$$

PQ 10.5

For the breeze-block wall to have the same thermal resistance as the cavity wall insulation, its thickness would have to rise to give an x/k value of 1.667. So the breeze wall would have to increase from its present 12 cm to $(1.667/0.15) \times$ 12 cm or 133 cm – an eleven-fold increase.

PQ 10.6

Exactly the same. The order of the materials, for the same face area and same overall temperatures, does not matter. Look at the denominator $x_1/k_1 + x_2/k_2 + x_3/k_3$ – the order makes no difference at all.

PQ 10.7

Without going into detail, you should arrive at

$$T_2 = - \dot{q} \; \frac{\ln(r_2/r_1)}{k_1 \, 2\pi \, L} + T_1$$

$$= \dot{q} \; \frac{\ln (r_3/r_2)}{k_2 \, 2\pi \, L} + T_3.$$

Thus

$$T_1 - T_3 = \frac{\dot{q}}{2\pi \, L} \left(\frac{\ln(r_2/r_1)}{k_1} + \frac{\ln(r_3/r_2)}{k_2} \right)$$

and

$$\dot{q} = \frac{2\pi \, L \, (T_1 - T_3)}{[\ln(r_2/r_1)]/k_1 + [\ln(r_3/r_2)]/k_2} .$$

PQ 10.8

This goes back to the simple pipe equation with only one term in the denominator.

The radii ratio is (2 + 0.2 + 0.2) cm external to 2 cm internal, i.e. 1.2. Thus

$$\dot{q} = \frac{2\pi\,(90 - 40)}{(\ln 1.2)/380} = 654.78 \text{ kW per metre.}$$

PQ 10.9

This is very similar to Example 10.5, so I will not go through every step. The one little error which may creep in is in assessing the radii ratios. For the copper pipe, it stays the same at 1.2 but take care with the scale.

There is a layer 1 mm on the inside of the copper pipe, so the outside diameter of the scale equals the inside diameter of the pipe which is 2 cm, but the inside diameter of the scale is (2 − 0.1 − 0.1) cm, i.e. 1.8 cm. The ratio (outer/inner) for the scale is thus 2/1.8 or 1.11. Thus

$$\dot{q} = \frac{2\pi\,(90 - 40)}{(\ln 1.2)/380 + (\ln 1.11)/1.25} = 3.74 \text{ kW/m.}$$

This problem is typical of car radiators which overheat – the fan cannot dissipate the heat because of the thermal resistance of the scale layer which is reducing the transfer of heat through the tube walls.

PQ 10.10

As noted, more surface means more heat transfer, so the bigger the area A, the greater the heat transfer rate.

Since temperature difference is the driving force, then the higher the difference $(T_2 - T_1)$, the higher will be the rate of convective heat transfer.

PQ 10.11

As $\dot{q}$ is in watts W, A is in m^2 and temperature difference in K, then the units of h are W/m^2 K.

PQ 10.12

Some are fairly obvious, like the fluid temperature and velocity – a fan heater running cool or warmer, slowly or quickly is a good example. Some points need a bit more thought. Specific heat capacity will govern how much heat a given size of splash may carry. The viscosity will count also – think of the difference between a hot water splash and a hot oil splash. We will meet the properties soon.

PQ 10.13

Without going through all the steps

$$T_2 = T_1 - \frac{\dot{q}}{h_1 A} = T_3 + \frac{\dot{q}}{h_2 A}.$$

Thus,

$$T_1 - T_3 = \frac{\dot{q}}{h_1 A} + \frac{\dot{q}}{h_2 A} = \frac{\dot{q}}{A}\left(\frac{1}{h_1} + \frac{1}{h_2}\right),$$

and

$$\dot{q} = \frac{A(T_1 - T_3)}{1/h_1 + 1/h_2}$$

PQ 10.14

By comparing the last two equations,

$$h = \frac{1}{1/h_1 + 1/h_2}.$$

TUTORIAL QUESTIONS

The following questions all refer to steady state and involve the application of the Fourier (conduction) or the Newton (convection) equations.

10.1 A solid furnace wall is made up of two layers of brick, the inner 0.1 m thick and thermal conductivity 1.04 W/m.K, the outer 0.15 m thick and thermal conductivity 1.83 Wm.K. If the inner face temperature is 500°C and the outer 150°C under steady conditions, what is the rate of heat loss per m² of wall and what is the intermediate (where the two brick layers join) temperature?   [1965 W; 311°C]

10.2 If the wall construction of the furnace in question 10.1 is reversed so that the 0.15 m thick (k = 1.83 W/mK) layer is on the inside and the 0.1 m thick (k = 1.04 W/m.K) layer is on the outside, but with the same inner and outer face temperatures, what is the new heat loss and the new intermediate temperature?   [1965 W; 339°C]

10.3 For the same furnace, the external surrounding atmosphere temperature is measured as 30°C and the internal furnace atmosphere is 800°C. What are the inner face and outer face convective heat transfer coefficients and what is the rate of convective heat transfer from the furnace outer wall to the external atmosphere?

[6.55 W/m²K; 16.4 W/m²K; 1965 W]

10.4 A central heating pipe of 2 cm outside diameter loses heat at the rate of 40 W/m length to the surrounding air from its surface which is at 65°C. What thickness of insulation, of thermal conductivity 0.125 W/mK, must be added to reduce the heat loss by 50%, assuming that the outside insulation face temperature is at 20°C? [4.85 cm]

10.5 For the insulated pipe of question 10.4, what is the external convective heat transfer coefficient if the local atmosphere temperature is 15°C? [10.9 W/m²K]

10.6 A thin steel plate of negligible thermal resistance separates two fluids, one a liquid at 100°C and the other a gas at 20°C. If the heat transfer rate through the plate is 480 W/m² and the liquid side heat transfer coefficient is ten times that of the gas side coefficient, what are the individual coefficient values?

[6.6 W/m²K (gas); 66 W/m²K (liquid)]

Heat transfer II – more on conduction and convection; radiation

11

11.1 COMBINED CONVECTION AND CONDUCTION

Although convection is important in ocean currents, in weather changes and in running a warm bath, by far the most common occurrence of it is in heat transfer between a fluid and a solid. The most common solid is either a wall (of a house or of a furnace) or a partition of some kind, such as the skin of a central heating radiator or an industrial heat exchanger. Almost always too the wall or partition is separating two fluids – the hot water in the radiator and the cooler air outside – so there is convection on both sides.

● **PQ 11.1** Think of another everyday example of heat being convected to a solid surface, passing through and then being picked up by another fluid.

Under steady state conditions, the convective flow of heat from one fluid to the wall or partition must equal the conduction heat flow through the solid wall which must equal the convection from the wall to the second fluid. That's what 'steady state' means (Figure 11.1).

> Convection from hot fluid to wall (Newton equation)
> = conduction through the wall (Fourier equation)
> = convection from wall to cool fluid (Newton equation).

● **PQ 11.2** Write this equation down in a mathematical form, using the notation of Fig 11.1 and a wall face area A, through which the heat is flowing.

That set of equations leads on to three important points. First, as it stands, we need to know the hot and cool fluid temperatures and the temperatures of the separating wall at either side. In practice it is fairly easy to measure fluid temperatures but not always so easy to measure

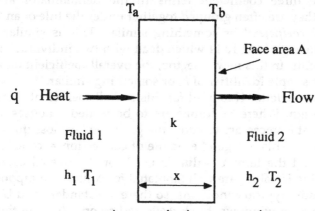

convection = conduction = convection

Figure 11.1 Combined convection and conduction.

the wall temperatures that are important here.

The problem is one of getting the thermometers in the right place and being quite sure that they are telling you what you need to know. It's alright for a simple case like a house wall but most industrial units are far less accessible. However, if the equations could be rearranged to eliminate the wall temperatures, just like the intermediate temperatures in conduction through composite walls, then this problem does not arise.

From the equations of PQ 11.2

$$\dot{q} = h_1 A (T_1 - T_a) = k A \frac{(T_a - T_b)}{x} = h_2 A (T_b - T_2),$$

or:

$$T_1 - T_a = \frac{\dot{q}}{h_1 A} \; ; T_a - T_b = \frac{\dot{q}}{k A/x} \; ; T_b - T_2 = \frac{\dot{q}}{h_2 A} .$$

Adding together the three left-hand side temperature terms and then the three right-hand side terms gives

$$(T_1 - T_a) + (T_a - T_b) + (T_b - T_2) = \frac{\dot{q}}{h_1 A} + \frac{\dot{q}}{k A/x} + \frac{\dot{q}}{h_2 A} .$$

Thus,

$$(T_1 - T_2) = \frac{\dot{q}}{A} \left(\frac{1}{h_1} + \frac{1}{k/x} + \frac{1}{h_2} \right)$$

and

$$\dot{q} = \frac{A (T_1 - T_2)}{1/h_1 + 1/(k/x) + 1/h_2} .$$

While the three coefficient terms in the denominator are truly individual, they are often grouped together under the title of an 'overall heat transfer coefficient' or something similar. This is similar to the overall coefficient of PQ 10.14 which dealt with two individual convection coefficients. In this new case, too, the overall coefficient may again be given the simple identifier of h, or something similar.

Sometimes the letter U is used for this overall coefficient, notably in building design. Where buildings are to be heated – houses, offices, stores – then it is necessary to know the probable heat loss through the building fabric. This is a good example of convection = conduction = convection and the term **U value** is used for the overall coefficient. These building U values are well tabulated for common components of buildings under typical conditions, so there are standardized U values for double-glazed windows, brick walls and so on. They are thus very useful design aids, reducing the amount of calculation required when working out the heating demands of buildings.

Note that this is only a little abbreviation or a handy design aid – it does not remove the need for finding the values of the individual components. The overall heat transfer equation can now be written quite neatly as

$$\dot{q} = h\,A\,(T_1 - T_2),$$

or, for buildings, $\dot{q} = U\,A\,(T_1 - T_2)$, where

$$\frac{1}{h} = \frac{1}{U} = \frac{1}{h_1} + \frac{1}{k/x} + \frac{1}{h_2}.$$

This leads on to the second point. There are three components of the overall coefficient and perhaps the immediate thought is that all three are important. While this is strictly true, it is unusual for all three to be equally important and it is commonplace to find that one factor exerts a **dominant** influence on the overall value. Refer back to Examples 10.8 and 10.9 where this was met for the two convection coefficients, and Examples 10.5 where a *rate controlling factor* arose.

Broadly speaking, for similar temperature differences (the driving force of heat transfer) and the same heat transfer area:

- rates of conduction tend to be faster than rates of convection;
- a dirty surface (like scale in a kettle) can upset this;
- rates of convection from liquids are faster than from gases, maybe by an order of magnitude or so;
- forced convection rates are higher than natural convection ones by an order of magnitude or so;
- convection from turbulent flow is commonly far higher than from laminar flow by a similar amount.

● PQ 11.3 A heat exchanger is a common industrial device which passes heat from a hot fluid through a separating wall to a cool fluid. In one design and using h_1, h_2 and k/x as above, it was found that h_1 was high, k/x was high and h_2 was low. Which if any of these would have the greatest influence on the overall coefficient? It may help to put some quite arbitrary numbers to the values.

So, if there are big differences between the components of the overall coefficient, the low value is the one that controls matters. It makes little difference if the high values are doubled or even more. The low value is the important one. As a reminder, you can think of the coefficients as gates. If the main gates to a football ground are open and 500 people a minute can walk through them but the turnstiles can only accommodate 50 people a minute, it doesn't matter how much wider the main gates are opened, the turnstiles will still control the rate at which people get on to the terraces.

Just to complete the picture, composite walls can be included in the new overall coefficient. It does not matter whether this is insulation of a house wall (a new tabulated U value) or a layer of scale on a heat exchanger surface. The overall coefficient would accommodate this by putting in another k/x term, exactly like the composite wall example:

$$\frac{1}{h} = \frac{1}{h_1} + \frac{1}{k_1/x_1} + \frac{1}{k_2/x_2} + \frac{1}{h_2}.$$

For the third point I have to say a few words about industrial **heat exchangers** – only an introduction, no detail. These devices are found throughout industry and are used for warming and cooling many different fluids. Designs are generally quite simple and the common ones consist of sets of tubes, through which one fluid flows, built into a container or shell through which the other fluid flows.

Figure 11.2 shows the idea of **parallel flow** exchangers, where both fluids flow in the same direction and **counterflow** exchangers, where the fluids flow in opposite directions. The choice for any particular application depends upon the properties of the fluids involved because the temperature characteristics of the two exchangers are different, as Figure 11.3 shows.

There is also a variation called a **cross-flow** heat exchanger, which is a descriptive name. The common example is the car radiator. Water flows through the radiator tubes from top to bottom, being cooled by air which is drawn or blown through the radiator from front to back. The flows are crossing each other rather than being parallel or counter to each other. Now back to that third point.

In the examples and in the determining of the overall coefficients, a single temperature has been taken for the hot fluid, another single

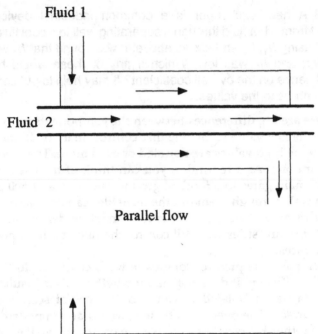

Parallel flow

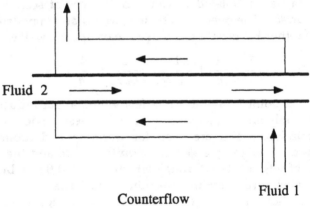

Counterflow

Figure 11.2 Parallel flow and counterflow heat exchangers.

temperature for the inner solid surface, another for the outer surface and another single value for the cool fluid. If you think of what the heat exchanger is trying to do though – take heat from one fluid and transfer it to another – then the temperatures must change as the fluids pass along the pipe. The hot one must get cooler and the cool one must get warmer. Fig 11.3 shows that.

At first, it was very convenient to consider just one temperature for each part of the transfer process. There are cases where this is valid – a small central heating radiator is very close to that; heat losses from a house can be calculated quite adequately by using one inside and one

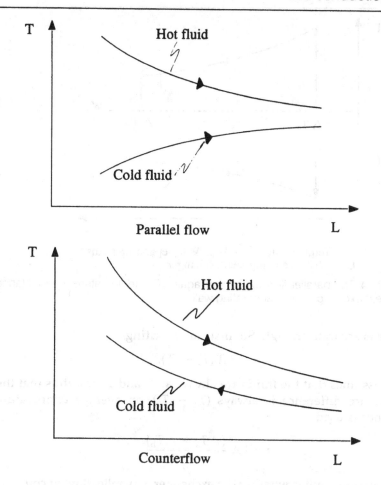

Parallel flow

Counterflow

T ... Temperature   L ... Distance along heat exchanger

Figure 11.3 Temperature characteristics of heat exchangers. T = temperature, L = distance along heat exchanger.

outside temperature. However, the chances of this being the usual industrial picture are limited – heat exchangers are definitely **not** single temperature devices! So something better than just the $(T_1 - T_2)$ term is needed for many real-life cases. Use the heat exchanger temperature profiles of Figure 11.4 to illustrate.

The temperature difference is the driving force for heat transfer, so a simple way round this might be to use an average temperature difference along the exchanger. In fact, providing that the difference at one end of the exchanger is no more than twice the difference at the other ($\triangle T_a$ not more than $2 \times \triangle T_b$), then a straightforward *arithmetic*

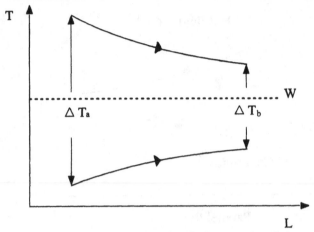

T ... Temperature   W ... Wall separating fluids
L ... Distance along heat exchanger

Figure 11.4 The parallel flow exchanger again. T = temperature, L = distance along heat exchanger; W = separating wall.

average is accurate enough. So, instead of writing

$$\dot{q} = hA\,(T_1 - T_2).$$

which assumes that the fluids are always at T_1 and T_2 and thus that the temperature difference is always $(T_1 - T_2)$, an average temperature difference is used

$$\dot{q} = hA\,\frac{(\Delta T_a + \Delta T_b)}{2}.$$

This equation applies whether the exchanger is parallel flow or counter-flow. For cross-flow ones, the same is used but with an empirical design factor. For instance, the cross-flow arithmetic average may be $0.8(T_a - T_b)$. This is a design factor and is not covered further here. Most standard texts on heat transfer discuss it further.

Example 11.1 In a parallel flow heat exchanger, Figure 11.4, the hot fluid enters at 140°C and leaves at 110°C. The cool fluid enters at 20°C and leaves at 40°C. The overall heat transfer coefficient is 15 W/m² K. What is the heat transfer rate per m²?

There are two ways of looking at the average temperature difference – either average the warm fluid temperature from one end to the other and then the cool, or look at the temperature differences ΔT_a and ΔT_b. Both give the same answer.

$$\Delta T_a = \text{(hot fluid in)} - \text{(cool fluid in)} = 140°C - 20°C = 120K$$

ΔT_b = (hot fluid out) − (cool fluid out) = 110°C − 40°C = 70K
Average temperature difference $(\Delta T_a + \Delta T_b)/2$ = (120 + 70)/2 = 95K
Average temperature of hot fluid T_h = (140 + 110)/2 = 125K
Average temperature of cool fluid T_c = (20 + 40)/2 = 30K
Average temperature difference $T_h - T_c$ = 125 − 30 = 95K.

Both ways give the same value (because they are both using the same numbers) of average temperature difference 95K. Thus the heat transfer rate per m² $(A = 1 \text{ m}^2)$ is

$$\dot{q} = 15 \times 1 \times 95 = 1425 \text{ W}.$$

This use of the arithmetic average holds good **provided** that the temperature difference at one end of the exchanger is no more than twice the difference at the other end. If it is more than twice the difference, then significant errors start to creep in. The reason is that using the arithmetic average says, in effect, that the temperatures are changing at a constant rate through the exchanger − a straight line temperature graph could be drawn. It is a simplification which is reasonable if the rule about twice the difference is observed. It is still a simplification, though, and the greater the temperature differences, the less accurate is the simplification.

Where the temperature difference at one end of an exchanger exceeds twice that at the other, then the **logarithmic mean temperature difference** (commonly LMTD) has to be used. The derivation is in Appendix B. This logarithmic mean, again using ΔT_a and ΔT_b as the temperature differences between the two fluids at either end of the exchanger, is given by

$$\frac{(\Delta T_a - \Delta T_b)}{\ln(\Delta T_a/\Delta T_b)}.$$

Thus, the heat transfer rate is given by

$$\dot{q} = hA\,\frac{(\Delta T_a - \Delta T_b)}{\ln(\Delta T_a/\Delta T_b)}.$$

The LMTD is usable at all times (except one, of course, where $\Delta T_a = \Delta T_b$ which gives an insoluble equation), it's just that the improved accuracy of the calculation at the lower temperature differences (one less than twice the other) is seldom justified in real circumstances.

Example 11.2 In a counterflow heat exchanger, the overall temperature difference at one end is 200 K and at the other 100 K. What is the average temperature difference and what error is introduced by using the arithmetic value? Just as a reminder, these are temperature differences, so there is no need to try to convert them to absolute values. The arithmetic average, $(\triangle T_a + \triangle T_b)/2$, is

$$\frac{200\text{K} + 100\text{K}}{2} = 150\text{K}.$$

The logarithmic mean temperature difference (LMTD) is given by

$$\frac{200 \text{ K} - 100 \text{ K}}{\ln(200/100)} = 144.3 \text{ K}.$$

This has been rounded from 144.27 because it is unlikely that real plant temperatures would be measured to 0.01 K and indeed it would be quite reasonable to round off the decimal also to 144 K. As in any other calculation though, the rounding off would otherwise be kept to the end of the calculation.

So the error is about 6.7 K, nearly 5%.

This is a measurable difference, so the LMTD is important. If the error was much less, then the arithmetic mean would be adequate. The reason for saying this is that while any basic design calculations have to be done accurately, they are often setting minimum sizes. The real-life interpretation of the design has to be done with cost and long service life in mind.

For example, standard size components, like tubes or nuts and bolts, are likely to be used in building refrigerators, guided by the design calculations; extra strength may be built in, like in bridges, or allowances for wear and tear, as in cars, again guided by the basic design calculations. For heat exchangers, extra heat transfer area may be designed in for when the exchanger starts to get dirty. Any minor errors incurred by using simplifications like the arithmetic mean are overwhelmed by the added sizes or strengths built in when standard sized components are used or long life is important.

- PQ 11.4 Why add extra area to a carefully designed exchanger?

For completeness it is worth noting that something parallel to the LMTD approach for temperatures can be used where areas change – the inside and outside areas of a pipe are not the same, for instance – and where heat transfer coefficients vary progressively, such as with change of temperature. That is for noting, however. Dedicated heat-transfer texts explore it fully but it is not particularly appropriate to this introductory treatment.

11.2 RADIATION

Since the word radiation has other connotations – X-rays, nuclear devices, for example – the proper title should be **thermal radiation**.

However, the simple title 'radiation' is used frequently. Thermal radiation is part of the electromagnetic spectrum and it covers the approximate wavelengths 0.1 to 1000 microns. Below the thermal radiation band is the visible spectrum (which is why you can see things that are red-hot) and above is the radio, radar and television band.

When looking at conduction and convection, there had to be some material present for the heat transfer to take place, a solid or a fluid respectively. Here is an important difference with radiation – nothing material (by way of a continuous solid or fluid) is needed for heat to be transferred by radiation from one body to another. In fact, anything between the source and sink is likely to be a barrier. Think of sitting in the sunshine – is a clear sky better than a cloudy one for you to receive the sun's heat? The heat you feel is radiated heat and, to all everyday intents, there is no significant material between us and the sun.

Just as you can feel the sun's heat even when clouds are present, so thermal radiation can pass through fluids (clouds, air) and solids (glass, clothing) to some extent. The extent depends upon the material properties and the wavelength of the radiation, (Figure 11.5). This latter fact is important when using radiation to heat treat various materials industrially, not unlike choosing the right wavelength for taking X-ray pictures.

There is another important difference when comparing the other heat transfer modes with radiation. In conduction and convection, the rate of heat flow depended on temperature difference in a simple way $(T_1 - T_2)$ is the sort of expression used, the temperature difference between source and sink. For radiation the rate depends upon the **fourth-power of absolute temperature,** i.e. T^4. The reasons for this fourth-power are bound up in the physics of electromagnetic radiation and are not going to be explored further here. Dedicated heat transfer texts cover the topic in depth.

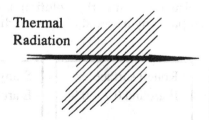

Figure 11.5 Radiation can pass through – it depends on the material and the wavelength.

● PQ 11.5 Without knowing any more about radiation, pretend there is a process where there is convection and radiation going on at the same time, from the same source to the same sink. At first both modes play an equal part. Which mode will dominate as source temperature rises? Why?

As before, early investigators' names are attached to the basic laws of thermal radiation, in this case the names are Stefan for the empirical or practical work and Boltzman for the theoretical. The fundamental law of radiation heat transfer is then the **Stefan–Boltzman law** and mathematically it is shown as something like

$$\dot{q} = \sigma \epsilon A T^4,$$

being careful to say 'something like' because that is the most basic form of the equation, indicating the heat radiated from a body at temperature T. There are several very valuable variations or sophistications.

It is worth having a look at the factors in that equation before going any further. The temperature raised to the fourth power T^4 has already been identified and the area A has the same significance as in other heat transfer calculations. The other two factors are commonly identified by Greek letters.

Epsilon, ϵ, represents the **emissivity** factor. For most practical purposes, it can be taken that thermal radiation is emitted from the surface of the source – the elements of an electric fire, for example. If there are several different surfaces, all the same size and all at the same temperature, they will radiate heat at different rates. In part, the rate will depend upon the material of the source. For instance, polished aluminium will radiate less than polished brass. Often far more important though is the condition of the surface – is it rough, smooth, coarse, fine? The rate of radiation from an oxidized copper surface – like an old copper tube that's been left outside – may be thirty times greater than from a shiny one of the same area and temperature. It is somewhat parallel to the property thermal conductivity – the actual value depends upon the material and its condition also (Figure 11.6).

Emissivity, then, has to take account of both material and condition and a simple way round this has been devised. A numerical value is essential for any calculation, and the solution used by Stefan and Boltzman was to compare any material and its condition to what might

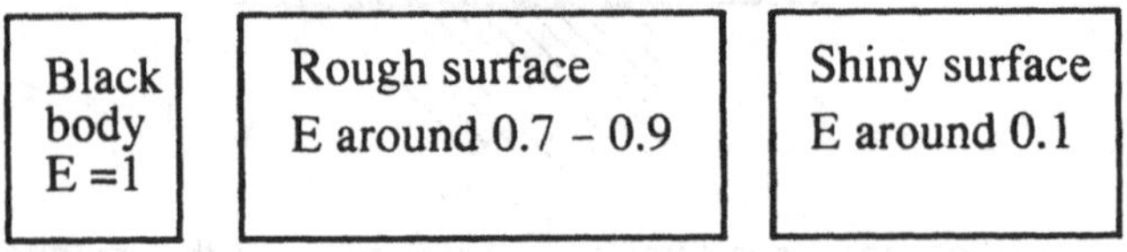

Black body E =1	Rough surface E around 0.7 – 0.9	Shiny surface E around 0.1

Figure 11.6 Emissivity (E) depends on surface conditions.

be called a **perfect radiator**. This ideal radiating surface – imaginary since it is ideal – is one which gives out the maximum theoretically possible amount of thermal radiation at any time. This perfect radiator is usually called a **black body** and it is given the ideal emissivity value of 1 or, sometimes, 100%.

● PQ 11.6 If a perfect radiator has an emissivity of 1, what in general can be said of the emissivity values of real (less than perfect) radiation sources?

While real materials have emissivity values less than 1, there are some very close approximations to a black body. The inside of a very hot furnace is one and deep space is another. There are plenty of everyday materials that have emissivities above 90%, but they are not all obvious. For instance, rough steel plate is about 0.95 or 95%, many painted surfaces exceed 0.9 or 90% as do both water and ice. You cannot guess the values – you must use the data tables.

Broadly, smooth shiny surfaces have low emissivities and rough surfaces have high ones. Incidentally, black body is a **thermal** name – it does not mean that the colour black is necessarily involved nor does it mean that every black-coloured surface has high emissivity. Emissivity values are usually temperature dependent but not always in the same fashion. For example, that of paper falls with rising temperature, whereas that of metals at large rises. The data tables are as useful here as in any other area of thermodynamics. Examples of emissivity values are shown below.

Temperature (K)	300	800	1200
Shiny metal	0.05	0.1	0.2
Dirty metal	0.7	0.8	0.9
Paper	0.9	0.7	0.5

Enough about emissivity for the moment. Compare the three modes of heat transfer again. The laws of Fourier and Newton say that the rate of heat transfer between bodies (source and sink) depends upon the temperature difference, something like $(T_1 - T_2)$. Stefan–Boltzman says that the rate of heat radiation from a body or source is proportional to T^4.

● PQ 11.7 Does that mean that the source will always radiate heat whatever the temperature?

This an important feature of thermal radiation – **all** bodies will emit radiation at **all** times. The opposite of emitting (giving out) is absorbing (taking in). Just as an emissivity value can be given to all materials or bodies, so an **absorptivity** value can be given, to indicate the ability to absorb radiation. The perfect absorber has an absorptivity of 1 (or

100%) – it will absorb all the radiation which falls on it – and real surfaces are compared to that. The absorptivity is usually identified by the Greek letter α, alpha.

Just as bodies can radiate at all times and all temperatures except absolute zero, all bodies can absorb at all times and all temperatures except absolute zero. The argument is exactly the same as for emission. Note carefully, though, an important difference. The emission is at the temperature of the emitting body – the energy source. The absorption also depends on the temperature of the emitting body. If there was no emitting body, there would be no absorption. Example 11.4 looks at this shortly.

● PQ 11.8 Is there likely to be any general relationship between absorptivity and emissivity? Clue – you know that shiny surfaces reflect light better than dirty ones.

For many purposes, the value of a body's or a surface's emissivity can be used as the value of its absorptivity. Both properties are temperature dependent, however, so errors can creep in because of differences between the emitter's and the absorber's temperature but, generally, the tabulated values of emissivity can be used for absorptivity.

The remaining factor in the basic Stefan–Boltzman equation, given the Greek letter sigma, σ, denotes the **Stefan–Boltzman constant**. What their work said was that the rate at which heat can be radiated from (or, of course, absorbed by) a surface was proportional to the surface area, the emissivity and (temperature)^4. Like any other equation which says something is proportional to something else, the introduction of a constant of proportionality makes this into something equals something else. The Stefan–Boltzman constant is one such and, in SI units, its numerical value is 5.643×10^{-8}.

● PQ 11.9 5.643×10^{-8} what? What are the units of the Stefan – Boltzman constant?

We can now have a closer look at the equation. If there is radiant heat transfer, which means transferring heat from one body to another, there must be an emitter and an absorber, a source and a sink. The source must be at a higher temperature than the sink otherwise there will be no net heat transfer – the Second Law tells us so.

Take careful note of this term *net heat transfer*. Previously, just heat transfer has been used when talking about conduction and convection. The heat has always flowed in one direction, from the higher to the lower temperature region. Here, though, we have to be careful with the words used because:

● all bodies or surfaces radiate at all times;

- all are capable of absorbing radiant heat at all times.

Refer back to PQs 11.7 and 11.8.

So, if bodies or surfaces can both emit and absorb all the time, there is not a one-way heat transfer, there is an *interchange*. The important factor is – how much more goes one way than comes back the other way? What is the net heat transfer? Suppose there is a high temperature surface, identified as M, and a low temperature surface identified as P, then M is radiating to P but also P is radiating to M, Figure 11.7. M will absorb heat from P and P will absorb heat from M. For simplicity, let them have similar emissivities and similar absorptivities.

However, as M is at the higher temperature (remember T^4), then M emits more than P. As their radiation properties are similar, then P will absorb more heat than M. There is a net heat transfer from M, the hotter body, to P, the cooler body, so the Second Law still holds good! It is important though to appreciate this **net** heat transfer idea.

Example 11.3 A flat plate of area 0.5 m² is at a temperature of 27°C, T_1, and has constant emissivity and absorptivity of 0.8. At what rate $\dot{q}_1$ is it emitting thermal radiation?

$$\dot{q}_1 = \sigma \epsilon A T_1^4 = 5.643 \times 10^{-8} \times 0.8 \times 0.5 \times (27 + 273)^4$$
$$= 182.8 \text{ W}.$$

Example 11.4 The same plate is put into a large furnace whose surface temperature is 900°C, T_2. At what rate $\dot{q}_2$ would it start to absorb heat? What is the initial net heat gain rate of the plate?

The important point to recall here is that the rate at which a body can absorb radiated heat depends upon the temperature of the emitter. To take a silly example, if the sun was cold, we would not feel warm in the sunshine. We only feel warm in front of an electric fire when it is switched on. We do not absorb radiant heat just because of our body temperature – the emitter has to be warm too!!

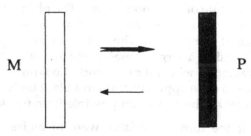

Figure 11.7 Radiant heat interchange.

Absorptivity means the proportion of incident radiation which can be absorbed. Recall that the inside of a large furnace is a good approximation to a black body, that is, it emits the maximum possible amount of thermal radiation. Thus the radiation which falls on the plate's 0.5 m² surface from the furnace's hot surfaces is

$$\sigma\, A\, T_2{}^4 = 5.643 \times 10^{-8} \times 0.5 \times (900 + 273)^4 \text{ W.}$$

but the receiving surface can only absorb the proportion that its absorptivity value allows,

$$0.8 \times 5.643 \times 10^{-8} \times 0.5 \times (900 + 273)^4 \text{ W.}$$

So, the Stefan–Boltzman equation for absorption still looks the same

$$\dot{q}_2 = \sigma\, \alpha\, A\, T_2{}^4,$$

but the T for absorption is the emitting or source temperature. Thus

$$\dot{q}_2 = 0.8 \times 5.643 \times 10^{-8} \times 0.5 \times (900 + 273)^4 \text{ W}$$
$$= 42732.9 \text{ W or, rounding, } 42.7 \text{ kW.}$$

The initial net heat gain rate is the rate of absorption minus the rate of emission:

$$\dot{q}_2 - \dot{q}_1 = 42732.9 \text{ W} - 182.8 \text{ W} = 42.55 \text{ kW.}$$

For this simple case then, an expression for net heat transfer rate $\dot{q}_3$ would be (absorptivity and emissivity have the same numerical value for this example, so don't be confused by the use of ϵ here!)

$$\dot{q}_3 = \sigma\, \epsilon\, A\, (T_2{}^4 - T_1{}^4).$$

● PQ 11.10 If the emissivity and absorptivity of the plate were not constant and equal, how would this equation change? Choose your own factors for the equation but identify them clearly.

In those examples the conditions were stated carefully – a large furnace, what were the initial rates, what is happening to the plate? As soon as the plate starts to warm up, its temperature changes and and thus so does the net rate of heat interchange. The plate will be radiating to the furnace too – but the numbers show that this is an insignificant amount, a fraction of one per cent. The T^4 effect is really dominant.

The plate was inside a large furnace, which means that the plate receives all the radiation which it can absorb. Looking back, the inside of a large furnace is a close approximation to a black body. The radiation available was therefore the maximum possible from the 900°C surfaces.

Example 11.5 If the furnace surfaces were of emissivity 0.9, what amount of radiation could the plate absorb from the furnace surfaces?

Again, absorptivity means the proportion of incident radiation which a surface can absorb. It cannot absorb heat if the heat does not fall on it. The amount available from these new furnace surfaces and falling on the plate is

$$\dot{q} = \sigma \epsilon A (900 + 273)^4 \text{ W}$$
$$= 48074.5 \text{ W}.$$

The plate has an absorptivity of 0.8, so it can only absorb 80% of the incident radiation, 0.8 × 48074.5 W or 38.46 kW.

So, what of relative sizes and temperatures?

● PQ 11.11 Put some progressively higher temperatures to the plate of Examples 11.3 and 11.4. Keep everything else constant. Suppose I can measure heat transfer rates to 1%. When, in your view, does the plate temperature become significant?

Now consider the relative size of the source (the furnace) and the sink (the plate). Really, it's not so much the size as the **overall system geometry**. Let us say that instead of a furnace at 900°C, the plate was exposed to a light-bulb filament at 900°C. Both sources are at the same temperature but common sense tells you that the plate will not get too warm from the light bulb.

● PQ 11.12 Why not?

Apart from the available energy, the temperature and the surface area, the amount of radiant heat which a surface receives depends upon the geometry. Think how hot the sun is – but you can't put a piece of steel in the sunshine and expect it to melt. It will get warm but that is all. The sun's radiant heat energy, like that of virtually all surfaces, is being spread out in all directions and it is a long way off. The proportion falling on the piece of steel is very small so, even though the sun is at a temperature which will melt steel, the distance apart of the source and sink means that it will not happen.

The same would happen with the plate in the furnace. Being inside the furnace, the plate would be able to 'see' all the radiating surfaces and thus receive energy from them. If, however, the plate was near a flat wall at 900°C, there would be no chance whatsoever of all the wall's radiant energy falling on the plate, as Fig 11.8 indicates. This being so, some sort of correction factor has to be introduced to allow for only a portion of the available radiation being received by the intended target.

Various so-called 'shape' or 'view' factors have been devised for many common radiant heat transfer applications and in this way the calculations can take account of system geometry. These factors are commonly given the capital letter F, with subscripts to identify individual ones if

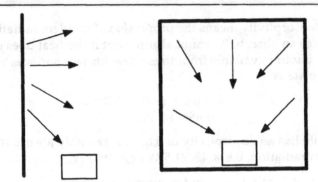

Figure 11.8 The geometry is important.

necessary. They are used as straightforward multiplication factors, just as emissivity and absorptivity are multiplying factors.

Example 11.6 For the conditions of Example 11.5, the furnace is replaced by another heat source such that the source-to-sink shape (or view) factor is 0.75. What now is the initial rate of radiation absorption?

The geometry of this new heat source/heat sink configuration will decide the shape factor – in fact, it is also called a 'configuration factor'. It is a fairly simple determination of how much of the emitted energy the plate could intercept, since radiant energy is emitted in all directions from most surfaces. Many dedicated heat transfer books either derive or list factors for many common shapes.

The new absorption rate, is therefore,

$$\dot{q} = F \sigma \alpha \epsilon A T^4$$
$$= 0.75 \times 5.643 \times 10^{-8} \times 0.8 \times 0.9 \times (900 + 273)^4 \text{ W}$$
$$= 28.84 \text{ kW}.$$

As with the combined modes of conduction and convection, so there arise combinations of radiation and other modes. Perhaps the most common is the domestic central heating radiator, where the heat transfer from the radiator to the room is around equal parts radiation and convection. While the combinations are not taken further here, they do occur together and, in applications of the laws and lessons of heat transfer, need to be considered together. There may well be a dominant mode but, until the calculations are completed, this may not emerge clearly.

Overall, then, the steady state rules of thermal radiation lay the ground for more advanced work when necessary, just as in conduction and convection. There are important differences to observe with radiation but there is much analytical and empirical information available.

11.3 MORE ON CONVECTIVE COEFFICIENTS

For many scientific and engineering applications, it is possible to write down an equation which links one measurement or factor with others quite accurately. The components of the equation can be measured readily and reliably. The steady state conduction equation

$$\dot{q} = -k A \frac{\Delta T}{\Delta x}$$

is a good example. Even allowing for the practical variability of k, the right-hand-side factors are straightforward everyday measurements and thus $\dot{q}$, the rate of flow of heat, is assessed with confidence.

Formulating other relationships with the same confidence may not be so easy. In a solid, with heat transfer by conduction, nothing moves very much and it is primarily a matter of common sense to find out what is important. With convective heat transfer in a fluid, anything that influences fluid motion may and most likely will play a part. When thermal gradients (the effects of temperature difference) are added, the picture becomes even more complex. Look back to PQ 10.12.

One way of dealing with this sort of difficulty is to prepare a set of experiments to investigate individual problems but the findings are likely to be limited to the individual study. For instance, experiments with a fan heater will give information which can be used with other fan heaters. It may be useless in heat exchanger design, even though both depend on convection. However problematical, there needs to be an analytical solution for widespread use.

One technique for this is called **dimensional analysis**. To illustrate, take forced convective heat transfer from a fluid to a solid surface. The Newton equation

$$\dot{q} = h A \Delta T$$

applies. In this equation, the area for heat transfer A and the temperature difference between the fluid and the solid surface ΔT can be measured. Recall that care has to be taken in respect of these temperatures – a local temperature? bulk fluid temperature? LMTD? but ΔT is easy to obtain in most practical circumstances.

As mentioned when we first met convection, however, the coefficient h is not always easy to fix. There are tabulated values for many common or routine applications but new designs or particular problems often arise. Thus the ability to calculate values which are not tabulated is important.

The question which arises when trying to analyse h is a simple one – what factors or properties may influence the flow of a fluid over a surface and the way in which heat is transferred to that surface? However, the answer is not so easy! Numerous possible factors can be listed and in the early days of heat transfer study, many experiments were done to eliminate the irrelevant ones or identify the important ones. The experiments are simple – set up a simple convection problem and vary each factor independently to see the effect.

• PQ 11.13 Suggest – no details, just the idea – a simple experiment to look at some of the factors.

For our present example of forced convective heat transfer from a fluid to a solid surface, the factors proved to be the fluid properties of thermal conductivity k, specific heat c_p, density ρ, dynamic viscosity μ and the fluid velocity V. Note **carefully** that the letter V is used both for velocity and for volume in various texts. Take care to use the right one. For this section, it is velocity.

As well as the fluid properties, some dimension which defines the solid surface, usually called a 'characteristic dimension' d, of the surface involved, such as length if the surface is a wall or diameter if a tube, is important. A general relationship can thus be written to show h as a function of these factors in the conventional form:

'heat transfer coefficient is some function of these properties'
$$h = f(k, c_p, \rho, \mu, V, d).$$

• PQ 11.14 What are the dimensions of each factor?

Experiments also showed that these factors are related in a relatively simple way and that the equation was quite valid if written

$$h = \text{constant } (k^a, c_p{}^b, \rho^e, \mu^f, V^x, d^y)$$

that is, the factors are each raised to some individual power. That simplifies matters but it does not provide a complete solution.

Any complete equation has to be dimensionally consistent, as was met in Chapter 1. For the present factors, the dimensions mass [M], time [T], length [L] and temperature [θ] are necessary and replacing the factors with their dimensions in the equation,

$$\left[\frac{M}{T^3\theta}\right] = \left[\frac{ML}{T^3\theta}\right]^a \left[\frac{L^2}{T^2\theta}\right]^b \left[\frac{M}{L^3}\right]^e \left[\frac{M}{LT}\right]^f \left[\frac{L}{T}\right]^x [L]^y.$$

• **PQ 11.15** Here is a more sophisticated use of dimensions than met previously. Bearing in mind what is meant by dimensional consistency, what may be a next step? Do not do it – just say what it might involve.

Comparing the indices on either side of the equation for [M], [L], [T] and [θ] will give four equations (one each for the four dimensions) but there are six unknowns – a, b, e, f, x and y. At first sight this seems insoluble because there are not enough equations to give discrete values for the indices. However, as we have come to expect in world of thermodynamics, much of the hard work has been done for us.

Early in the 20th century, a researcher called Buckingham showed a simple link between the number of variables in an equation like this and the number of dimensions involved. It uses the concept of the **dimensionless group** – that is, several variables arranged together so that their dimensions cancel out in that group.

For instance, group together velocity V, characteristic dimension d, density ρ and viscosity μ as $(V\,d\,\rho\,\mu)$. Applying the dimensions,

$$\left[\frac{L}{T}\right] \times [L] \times \left[\frac{M}{L^3}\right] \times \left[\frac{LT}{M}\right]$$

and they all cancel out. The group is said to be **dimensionless**. This particular group of properties is important in fluid flow studies and is named after a researcher, being called the **Reynolds number**. It is traditional to give important groups the title 'number', usually with an investigator's name attached.

The link that Buckingham showed was that in any problem, if there are X variables (density or specific heat for instance, to a total of X of these properties) and there are Y dimensions, then they can be tied together firmly by $(X - Y)$ dimensionless groups. For this case of forced convection, there are seven variables $(h, k, c_p, \rho, \mu, V, d)$ and four dimensions $([M], [L], [T], [\theta])$ so three dimensionless groups are sought. It is now a matter of putting the variables into three usable or recognizable groups and again, the hard work has been done by past researchers.

In the present case, the groups of value are called the Reynolds number, the Nusselt number and the Prandtl number, which are in turn usually abbreviated to Re, Nu and Pr. These groups are

$$Re\text{: } \frac{V\,d\,\rho}{\mu}, \quad Nu\text{: } \frac{h\,d}{k}, \quad Pr\text{: } \frac{c_p/\mu}{k},$$

and they are used widely in aerodynamics, chemical engineering and so on.

- PQ 11.16 Check that these are truly dimensionless.

One of the consequences of Buckingham's work was that the required number of dimensionless groups could then be arranged as a simple equation between the groups. For turbulent convective heat transfer, the Reynolds, Nusselt and Prandtl numbers are usually arranged in the form

$$Nu = \text{constant } (Re)^p (Pr)^q.$$

So, an analytical relationship for h is derived. The value of the constant and the indices p and q have to be found by experiment but this has already been done for a wide range of operations. For instance, the expression for turbulent convection in a pipe is

$$Nu = 0.023 \, (Re)^{0.8} \, (Pr)^{0.4},$$

and this applies very closely for gases and many liquids over a wide range of pipe sizes.

Example 11.7 Given these properties for water – density 1000 kg/m³, specific heat capacity 4.18 kJ/kg K, dynamic viscosity 4.8×10^{-4} kg/m s and thermal conductivity 0.65 W/m K – which is flowing at 2 m/s through a pipe of 0.15 m diameter, determine the convective heat transfer coefficient from the foregoing equation.

Putting the individual factors into the equation,

$$(h \, d/k) = 0.023 \times (V \, d \, \rho/\mu)^{0.8} \times (c_p \, \mu/k)^{0.4}$$

$$\left(\frac{h \times 0.15}{0.65}\right) = 0.023 \left(\frac{2 \times 0.15 \times 1000}{4.8 \times 10^{-4}}\right)^{0.8} \left(\frac{4180 \times 4.8 \times 10^{-4}}{0.65}\right)^{0.4}$$

from which $h = 6.777$ kW/m² K.

There are plenty of similar equations in the reference texts to cover flow over banks of tubes, across flat plates and so on. It is even simpler than that – most of the expressions have either been simplified for very common operations or converted to nomograms for general use.

PROGRESS QUESTION ANSWERS

PQ 11.1

The house again – warm air inside, cool air outside; a jug of water cooling down; a pan of milk warming on a gas ring.

PQ 11.2

This needs you to think very clearly about the minus sign in the Fourier equation. To be consistent, I have eliminated it by putting all the temperature differences as: warmer − cooler. If you have any doubts, revise now. Thus,

$$\dot{q} = h_1 A (T_1 - T_a)$$

$$\dot{q} = k A \frac{(T_a - T_b)}{x}$$

$$\dot{q} = h_2 A (T_b - T_2).$$

PQ 11.3

The overall coefficient is made up of the three individual ones as

$$\frac{1}{h} = \frac{1}{h_1} + \frac{1}{k/x} + \frac{1}{h_2}.$$

Using arbitrary numbers – think some up – say the two big ones h_1 and k/x are 100 W/m^2 K, and the little one h_2 is 10 W/m^2 K. Thus,

$$\frac{1}{h} = \frac{1}{100} + \frac{1}{100} + \frac{1}{10}$$

and h = 8.33 W/m^2 K. If h_1 and k/x were 1000 W/m^2 K, h would only rise to 9.8 W/m^2 K. This is exactly the same as in Examples 10.8 and 10.9 – the low value controls the overall coefficient.

PQ 11.4

The extra is of course extra heat transfer surface area. The overall heat transfer coefficient can be estimated quite accurately for clean surfaces. If the surface starts to get dirty or attracts a layer of scale (like a kitchen kettle), then the overall coefficient will change. Look back at PQ 10.9. Since dirt and scale are good insulators, the overall coefficient will fall. Recall PQ 11.3 – the overall coefficient governed by the poorest. If the coefficient falls, then the heat transfer rate falls. This can be offset by having more area available.

PQ 11.5

As temperature rises, radiation will play a progressively greater part. T^4, the radiation driving force, rises much quicker than T, the convection driving force.

PQ 11.6

Real values must be less than 1.

PQ 11.7

Yes, because the law says (absolute temperature)^4. The only time that the source or body will not radiate then is when it is at absolute zero, therefore.

PQ 11.8

Light and thermal radiation are both parts of the electromagnetic spectrum so they follow the same general rules. If a shiny surface reflects some light, then it is likely to reflect some thermal radiation. That is, it will have a lower absorptivity than a poorly reflecting dirty surface. Emissivity is less for a shiny surface also, so it looks as though emissivity and absorptivity are linked. Low absorptivity, low emissivity and vice versa.

PQ 11.9

The simple equation says

$$\dot{q} = \sigma \epsilon A T^4.$$

The units are: $\dot{q}$ is in W; ϵ is a ratio, no units; A is in m^2; T^4 is K^4. Thus the units of σ are W/m^2 K^4 and the constant is 5.643×10^{-8} W/m^2 K^4.

PQ 11.10

Emissivity and absorptivity both vary with temperature, like so many thermodynamic properties. As emission of thermal radiation from a surface is to do with the temperature of the surface, say T_1, then the emissivity must be the value at that temperature, say ϵ_1.

Absorption by a surface is to do with the temperature of the emitting source, say T_2, not the receiving surface primarily (sunshine!) so the absorptivity value must be the value relevant to that source temperature, say α_2.

So, the net heat gain equation becomes

$$\dot{q} = \sigma A (\alpha_2 T_2^4 - \epsilon_1 T_1^4).$$

PQ 11.11

If I can measure to 1%, then the plate temperature may be significant when the plate radiation is about 1% of the furnace radiation, the big one. I estimate that the plate has to reach about 390°C which again underlines the domination of T^4. Note this value is only to do with this example. It is not a general figure in any way.

PQ 11.12

With the furnace, there is a lot of heat and a lot of radiating surface. With the filament, even at 900°C, there is not much energy being put in (say a 60 W bulb) so there is not much to give out. The surface is very small too – the emission rate depends upon the surface area A as well as the temperature.

PQ 11.13

One idea would be to take a clean radiator from a car with an electric fan blowing air through it. Now flow through its pipework various fluids – air, oil, water – at various temperatures. Check the rate of heat transfer and see how these fluids influence it. The fluid properties are known, so their influence can be deduced.

PQ 11.14

h: $[M/T^3\theta]$; k: $[ML/T^3\theta]$; c_p: $[L^2/T^2\theta]$; ρ: $[M/L^3]$; μ: $[M/LT]$; V: $[L/T]$; d: $[L]$.

PQ 11.15

It might involve trying to find the indices a, b, e, f, x, y so that the equation has the same dimensions on either side – so that it is dimensionally consistent.

PQ 11.16

$$\frac{hd}{k} : \left[\frac{M}{T^3\theta}\right] \times [L] \times \left[\frac{T^3\theta}{ML}\right],$$

$$\frac{c_p\mu}{k} : \left[\frac{L^2}{T^2\theta}\right] \times \left[\frac{M}{LT}\right] \times \left[\frac{T^3\theta}{ML}\right],$$

and the Reynolds number has already been discussed.

TUTORIAL QUESTIONS

These questions refer to steady state operation and are applications of the fundamental heat transfer equations. Take care to understand whether temperatures apply to solid faces or to surrounding atmospheres.

11.1 Air on one side of a wall is at 150°C and on the other side at 20°C. The convective heat transfer coefficient is identical on either side at 5 W/m² K. The wall is 12 cm thick and of thermal conductivity 2 W/m K. What is the steady state rate of heat transfer through the wall?

[282.6 W/m²]

11.2 Water at 90°C flows through a steel pipe of inside diameter 5 cm, outside diameter 6 cm and thermal conductivity 70 W/m K. There is a 3 cm thick insulation layer, thermal conductivity 0.2 W/m K, around the pipe and the surrounding atmosphere temperature is 15°C.

Given that the heat loss from the pipe to the surroundings is 60 W per metre run and that the outer face (insulation to air) convective heat transfer coefficient is 5 W/m² K, what is the value of the convective heat transfer coefficient from the water to the inner surface of the pipe?

[38 W/m² K]

11.3 A parallel flow heat exchanger has the hot fluid entering at 250°C and leaving at 60°C, while the cold fluid enters at 30°C and leaves at 50°C. What is the log mean temperature difference for this exchanger?

If the overall heat transfer coefficient was 21 W/m² K, what would be the rate of heat exchange?

[67.9K; 1.427 kW/m²]

11.4 If a counterflow heat exchanger worked with the same temperature limits and overall coefficient as the fluids in question 11.3, what would be the LMTD and heat transfer rate in this case?

What error would be introduced in either case by the use of arithmetic mean temperature differences?

[89.6K; 1.882 kW/m²; parallel, 69%; counterflow, 28%]

11.5 A large source of thermal radiation at 800°C is heating a 10 cm × 10 cm ceramic plate which is initially at 30°C. If the source emissivity is 0.8 and the plate absorptivity is 0.75, what is the initial net rate of radiant heat transfer to the plate?

[446 W]

11.6 One face of a flat steel plate, whose emissivity and absorptivity are equal at 0.6, is exposed to a large source of thermal radiation of emissivity 0.9, temperature 400°C. The plate can emit radiation from both its faces to its surroundings and water is sprayed onto the rear face (that is, the one not receiving the radiation) to remove heat by convection. The plate's thermal resistance may be ignored.

If the water is at 10°C and the spray's convective heat transfer coefficient is 50 W/m² K, estimate the steel plate temperature at steady state.   [About 120°C]

Reacting mixtures **12**

12.1 WHAT IS A REACTING MIXTURE?

Up to now, we have met gases and liquids and vapours which have been heated or cooled or compressed or expanded. They have been treated as working fluids in that they can be agents for energy usage or energy transfer, that is for doing something useful. If a gas is heated, it may be used to do some work. If water is boiled, the steam produced can be used to drive a turbine and so on. They have, however, all been treated as though they were inert – incapable of much other than doing something physical (heat or work transfer) with the energy that has been supplied to them.

There are plenty of engineering demands for an inert working fluid. It is necessary for the electricity industry – steam from boilers being used to drive alternators. It is necessary in most manufacturing industries – compressed air being the inert fluid for paint spraying or for some automatic machines. However there are also many examples where it is essential for the working fluid to react chemically as well as physically (Figure 12.1).

A **reacting mixture** then is a mixture of two or more substances that will interact chemically. While these substances can be as straightforward as say an acid and a base as in school chemistry, the possibilities are wide ranging. For proper use their interaction needs to be controlled and the outcome applied to some sort of process or operation.

Combustion of fuels is the obvious reacting mixture example for thermodynamics – as in using a fuel for firing the electricity industry boilers to make steam, or as in using petrol to make a car engine work. It is certainly the most widespread and largest user of the principles of reacting mixtures. Since combustion covers the general principles of the topic and has such important real life applications, **fuels and combus-**

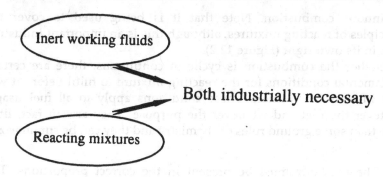

Figure 12.1 Inert and reacting fluids are needed.

tion will be used to illustrate the the whole topic.

● **PQ 12.1** To set the scene, list three useful applications of a fuel and say what other common substance is in the reacting mixture.

Strictly, the word 'combustion' can be applied to reactions between many materials which result in significant heat release but commonly the word is used for reactions between fuels and air. There are some quite exotic fuels, such as those used in rocketry or in some speed record attempts, but here we are staying with everyday fuels. Thus the word fuel, from here on, means a routine one such as natural gas, petrol or coal which needs air to burn. There is no need to go into their properties in any great depth nor to study the equipment in which they are used but it is worthwhile looking at the background to their use.

12.2 THE ELEMENTS OF COMBUSTION

Combustion is the general name given to an overall reaction between fuel and air or, more correctly, the oxygen in the air. It is often subdivided into **continuous combustion** and **cyclic combustion**, these names being descriptive. In the former, fuel is supplied continuously at a steady rate, to burn steadily and continuously. A gas-fired central heating system would be a good example, where the gas burns steadily in the boiler until the house thermostat tells it to stop.

The motor car engine is the most common example of cyclic combustion, where individual amounts of fuel and air are fed to a cylinder. There they react, do work and are ejected and then the whole process is repeated. There is a **cycle of operations**.

● **PQ 12.2** Why is this cyclic combustion? If I am on a journey, my car engine usually runs continuously, so what is cyclic about it?

Since thermodynamic cycles – cyclic or repeated sequences of operation – are in later chapters, most of what follows here will deal with continuous combustion. Note that it is being used to cover the principles of reacting mixtures, although this is an important industrial topic in its own right (Figure 12.2).

Whether the combustion is cyclic or continuous, there are certain fundamental conditions for the reacting mixture to fulfil before it will react in a usable fashion. These conditions apply to all fuel usage, whatever the fuel and whatever the purpose. They are, in fact, little more than some ground rules of chemistry and they can be summarized as:

(a)   The reactants must be present in the correct proportions. Too much or too little of one and either they will react or something is going to be wasted.

(b)   They must be mixed together thoroughly. If they are not mixed, then they cannot react at all. If they are not mixed thoroughly, then they cannot react completely.

(c)   The correct proportions well mixed does not ensure reaction. There has to be the correct burning conditions. Something has to start the reaction.

● PQ 12.3 Think of everyday fuel usuage. Name two ways of starting the combustion reactions.

(d)   Time is necessary for the reactions to go to completion. Most burning reactions seem very quick but they are progressive and they do take a finite time.

That is very much a summary and each of these factors needs to be studied a little further (Figure 12.3).

For the first, the **correct proportions**, there are two aspects to 'correct'. At its simplest, this means the proportions that could be represented by a complete chemical equation, displaying the reactants and the products. Each reactant is consumed completely, there are no spare bits and the products are predictable from the rules of chemistry. Methane and oxygen burning together is a good example:

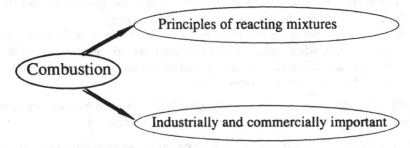

Figure 12.2 Combustion illustrates reacting mixtures.

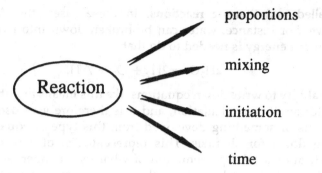

Figure 12.3 Reaction conditions.

$$CH_4 + 2O_2 \rightarrow CO_2 + 2H_2O.$$

● **PQ 12.4** In simple words, no technicalities, what does this equation mean? Noting that this is an example of a combusting or burning mixture, is something missing?

The reaction as written shows the proportions of reactants and of products. The proportions in the reaction are **molecular proportions** because this sort of reaction representation is of a molecular reaction. It is a theoretical statement and there must be the unspoken words 'If everything is perfect then this would be the reaction.' It is a model of what might be.

Just to digress a little, even the simplest reaction is actually a chain of events but it is normal to show only the starting and finishing points. Reactions do not go to completion but actually reach a state of equilibrium. If the reaction conditions have been controlled properly, however, there will be almost complete conversion. For our purposes, the molecular reaction is adequate.

Since the reaction is an example of a combusting mixture, then a complete statement should say something about an energy release. Where energy is given out (heat released) this is called an **exothermic reaction**. Combustion reactions are exothermic and may be shown by the inclusion of an enthalpy term, often identified as ΔH, at the end of the reaction,

$$CH_4 + 2O_2 \rightarrow CO_2 + 2H_2O - \Delta H.$$

The value of ΔH has been determined for most common fuels and is usually reported on a mass basis, such as an oil fuel having a value of 42 MJ/kg. Notice the negative sign. The enthalpy term is negative in exothermic reactions, $-\Delta H$. It is a convention showing that energy is leaving the system to be used elsewhere.

Some reactions (not combustion ones) may take in energy and these are called **endothermic reactions**. In these cases, the ΔH term is positive. For instance, water can be broken down into hydrogen and oxygen but energy is needed to do that:

$$2H_2O \rightarrow 2H_2 + O_2 + \Delta H.$$

The ability to write down equations like this means that there is some basis for any other calculation, and it is therefore very useful. Similar equations or something developed from this type of equation can be written down for all fuels. This representation of a reaction is the theoretical or ideal representation of what may happen and the word **stoichiometric** is used for these chemically correct proportions. However, this is the ideal picture and it is not completely true-to-life.

Various real-life demands bring about modifications to the proportions shown stoichiometrically. One example is that a motor car engine requires a **fuel-rich mixture** for maximum power, such as when accelerating or overtaking. Another is the use of **excess air** (that is, more than the stoichiometric demand) in boiler or furnace combustion to counteract the practical limitations to air and fuel mixing, thus to ensure complete use of the fuel. These are very practical demands. In no way do they make the combustion equations incorrect but they do show that there is more to the real problem than an ability to write down equations.

That first ground rule of **correct proportions** therefore has a theoretical and a practical part to it (Figure 12.4). The common way of representing real life proportions is by ratio rather than absolute amount. Thus you will meet terms like 'air/fuel ratio', 'stoichiometric air/fuel ratio' and so on. For instance, the stoichiometric ratio for burning kerosine (paraffin) with air is about 14.7 kg air to each 1 kg kerosine. This is usually written as 'A/F ratio 14.7:1 w/w', meaning an air-to-fuel ratio of 14.7 to 1 on the weight-to-weight basis. The practical ratio is higher, for instance around 16:1 to 19:1 if used in a boiler, the actual value depending upon operational conditions.

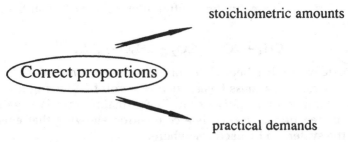

Figure 12.4 The first ground rule.

For the second factor, that of **mixing the reactants**, the ideal would be the most intimate contact possible. How well this can be done for common fuels and air depends mainly on the type of fuel.

For a gaseous fuel, this intimate mixing can be approached very closely – all gases are mutually miscible in theory and very nearly so in practice. For a liquid fuel this is a bit more difficult unless the fuel is very volatile and can be made to evaporate into and then mix with the combustion air.

● PQ 12.5 If the liquid fuel does not evaporate too well, like a heavy oil, what else can be done to the fuel to aid mixing with the air? Think of our model of wet steam.

This process of breaking the fuel up into tiny droplets (actual size a few tens of microns) is called **atomization**. It is a trade type title – there are no atomic sizes involved!

With a solid fuel, evaporation can be ignored other than the escape of the volatile matter which forms part of most solid fuels. Generally, the best that can be achieved is to make sure that the air can get to the fuel surface as effectively as possible, either by breaking the fuel up finely or by careful control of the combustion air supply.

Mention was made earlier of the need for excess air to ensure the complete combustion of fuel in a boiler or a furnace. The mixing of materials, whether it is the ingredients of a cake, a pot of paint or fuel and air for burning, needs energy. It is unlikely practically that complete mixing can ever be achieved – the last dregs of the ingredients are going to be trapped in the mixture somewhere, however turbulent it is. There has to be a practical and economic balance, therefore, of how much mixing is done to achieve acceptable results – a balance between the energy used for mixing and the end result.

● PQ 12.6 If perfect mixing of fuel and air is unlikely and fuel wastage is to be avoided, what can be done?

The skill is in putting in enough extra air – generally called '*excess air*' because it is excess to stoichiometric requirements – to mop up the fuel but no more. In practice, this depends upon the fuel and the equipment quality but the extra proportions range from virtually nothing to around 50% of stoichiometric (Figure 12.5). The same sort of conclusion applies to most reacting mixtures, not just fuel combustion.

The third factor is that of **proper conditions** for reaction. We think of petrol burning very easily but if I have a jar of petrol standing in the open air, all that will happen is that the petrol will start to evaporate. It is very unlikely to burn spontaneously. If I took that jar of petrol and bubbled cold air through it so that the two were mixed thoroughly, little other than yet more evaporation would occur. If, however, I put a

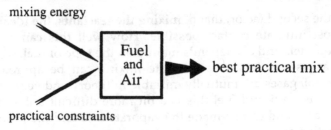

Figure 12.5 The second ground rule.

lighted match to the evaporating mixture – not a wise thing to do! – then it would be likely to start to burn.

Some mixtures react more or less spontaneously, like acid + base mixtures which you will have met in school chemistry. Others need specific starting conditions – an energy boost – like a Bunsen burner needs a match or a spark to light its gas flame. The starting amount of energy is often called the **activation energy** – it activates the reaction – and the Bunsen burner match or the car spark plug are the suppliers of this. The activation energy demand of a reaction is a precise, measurable and tabulated amount. In real life it is not metered into the reaction but satisfied by whatever is a sufficient source – the spark plug or the match.

Generally, in fuel combustion, once a reacting mixture has been given a start, some of the energy released by the initiated reaction is used to maintain the reaction. A steady flame in a central heating boiler ignites the fresh incoming fuel continuously. Once a spark in a car engine cylinder has ignited the fuel air charge, the flame will pass through the rest of the mixture in the cylinder. That is – it will if the proper reaction conditions are maintained. This is achieved through the design of the boiler or furnace or engine, primarily by keeping up the temperature in the combustion zone. So the 'proper conditions' means those for starting the reaction and those for keeping it going to the very end. Pressure does have an effect but, rather like most thermodynamic properties, reactions are temperature dominant (Figure 12.6).

We have already touched upon the fourth factor in looking at the third. The flame (recall that continuous combustion is being used as an

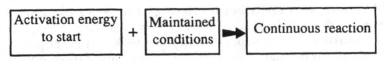

Figure 12.6 The third ground rule.

example) has to be kept under the proper working conditions until it is completely burnt out. **Time** is needed for this to happen and the time required for any reaction to complete varies with conditions.

For instance, if a tube of about 5 cm diameter is filled with a stoichiometric mixture of natural gas and air and this mixture is ignited at one end of the tube, the resultant flame will travel at a brisk walking speed. If a lean mixture (excess air) was used, the speed would be less. If a wider tube was used, the speed would be higher. There is, incidentally, a proper apparatus for doing this experiment – do not use any old bit of pipe!

Different conditions will give different results but this shows that combustion reactions are certainly not instantaneous. In fact, no reaction is instantaneous, whatever the reactants (Figure 12.7).

● PQ 12.7 Assuming that the reactants are in their correct proportions, there is a 'golden rule' in combustion called 'the three Ts'. Each *T* is an initial drawn from the ground rules. What are they? You have met two by name but you will have to think about the third.

That then is an introduction to combustion, an example of reacting mixtures. The theoretical aspects form a good basis for design because calculations can be attached to them – the stoichiometric amounts, the activation energy. The practical aspects, in general, have to be assessed empirically – how much excess air, for example. Whatever the reacting mixture, combustion or otherwise, the proportions have to be correct for the demand, the reactants have to be mixed as well as is economically possible, the conditions for good reaction have to be determined and maintained and then the reaction has to be given time to complete.

While these points are important to the practice of reacting mixtures, the basic laws of thermodynamics still apply and an old term, enthalpy, has been met in a new context – **enthalpy of reaction**. It is still enthalpy and its interpretation is unchanged.

● PQ 12.8 Just remind yourself about the definition and the interpretation of enthalpy.

12.3 STOICHIOMETRY – THE REACTANTS

Stoichiometry covers the calculation of the theoretical (stoichiometric) proportions of reactants and the theoretical products and their quanti-

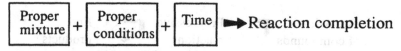

Figure 12.7 The fourth ground rule.

ties. While it may seem that dealing solely with the theoretical quantities limits its value, there are two very clear advantages to having the
stoichiometric information.

● **PQ 12.9** As much of real-life engineering requires experiment to
produce reliable designs, what are these two advantages?

Recall that continuous combustion of an everyday fuel and air
mixture, as used in boilers and furnaces, is being used to illustrate
reacting mixtures. Most of that which follows is to do with the
stoichiometry of combustion but the lessons can be applied widely.

Calculations for practical combustion in boilers and furnaces – the
main users of continuous combustion – cover two particular areas.
These are the determination of stoichiometric air/fuel ratios upon
which to base real or practical ratios and the prediction of exhaust or
product properties.

The calculations are based on the reactions of the fuel constituents
with oxygen, which in turn is supplied almost exclusively as atmospheric air. There are some very specialized processes where neat
oxygen or oxygen-enriched air is used but they are few. The important
energy-providing elements in fuels are carbon and hydrogen, which all
common fuels contain. Sulphur is usually present also but quantities
vary from negligible to a few per cent.

These elements, mostly present in fuels as components of compounds, release energy by the compounds which contain them being
broken down and then the available elements being converted ultimately into their **oxides**. The fuel's carbon ends up as carbon dioxide
CO_2, and the fuel's hydrogen ends up as water substance H_2O. Any
other reactable elements, sulphur for instance, also end up as their
oxides. There are good thermochemical reasons for this and it is the
most practicable route for usable energy winning.

Combustion then is actually a **chain of reactions** but the chain can be
summarized very simply by considering what goes into the combustion
chamber and what comes out – the start and finish of the chain. This
does not mean that the intermediate steps are not important, only that
we can omit them for our purposes (Figure 12.8).

The amount of energy which any given fuel can provide is calculable
for simple fuels – methane, for example – but is usually determined
experimentally by burning a small quantity under highly controlled

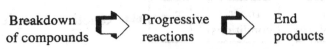

Figure 12.8 Combustion is a chain of reactions.

conditions. This gives the fuel's **calorific value** and it is referred to some arbitrary datum temperature – it is in effect the amount of heat available per unit mass of fuel above the datum. 25°C is a common datum. It is a very practical measure and it has a very practical application.

● PQ 12.10 What, apart from the elements that the fuel contains, will have a major influence on the calorific value? Remember that the elements are usually present as compounds which have to be broken down.

The common fuels – coal, oil, gas – are **hydrocarbon** fuels, so called because their main elements are hydrogen and carbon. However, they are combined in different ways to give the various fuels their different properties. Energy is involved in the formation of compounds – think of burning hydrogen to form water:

$$2H_2 + O_2 \rightarrow 2H_2O - \Delta H.$$

As the hydrogen burns, it releases heat – the enthalpy of reaction or, for the formation of this compound, the **enthalpy of formation**. If I now want to break this compound down, I will have to supply the same amount of energy because the reverse reaction will be endothermic:

$$2H_2O \rightarrow 2H_2 + O_2 + \Delta H.$$

The endothermic energy to break down the fuel's compounds has to come from the fuel itself, hence the compounds' influence on the calorific value, the externally usable heat release.

At first sight then it would seem a virtually impossible task to prepare a series of chemical equations to represent the process of combustion unless every compound in every fuel was known. An estimate some time ago gave around 3000 different compounds in coal alone. However, there are ways round the problem.

First, gaseous fuels, such as North Sea gas. Common gases have simple formulations – methane CH_4, propane C_3H_8, carbon monoxide CO for instance. The reason is straightforward – the more complex the molecule, the more likely the material is to exist in liquid or solid form at everyday temperature and pressure. So gaseous fuels containing these are easy to analyse and the gaseous fuel reaction can usually be represented in its molecular form quite accurately – for example,

$$CH_4 + 2O_2 \rightarrow CO_2 + 2H_2O.$$

● PQ 12.11 Recall Avogadro's hypothesis and explain how the stoichiometric air/fuel ratio can be assessed from that equation.

So, any individual gas can be written in equation form and its **volumetric** air/fuel ratio found. Since this is so simple and since gaseous fuels are normally measured by volume, air/fuel ratios for them are almost always recorded by volume. A little practice may be needed in balancing the equations – the same number of atoms of each sort on both sides – but the ones for gas fuels are quite simple.

Example 12.1 Prepare the balanced equation for propane, C_3H_8, reacting stoichiometrically with oxygen. Assume complete reaction.

Start by writing down the reactants and products, without proportions:

$$C_3H_8 + O_2 \rightarrow CO_2 + H_2O.$$

I know that the starting materials are propane and oxygen and that the products will be carbon dioxide and water substance. These latter are the end products of any hydrocarbon material reacting with oxygen, as already noted.

Now look at the carbon alone. There are three atoms on the left, C_3, so these must appear on the right. The only place where they can appear is in the carbon dioxide, CO_2, and the only way therefore that three carbon atoms could appear is if there were three molecules of carbon dioxide, $3CO_2$. Similarly, there are eight atoms of hydrogen, H_8, on the left which must appear on the right. Water substance, H_2O, is the only home for them, with two atoms of hydrogen per water molecule, H_2O. There has to be four molecules of water, $4H_2O$, to cover the eight atoms of hydrogen. The equation with the amounts of products can now be written

$$C_3H_8 + O_2 \rightarrow 3CO_2 + 4H_2O.$$

To be able to write the right hand side like this, there has to be supplied the oxygen of the product molecules, $3O_2$ in the CO_2 (thus six oxygen atoms) and $4O$ in the H_2O (thus four atoms).

The total oxygen demand is therefore ten oxygen atoms, that is five oxygen O_2 molecules. The complete equation can now be written:

$$C_3H_8 + 5O_2 \rightarrow 3CO_2 + 4H_2O.$$

The same technique can be used on all equations of this type.

Where the fuel is a mixture of gases, then the easy way is to write individual reactions for each individual gas and multiply the whole by its proportion. For instance, if a fuel was 35% methane and 65% propane by volume, then the reactions could be written

$$0.35 \, (CH_4 + 2O_2 \rightarrow CO_2 + 2H_2O)$$
$$0.65 \, (C_3H_8 + 5O_2 \rightarrow 3CO_2 + 4H_2O).$$

Example 12.2 What is the stoichiometric oxygen demand, on a volume basis, for that gas mixture?

We need an arbitrary basis, such as one volume or 100 volumes. Using 100 volumes avoids a lot of decimal places and the answer can be converted to any other basis quite easily. In 100 volumes of the mixture, there are 35 volumes of methane and 65 volumes of propane.

The individual equations show that one volume of methane requires two volumes of oxygen for stoichiometric reaction, so 35 volumes of methane require 70 volumes of oxygen. One volume of propane requires five volumes of oxygen, so 65 volumes of propane require 325 volumes of oxygen.

Thus, 100 volumes of the mixture require 395 volumes of oxygen for the **overall** reaction. This would normally be reduced to a simple ratio such as 3.95:1, rather than 395:100.

Example 12.3 Now determine the stoichiometric air/fuel ratio for that mixture. Air is 21% by volume oxygen.

Use the arbitrary 100 volumes again for the air also. As air is 21% by volume oxygen, so 100 volumes of air contain 21 volumes of oxygen. The mixture requires 395 volumes oxygen per 100 volumes mixture, so the air demand is 395 × (100/21) volumes air per 100 volumes of fuel, that is 1881 volumes air per 100 volumes of fuel, which would be written as 18.81:1 v/v, where v/v means volumes per volume, all these volumes being measured at the same temperature and pressure.

That is all there is to calculating gaseous reactant ratios!!

For liquid and solid fuels generally, this generation of simple equations for the compounds in the fuel does not apply. The compounds which they contain are quite complex but, while they could be analysed, it is not necessary. More complex fuels can be treated as though their compounds had collapsed into separate heaps of carbon, hydrogen and so on, with these elements then reacting separately.

● PQ 12.12 How is it that this simple technique can be valid?

Any common fuel can be analysed for its elements very simply, however those elements may be bound into compounds. The term *'ultimate analysis'* is used for this, since the end product is the finding of the ultimate contents – the carbon, hydrogen, sulphur – rather than their compounds. Most common fuels have their ultimate analyses well tabulated in reference texts or available from the major fuel suppliers.

Therefore, instead of writing an overall equation to represent the fuel's compounds reactions with air (oxygen), a series of reactions is written to deal with each important constituent element, such as

$$S + O_2 \rightarrow SO_2$$
$$H_2 + \tfrac{1}{2}O_2 \rightarrow H_2O$$
$$C + O_2 \rightarrow CO_2,$$

just as though they were in their simplest everyday elemental form. The quantities of each element in a given amount of fuel are then built into each equation to determine the amounts of oxygen necessary to complete the calculations.

● PQ 12.13 Strictly, what are these equations saying?

Now, trying to use one atom or one molecule is unrealistic, so for these equations to be really useful, they need adapting for the amounts of fuels used in real life. This can be done through the concept of the *kilogram-mol*, sometimes called just 'mol' or 'kmol' or 'mole'. The titles vary a little but here is a reminder of the concept.

The molar or atomic mass (molecular weight or atomic weight in some texts) of a compound or element is a relative measure of the mass of the smallest identifiable or free-standing portion of that material. Elements are given atomic masses which reflect the mass of one of their atoms compared to an atom of any other element. In round numbers, for example, the atomic mass of hydrogen (the lightest material) is 1, that of carbon is 12, of oxygen 16, nitrogen 14, sulphur 32. There are progressive reviews of these values as measurement techniques improve but for our purposes the round numbers will suffice for the present. Also, for the present, take it that these are relative masses or ratios, so no units are attached.

Molecules, being agglomerations of atoms, have relative masses which are additions of those of the constituent atoms.

● PQ 12.14 What are the molar masses of methane CH_4, oxygen O_2, water substance H_2O and sulphur dioxide SO_2?

Notice that these are **relative** masses – the mass of the molecule compared to others, not the mass in kg or any other measure – and that no units have been put to the masses. Now take the reaction of carbon being converted to carbon dioxide and add these relative mass figures to it:

$$C + O_2 \rightarrow CO_2$$
$$12 \quad 32 \qquad 44 \text{ unspecified units.}$$

The equation is now far more useful – it gives the relative mass proportions of the reactants and the product! For every 12 relative mass units (whatever they are) of carbon, there will be needed 32 relative mass units of oxygen and the reaction will yield 44 relative mass units of carbon dioxide.

Instead of having unspecified mass units, the whole equation can be multiplied through by the term kilograms – so that everything changes from unspecified units of relative mass to kilograms. Thus it follows that

$$C + O_2 \rightarrow CO_2$$
$$12 \text{ kg} \quad 32 \text{ kg} \quad 44 \text{ kg.}$$

The kmol (mol, mole, kg mol) concept is then that atomic or molar mass is expressed in kilograms. Now, any equation can be written down for materials which are reacting together and it gives the reacting mass proportions – it has real use. The appropriate molar or atomic masses are given as kilograms instead of unspecified relative values and the real amounts can thus be calculated.

Example 12.4 Calculate the theoretical amount of oxygen needed and the masses of the products for a liquid fuel whose ultimate analysis by mass is 85% carbon, 15% hydrogen.

Start with 100 kg of fuel as a convenient number, just like the 100 volumes for gases. The two reactions, with their kmol concept masses, are

$$H_2 + \tfrac{1}{2}O_2 \rightarrow H_2O$$
$$2 \text{ kg} \quad 16 \text{ kg} \quad 18 \text{ kg}$$
$$C + O_2 \rightarrow CO_2$$
$$12 \text{ kg} \quad 32 \text{ kg} \quad 44 \text{ kg.}$$

The fuel contains 15 kg hydrogen and 85 kg carbon rather than 2 kg and 12 kg, so the equations are arithmetically adjusted to take account.

$$H_2 \quad + \quad \tfrac{1}{2}O_2 \quad \rightarrow \quad H_2O$$
$$15 \text{ kg} \quad 16 \times (15/2) \text{ kg} \quad 18 \times (15/2)$$
$$15 \text{ kg} \quad 120 \text{ kg} \quad 135 \text{ kg}$$

● PQ 12.15 And for the carbon?

So, the total oxygen demand for this fuel is 346.7 kg per 100 kg fuel, which would be written something like 3.467:1 w/w, meaning weight for weight, the normal way of measuring the needs of solid or liquid fuels because of the way in which they are supplied and/or analysed. Yes, I know it should be mass but the commonplace way is now that which it has been for many years, showing the ratio as w/w. Strictly then, this sort of analysis or result for solid and liquid fuels is on the **mass basis**.

The total masses of products are 135 kg water substance, or 1.35 kg per kg of fuel and 311.7 kg carbon dioxide, or 3.12 kg (rounded) per kg fuel.

Example 12.5 Given that air contains 23.2% by mass oxygen, now find the stoichiometric air/fuel ratio for the mass of air needed for 20 kg of fuel, and the masses of products from the stoichiometric combustion of 20 kg fuel.

Compare this to the gaseous fuel ratio found in Example 12.3. The air/fuel ratio is 346.7 × (100/23.2) kg air per 100 kg fuel or 1494.4 kg air per 100 kg fuel. This would normally be rounded and presented as a ratio per unit mass rather than per 100 kg, 14.9:1 w/w.

The mass of air required for 20 kg fuel then is 20 × 14.9 or 298 kg but if the rounding had been done now, the answer would be 299 kg! Do the rounding at the end.

- **PQ 12.16** Are the products the same?

Note that throughout, we have stoichiometric proportions and have assumed complete reaction. Whether reactant proportions are given on a volume or a mass basis (not just fuels) depends upon their use, their usual way of being measured and their usual way of being analysed. This has been an example of a way of doing the calculation. It is not the only way but the same numbers are used whichever way it is done. For instance:

100 kg fuel contains	85 kg Carbon	15 kg Hydrogen
In kmol:	7.08	7.5
reactions	$7.08C + 7.08O_2 \rightarrow 7.08CO_2$	
	$7.5H_2 + 3.75O_2 \rightarrow 7.5H_2O$	
total oxygen, kmol	10.83	
total oxygen, kg	346.56	
products, kmol	$7.08\ CO_2 +$	$7.5\ H_2O$
products, kg	311.52	135.0

For 20 kg of fuel rather than 100 kg, the oxygen demand and products are oxygen 69.3 kg, carbon dioxide 62.3 kg; water 27 kg.

12.4 STOICHIOMETRY – THE PRODUCTS

One useful by-product of the combustion calculations so far is that they automatically give the quantities of the products of reactions. As the equations have been laid out to find stoichiometric proportions of reactants, the **products** shown must also be in stoichiometric proportions. This makes the clear assumption that the equation is correct in terms of what is produced by the reaction, of course, but most combustion reactions are well established and as reliable as any (Figure 12.9).

Take the combustion of propane, a common industrial and commercial fuel gas, as an example and refer back to PQ 12.11, Example 12.1 and Example 12.2.

The stoichiometric, perfect reaction is

$$C_3H_8 + 5O_2 \rightarrow 3CO_2 + 4H_2O.$$

The products of this reaction are three volumes of carbon dioxide and 4 volumes of water substance per volume of propane.

● PQ 12.17 Is this so? Will it be four volumes of water substance? If 1 m³ of propane was burnt, would it produce 3m³ of carbon dioxide? Must some conditions attach to this recorded volume?

That theoretical reaction only takes account of the oxygen needed. The oxygen will almost certainly have been provided as a constituent of air, so the complete equation **must** include anything else which is brought into the system by the combustion air. While atmospheric air contains a wide variety of gases, for instance small amounts of argon, the dominant ones by a very large margin are the oxygen which we need, about 21% by volume and 23.2% by mass, and nitrogen about 79% by volume and 76.8% by mass. The word 'about' allows a little for the various other constituents.

In round numbers then, for each portion of oxygen put into the reaction, something approaching four portions of nitrogen will accompany it. While there are reactions which involve nitrogen – the production of so-called nitrogen oxides for instance – these are on the 'parts per million' level. These nitrogen oxides can have important side-effects but here they are ignored as being too small to influence the overall combustion outcome.

Going back to the reaction and adding the nitrogen, it will appear in either side of the equation as we are quite realistically taking it as inert. As volumes are being considered for this gaseous fuel, the volume ratio of 79:21 nitrogen:oxygen in air applies. There are 79 volumes of nitrogen in the air to every 21 volumes of oxygen, so

Figure 12.9 Equations give products proportions too.

$$C_3H_8 + 5O_2 + (5 \times \frac{79}{21})N_2 \rightarrow 3CO_2 + 5H_2O + (5 \times \frac{79}{21})N_2$$

This then, ignoring the minor constituents of air and the minor reactions, is the complete stoichiometric equation. While combustion is being used here to illustrate the important features in this context of reacting mixtures, a similar approach would have to be taken whatever the reactants. It is insufficient just to consider those materials which are chemically active – anything else in the mixture may have an important contribution to make somewhere. In this case the nitrogen may not take part chemically but it will be heated up during the burning reaction, so it must influence the final temperature (Figure 12.10).

The complete theoretical equation now gives the complete list of ideal products and these can be expressed as percentages on a volume basis – Avogadro says so. Since these products are the products of an ideal reaction, any real reaction products can be compared to them to see how close to the ideal is the real operation. On boiler plant for example, a routine analysis of the exhaust products is a very useful tool when investigating the overall efficiency of operation of the boiler. As an aside, these exhaust products of combustion are called variously exhaust gases, flue gases, chimney gases or stack gases in boiler and furnace operation.

Since the product analysis is so useful industrially and commercially, many pieces of analytical equipment have been designed and used. Some use the physical properties of the exhaust gas mixture and some the chemical properties but in virtually all the practical plant analysers, a sample of the exhaust gas is taken from a suitable point in the system and then analysed at **ambient** (that is, surrounding atmospheric) temperature.

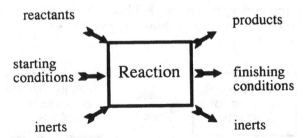

Figure 12.10 Consider everything.

● PQ 12.18 Of the products of the combustion of propane, which will appear in the exhaust gas analysis? Put some percentages to that gas analysis from the complete equation.

So, while the term 'flue gas analysis' (the common industrial term for the analysis of combustion products) is used, it should strictly be *'dry flue gas analysis'* as the water substance condenses out of it and is not counted in the analysis. The reporting of dry flue gas analysis is usual.

That analysis from the reaction equation is straightforward for gaseous fuels but they are only part of the total fuel usage of the world. A technique has to be developed for solid and liquid fuels which make up the rest. As a starting point, take the combustion of solid carbon (coke is nearly all carbon) and applying the same allowance for nitrogen, the reaction without numbers is

$$C + O_2 + N_2 \rightarrow CO_2 + N_2.$$

If the volume proportions of $N_2:O_2$ in air are 79:21, then the molar proportions must also be 79:21, Avogadro again. Thus for every mol (kmol, kilogram-mol) of O_2, there must be 79/21, that is 3.76 of N_2:

$$C + O_2 + 3.76N_2 \rightarrow CO_2 + 3.76N_2.$$

Even though the reactants are a mixture of solid (carbon) and gas (air), the products are both gases and so their molecular proportions must be the same as their volume proportions at the same temperature and pressure.

Thus there is one molecule of carbon dioxide to every 3.76 molecules of nitrogen – the reaction equation tells us that – so there will be one volume of carbon dioxide to every 3.76 volumes of nitrogen. The equation immediately gives the volume basis exhaust gas analysis.

The exhaust or products or flue gas analysis from the combustion of any solid or liquid fuel is simply an addition of the products from each of the constituents of that fuel. The important first step is working out how many mols of each constituent there are from the ultimate mass analysis. You will recall that ultimate analysis is the usual way of dealing with liquid and solid fuels.

Example 12.6 The ultimate analysis of an oil fuel is 86% carbon, 14% hydrogen. Determine the stoichiometric dry flue gas analysis for the fuel when burnt with air.

Since there is no stated amount of fuel being burnt, take a quantity such as 1 kg or 100 kg to work on. These are simple round numbers and they in no way affect the outcome. 100 kg avoids too many decimal places.

100 kg of the fuel contains 86 kg carbon and 14 kg hydrogen – that is what the ultimate analysis means. Change these mass quantities to kg mol quantities – 1 kg mol (kmol, mole and so on) contains the atomic or molar mass in kilograms.

A simple rule when dealing with the elements in this sort of calculation – if the element is solid at room temperature then you need its atomic mass; if the element is gaseous at room temperature then you need its molar mass for this kilograms-to-mols change.

● **PQ 12.19** Change the ultimate analysis, kg element per 100 kg fuel, into mols of element per 100 kg fuel.

Now we can write the molecular equations for 100 kg fuel plus oxygen

$$7.167C + 7.167O_2 \rightarrow 7.167CO_2$$
$$7.00H_2 + 3.50O_2 \rightarrow 7.00H_2O$$

and with 3.76 mol N_2 for every mol O_2,

$$7.167C + 7.167O_2 + 26.95N_2 \rightarrow 7.167CO_2 + 26.95N_2$$
$$7.00H_2 + 3.50O_2 + 13.16N_2 \rightarrow 7.00H_2O + 13.16N_2.$$

So the products are 7.167 mol CO_2, 7.00 mol H_2O and 40.11 mol N_2, of which the water substance condenses out from the dry flue gas analysis to give only CO_2 and N_2 proportions. The stoichiometric dry flue gas analysis of this fuel is thus 15.16% CO_2 and 84.84% N_2.

In practice, it is not easy to measure nitrogen with typical plant analytical instruments, so nitrogen is usually reported **by difference** – that is, measure everything and take it away from 100%. It follows that the nitrogen percentage will include anything not measured plus all the errors! It is not quite as bad as it sounds because the readily measured items are the important ones for any plant work (Figure 12.11).

The final point to be covered here on the products moves away from the stoichiometric mixtures towards real ones. Refer back to PQ 12.6 and its discussion. For real industrial reacting mixtures, it is very

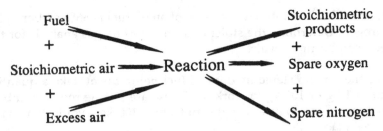

Figure 12.11 Excess air leads to extra products.

unlikely that the stoichiometric proportions will be met. There is usually an excess of one or other, so that the more expensive reactant is used as fully as possible. In continuous combustion, this is seen as excess air, air put into the combustion process as an extra percentage above the stoichiometric amount. The amount is determined by experience.

• PQ 12.20 If excess air is used and assuming complete reaction, how will this affect the products analysis?

Example 12.7 For the fuel oil of Example 12.6, what will be the actual air/fuel ratio and the dry flue gas analysis if 15% excess air is used?

First the air/fuel ratio. The hard work has been done, in that the stoichiometric amount of oxygen is known – 7.167 mol for the carbon and 3.50 for the hydrogen, a total of 10.667 kg mol.

• PQ 12.21 If the relative molar mass of oxygen is 32 (thus 32 kg per kg mol) and there is 23.2% oxygen in air by mass, what is the stoichiometric air/fuel ratio for this fuel?

Therefore, with 15% excess air, the total air supply is 115% of stoichiometric, or 16.92 : 1. This extra 15% of air means an extra 0.15 × 10.667 kg mol of oxygen and an extra 0.15 × 40.11 kg mol of nitrogen, so the overall reaction now, combining the carbon and the hydrogen reactions, is

$$7.167C + 7.167O_2 + 7.00H_2 + 3.50O_2 + 0.15$$
$$\times 10.667O_2 + 40.11N_2 + 0.15 \times 40.11N_2$$
$$\rightarrow 7.167CO_2 + 7.00H_2O + 1.6O_2 + 46.17N_2.$$

The oxygen in the products is from the excess (unreacted) air and the nitrogen is the total from the stoichiometric and excess air. The dry flue gas analysis eliminates the water substance because it condenses, so the products which contribute to that analysis are carbon dioxide, excess oxygen and the total nitrogen.

Thus the dry flue gas for the arbitrary 100 kg fuel is 7.167 kmol carbon dioxide, 1.600 kmol oxygen and 46.127 kmol nitrogen. Changing to a percentage, the analysis is 13.06% carbon dioxide, 2.91% oxygen and 84.03% nitrogen. All the rounding-off has been left to this point. It is common practice to report only the active, measured constituents since the nitrogen percentage would automatically include all the errors, as mentioned earlier.

This large amount of nitrogen in the exhaust gases from combustion of fuels is commonplace. Typical real air/fuel ratios are in the broad range 10:1 to 20:1. About four-fifths of air is nitrogen, the air is ten to twenty times the fuel amount, so about four-fifths of everything going into a combustion zone is nitrogen and about four-fifths of everything coming out is nitrogen.

Make sure that you understand that **only** the stoichiometric amounts can be calculated from first principles. Anything to do with practical constraints – such as here the need for excess air – has to be found **empirically**, from experience that is. However, the calculable, ideal amounts are a very good point of comparison, showing how well any real process is performing compared to the ideal.

One last point – just words, no calculations – about the release of heat which accompanies an exothermic reaction.

● **PQ 12.22 Think of the NFEE. Can the reaction enthalpy tell what temperature may develop as a result of the reaction?**

When nothing is known about the reaction conditions for any reacting mixture, little can be said about temperature changes. If however, it is assumed that all the energy released goes into raising the temperature of the products of reaction, so-called **adiabatic temperature rise** can be calculated. This would give the temperature of the products above the temperature of the reactants.

At its simplest, this temperature rise would be found from

$$\Delta T = \frac{\text{enthalpy of reaction}}{\text{mass of products} \times \text{specific heat capacity}} .$$

However, other thermochemical effects become involved, which are beyond the scope of this text. The point is, though, that an ideal value, which may be used as a basis of comparison with real values and as a design guide, is calculable. Anything better than that can only be estimated when all the operational conditions are known.

Finally, a reminder that the topic of this chapter has been some principles relating to reacting mixtures. Fuel combustion has only been used as a convenient and common example of reacting mixtures. The principles apply to all and, perhaps more than in any other section of thermodynamics, practical influences are as important as the theoretical calculations.

PROGRESS QUESTION ANSWERS

PQ 12.1

There are many everyday uses for fuels and here are three common ones.

(a) A house central heating system, where perhaps gas is burnt in a domestic boiler to heat water, the water then circulating through radiators to warm the house.
(b) A car engine, usually gasoline (petrol) which is burnt in the cylinders of the engine, the energy released then appearing as work to drive the car along the road.
(c) A power station, where fuel is burnt in a large boiler to raise steam, which is then used to drive an alternator for electricity generation.

In each of these cases, the fuel is burnt and in each case air is needed for the combustion. More accurately, oxygen is needed but it is almost exclusively supplied as a constituent of air. The other common substance then is air and it is the oxygen of the air which is used.

PQ 12.2

It all comes down to system boundaries, those pretend boxes which surround any working system. If I draw a boundary round the car engine, then fuel and air flow steadily across the boundary into the engine, and work – the propulsion of the car – flows steadily out across the boundary also. However, we are now looking at combustion rather than the steady flow energy equation. If I now draw the boundary round just at one of the cylinders of the engine, then the picture is changed.

Some fuel comes into the cylinder, with the necessary air. They are fixed amounts according to the engine design. The flow of fuel and air stops, the fuel/air mixture is burnt to do work and the combustion products are then pushed out of the cylinder. The whole process is repeated in this and the other cylinders as long as the engine runs, so there is a repeated cycle of operations. The combustion itself is thus definitely cyclic, even though the car progresses smoothly.

PQ 12.3

A spark from a spark plug in a car. A match to light a fire.

PQ 12.4

At its simplest, we can say that one portion of methane reacts with two portions of oxygen to give one portion of carbon dioxide and two portions of water substance.

As this is combustion, heat will be released but it is not shown in this equation.

PQ 12.5

The model of wet steam was of many tiny liquid water droplets being carried on the gaseous water substance. The liquid and gas were very much mixed together. So if the liquid fuel is broken down into lots of small droplets, they can mix intimately with the air.

PQ 12.6

If I am trying to mix (practically, not theoretically) fuel and air, there are three things I can do:

- put in the stoichiometric proportions and accept what comes out, knowing that mixing cannot be perfect;
- put in a bit more fuel to use up all the air;
- put in a bit more air to use up all the fuel.

With the first two, some fuel is inevitably lost, either by accepting incomplete mixing or by deliberately putting more into the mixture. Fuel is expensive and cannot be thrown away, so those two are out. The third option is best, not because the air is free (for someone has to pay to move it around, such as by electric fan) but because the air is far cheaper than the fuel.

PQ 12.7

(a)   Turbulence – the thorough mixing of the reactants.
(b)   Temperature – to maintain good operational conditions of which that is the dominant one.
(c)   Time – so that the reactions may go as close to completion as is practically possible.

Just as a little addition to that last point, no reactions go to total completion. They achieve a position of equilibrium, but, with care, that equilibrium in industrial combustion is as near to completion as practical circumstances allow.

PQ 12.8

Refer back if necessary. The equation $Q = mc_pT$ applies and enthalpy was taken as meaning the total ability to supply heat or to do work.

PQ 12.9

First, calculating the theoretical values will give the ideal picture, so that practical shortcomings can be recognized. Second, the calculations are a good starting point for real designs. Without such information, the designer would either have to think of a number or use all the old designs – which is no way forward!

Most real designs are based on some starting calculations, leading to a prototype from which the practical lessons are learned.

PQ 12.10

The elements are only the building bricks of a fuel. They are present as compounds, so it is the way in which the elements are combined that also influences the calorific value. In the chain of reactions, there will be some energy used to break down the compounds. This energy, part of the fuel's energy, is not available for outside use therefore.

PQ 12.11

Avogadro tells us that for gases at the same temperature and pressure, the same volume of gas contains the same number of molecules. So if there is 1 m^3 of hydrogen at 20°C and 1 bar in one container and, in another container, 1 m^3 nitrogen at the same temperature and pressure then there will be as many molecules of hydrogen in the one as there are molecules of nitrogen in the other. That is, one molecule of hydrogen takes up as much space as one molecule of nitrogen, under the same temperature and pressure conditions.

Thus if a molecular reaction for gases is written down and the gases are all measured at the same temperature and pressure, the molecular reaction must also represent a volume reaction. That reaction gives the stoichiometric volume ratio of oxygen to methane and, as oxygen is a fixed percentage in air, it also gives the volumetric air/fuel ratio.

PQ 12.12

Those compounds are still made up of the building bricks of carbon, hydrogen or whatever, so it is reasonable to say that the compounds could be dismantled. Having taken the compounds back to their elements, then the elements can react afresh to their oxides. It is rather like knocking down a house and using the bricks to build a wall.

The compound formulation will affect the calorific value, as noted earlier but not (other than in proportions) the amount of oxygen to burn the constituent elements. As combustion is a chain of reactions, so the compounds of the fuel will undergo a progressive change as they burn which is rather like the compounds being dismantled.

PQ 12.13

All these equations, whether the complete compound as in gaseous fuels (CH_4, H_2, C_3H_8) or in the elemental form for solid and liquid fuels, are really indications of what would happen ideally at the molecular or atomic level. The last one, for example, is saying that one atom of carbon would react with one molecule of oxygen to generate one molecule of carbon dioxide.

PQ 12.14

As an example, the molar mass of methane is the sum of the relative atomic masses of the constituent atoms per molecule – one carbon, $C = 1 \times 12$, plus four hydrogen, $H_4 = 4 \times 1$, total 16. Similarly, oxygen is 32, water substance 18 and sulphur dioxide 64.

PQ 12.15

Similarly

$$
\begin{array}{cccc}
C & + & O_2 & \rightarrow & CO_2 \\
85 \text{ kg} & & 32 \times (85/12) \text{ kg} & & 44 \times (85/12) \text{ kg} \\
85 \text{ kg} & & 226.7 \text{ kg} & & 311.7 \text{ kg}
\end{array}
$$

PQ 12.16

The same amount of H_2O and CO_2 are produced but the air brings with it nitrogen, which was absent when only using oxygen. Taking air to be oxygen and nitrogen for simplicity (there are other components), then there is $(100 - 23.2)\%$ nitrogen in air, 76.8% by mass.

The products will thus contain 229.5 kg nitrogen as well as the carbon dioxide and the water substance.

PQ 12.17

Remember that this applies to a gaseous fuel and Avogadro's hypothesis refers to gases. Therefore, it can only be four volumes of water substance if that water substance is gaseous at the same temperature and pressure that the other gases are measured. If the products of the reaction are cooled down so that the water condenses, then it is not in the gaseous state and the rule cannot apply.

The carbon dioxide volume must also be measured similarly. At the time of the reaction, the carbon dioxide will be very hot because of the enthalpy release. The 3 m^3 can only be so if all the gaseous volumes are measured at the same temperature and pressure, which gives us the necessary attached conditions.

PQ 12.18

As the reaction is written down, the products of combustion are carbon dioxide, water substance and nitrogen. When an exhaust gas is sampled and analysed at ambient temperature, the water substance will condense out, so the analysis here will only show the carbon dioxide and the nitrogen.

For the percentages, the molecular ratios are the same as the volume ratios provided everything is measured under the same conditions, Avogadro's hypothesis.

So, there are 3 volumes of carbon dioxide and $(5 \times 79)/21$ or 18.81 volumes of nitrogen. Converting these to percentages in a mixture of the two gives 13.76% CO_2 and 86.24% N_2.

PQ 12.19

There are $(86/12)$ kmol of carbon, 7.167 kmol and $(14/2)$ kmol hydrogen, 7.00 kmol, taking the round number molar masses for these materials.

PQ 12.20

Excess air is added for physical purposes – making sure that little or no fuel avoids air contact. However, the fuel cannot use any more air than the stoichiometric amount, so there must be spare air at the end of the reactions.

The products will thus contain the unused oxygen and the extra nitrogen which accompanied it in the excess air.

PQ 12.21

10.667 kg mol of oxygen is 10.667×32 kg $= 341.34$ kg. This is supplied by $341.34 \times (100/23.2) = 1471.3$ kg air. The required ratio is therefore 14.71:1.

PQ 12.22

No. The energy release may be retained adiabatically, may all go to do work or anything in between. Thus temperature rise cannot be forecast without knowing more about the reaction conditions.

TUTORIAL QUESTIONS

Recall that gaseous fuels' air/fuel ratios and all product gas analyses are usually by volume. Liquid and solid fuel air/fuel ratios are by mass.

When reporting exhaust or flue gas analyses, it is commonplace to include only the 'active' constituents. The remainder is nitrogen plus accumulated errors plus minute constituents.

12.1 What is the mass of water substance produced if 1 kmol of methane, CH_4, is burnt completely? [36 kg]

12.2 Butane, C_4H_{10}, burns with 10% excess air. What is the carbon dioxide, CO_2 percentage content of the dry exhaust gases? [12.68%]

12.3 A solid rocket propellant is 80% carbon, 12% hydrogen and 8% sulphur by mass. What is its stoichiometric oxygen/fuel ratio? [3.17:1]

12.4 What is the stoichiometric exhaust gas analysis by volume from the rocket fuel of question 12.2, with the water in the gaseous phase? [$CO_2 = 51.61\%$; $H_2O = 46.45\%$; $SO_2 = 1.94\%$]

12.5 A gas mixture of 94% methane, CH_4, 3% propane, C_3H_8 and 3% nitrogen, N_2, is burnt to give a dry flue gas analysis of 10.76% CO_2, 2.11% O_2, the remainder being nitrogen. What is the air/fuel ratio being used and the excess air percentage? [10.55:1; 10% excess air]

12.6 A heating oil has the approximate analysis by mass 85.7% carbon, 13.4% hydrogen, 0.9% sulphur. What is its stoichiometric air/fuel ratio and its dry flue gas analysis if burnt with 20% excess air? [14.85:1 bymass; 12.32% CO_2; 0.05% SO_2; 3.72% O_2].

Cycles of operation – gases

13

13.1 EVENTS IN SEQUENCE

Up to now, most of the thermodynamic processes which we have met have been single processes or substantially so – a gas has been compressed, a liquid boiled, a reaction followed through. In the majority of real thermodynamic operations, however, there is a repeated sequence of events, often called a **cycle of operations**. It is possible to have both open (where something is used and then ejected) and closed (where something is used repeatedly) sequences of events. Incidentally, take care not to confuse that with open and closed systems.

Think of an oil-fired power station boiler, making steam for electricity generation. In the boiler, fuel is burned with air to release heat. This heat is then transferred to the water substance contained within the boiler, to boil the liquid and to then produce superheated steam. The products of combustion have as much heat as possible taken from them before being discharged from the system. The superheated steam which has been generated is used to drive a turbine which in turn drives an alternator to make electricity (Figure 13.1).

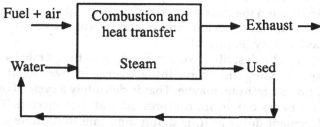

Figure 13.1 Sequences of events in the boiler.

First the fuel and air. These are fed into the combustion zone of the boiler where they burn and where some of the heat released is transferred to the surrounding water. The products of combustion leave that zone and pass through other heat removal or heat transfer zones before being discharged to the atmosphere as waste gases.

This is not a single dose of oil and air – the whole is a continuous process with all the events happening at all times in different parts of the boiler plant. However, if a single atom of carbon from the oil could be followed through the system, it would be involved in a *sequence of events*, the combustion, the heat transfer and the ejection to atmosphere. Since each stage of the process is repeated continuously, with new fuel and air throughputs, any later atom of carbon would follow the same sequence. The process is an **open sequence** or **cycle** of operations.

The water substance on the other hand goes round and round in a circuit. Starting from cold water, this is warmed or preheated by heat exchange from the some of the outgoing last dregs of heat in the exhaust gases of combustion. The heat exchanger used is commonly called an *economiser*. It is then boiled in the main part of the boiler, superheated in another section and then fed to a turbine to do work. At the end of its work, the steam condenses to liquid and is returned to the beginning of the boiler circuit.

This is very much a **closed sequence** or **cycle** of events, with each particle of water substance being used numerous times. Commonly but not exclusively, the term 'cycle' tends to be used in the latter context, where a repeated operation on a working fluid or a repeated mechanical motion can be identified. We will be concentrating on this type of cycle mainly but not exclusively.

● PQ 13.1 Will the laws of thermodynamics apply to both sorts of sequence just as they do to single operations? Why?

As another example a reciprocating air compressor feeds air to a paint spray. This compressor is a simple engine with a piston moving up and down a cylinder. There may be multiple pistons and cylinders in bigger units. Valves at the head of the cylinder allow air in and out. The piston, driven by an external motor via a crank, moves down the cylinder to draw air in through the inlet valve and then back up to compress the air. The compressed air is then fed through the exhaust valve to its intended paint spraying duty (Figure 13.2).

The cycle can be described from two different but parallel views. The mechanism is doing the same thing over and over again – a few thousand times a minute, maybe. That is definitely a cycle of operations with the air being drawn in, compressed and then ejected. The air, at least replacement doses of it, is doing the same thing time and again. Some is drawn in from the atmosphere, work is done on it and it flows

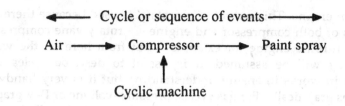

Figure 13.2 Sequence and cycle.

out at high pressure. The next dose of air follows the same sequence.

● PQ 13.2 Air from a compressor is also usually warm. Why?

It is convenient to study cycles of operation by looking at two areas – those where a **gas** is the working fluid and those where a **vapour** is involved. These cycles are similar but there are some differences. It is equally convenient to use the word gas to mean the dominant working fluid in the cycle, not the only working fluid. The electricity generating boiler was oil-fired but there would be far more air than oil (Chapter 12, air/fuel ratios). Once the fuel burns, then the gaseous products of combustion are the working fluid. This whole process would qualify as a gas cycle.

● PQ 13.3 Identify one of each type of cycle, gas and vapour, from domestic life.

That latter example in the answer to PQ 13.3, incidentally, is a good example of the Second Law in action. External work (the compressor) is needed to drive heat uphill against the temperature gradient. The heat is taken from the cold inside of the fridge and passed to the warm heat exchanger. That could not happen without an external energy source.

13.2   GAS CYCLES AND WORK

Remember that the word 'gas' here is a handy receptacle for all sorts of things and the careful wording of '*gas is the dominant working fluid*' is intentional. We have already met, briefly, the reciprocating air compressor and a gas is certainly the dominant working fluid – the air. In the same bracket, however, is usually included internal combustion engines – petrol engines as in most cars, diesel engines as in most lorries for instance. The fuel may well be admitted as a liquid but it is vaporized and burnt rapidly to generate hot gases and the combustion air is present in far larger quantity than the liquid fuel. Gas in various forms is thus the dominant working fluid in these engine cycles.

Much gas cycle study can be done very easily by reference to these two, the reciprocating air compressor and the reciprocating internal combustion engine. This word *'reciprocating'* is used because there are other sorts of both compressor and engine – a rotary vane compressor and a gas turbine engine for example – but from here on the word reciprocating will be assumed. It is useful to describe cycles and operations in words to ensure understanding but it is very handy to show cycles graphically. For gas cycles, pressure–volume or P–V graphs are usual. Where a set of graphs (perhaps more precisely called graph traces) join together to show a cycle, the resulting picture is often called a *diagram*, as in p–V diagram.

● PQ 13.4 Remind yourself – sketch p–V curves for reversible adiabatic, isothermal, constant pressure and constant volume changes of an ideal gas.

For each of these four, an expression was derived for the external work done, in terms of pressure and volume changes. If you have doubts, refer back to Chapters 2, 3 and 4. Calculus derivations, Appendix B, use the relationship $dW = p\, dV$, or $W = \int p\, dV$. On any p–V curve then, this external work is also shown by the area under the curve – the usual calculus interpretation of such relationships (Figure 13.3).

If a fixed mass of air is compressed and expanded in a series of changes, say a constant volume, constant pressure and a pV^n in that order, to return to the starting pressure and volume, then this could be shown on a p–V diagram. The sequence is ABCA (Figure 13.4).

● PQ 13.5 Identify the changes and, using the letters ABCDE, write down the areas which represent the external work associated with each step.

There is no work associated with AB but there is with both BC and CA. For BC, the gas is being compressed, so external work is done on it.

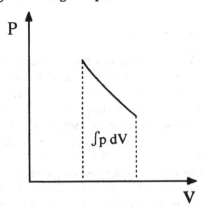

Figure 13.3 The area under the curve is the work done, $\int p\, dV$; p = pressure; V = volume.

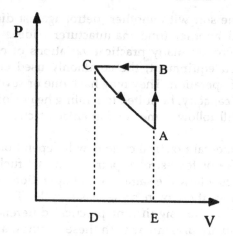

Figure 13.4 Changes in sequence.

For CA, the gas is expanding and and must therefore be doing external work on something. Thus some of the work done on the gas is recovered by the gas doing some work, but there is still a net expenditure of work on the gas – the area BCDE (work done on the gas) is bigger than the area CAED (work done by the gas).

- PQ 13.6 Which area in Figure 13.4 shows the net work?

This is an important conclusion for any sequence or cycle of events which can be drawn to form a closed loop – that is, it gets back to its starting point – on a p–V diagram. The **area** *enclosed* by the loop shows the **net external work** done on or by the gas. It applies to air compressors, engines, anything which can be fitted on a p–V diagram.

13.3 ENGINES AND AIR STANDARD CYCLES

Just looking around the roadways shows that there are several possible types of engine which can be used to drive a car or lorry, bus or motor cycle. Using the everyday terminology, there is a two stroke petrol one common amongst motor cycles, the four stroke petrol engine usual in family cars, and both two- and four-stroke diesel ones found in the range of commercial vehicles for instance. There are good reasons, mostly economic rather than technical, for choosing one or other of these for any particular application. It will be helpful, if you are unfamiliar with engines, to take a quick look now at one of the many elementary mechanical engineering books which describe them and their operation.

For both economic and technical design improvements, engines need to be compared, one sort with another (petrol against diesel) and one interpretation with another (one manufacturer's ideas compared to another). While there are many practical variations of engine design and their associated equipment, the commonly used engines follow some basic cycle of operation. They may have one or several cylinders, be of large or small capacity, be suited to hauling heavy loads or to high speeds but they still follow some fundamental practical cycle (Figure 13.5).

What this fundamental practical cycle is will depend on whether the engine is two-stroke or four-stroke, petrol or diesel fuelled. All four-stroke petrol engines follow the same cycle of operation; all four-stroke diesel engines follow their own same basic cycle. There are strong theoretical similarities but significant practical differences. As mentioned earlier, if you are unfamiliar with these engines, a quick review now will help your understanding later. Much of what follows assumes that you now have that engine background.

As we are looking at cycles of operation, then actual engine designs and design differences will be avoided. We can do all that is necessary by taking two engine cycles and analysing them.

● PQ 13.7 What is the value of analysing engine cycles? A clue – recall absolute and relative efficiencies.

The two engine cycles are the **spark ignition** (petrol engine) cycle and the **compression ignition** (diesel engine) cycle. Note clearly that we are now talking about the fundamental cycles of operation, not any of the numerous practical interpretations. These cycles tell the order of events and show the important features of operation. They say nothing about how big the engine is, which fuel pump is used and so on.

● PQ 13.8 If I want to look at the basic cycle of operation of a spark ignition (petrol) engine, list a few things on real engines which I should try to ignore.

When looking at engine fundamentals, the facts that different fuels or different arrangements of inlet and outlet valves are employed may well mask the more important features of comparison. One way of overcoming this to a great extent is to use the so-called *'air standard cycles'*.

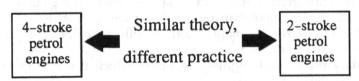

Figure 13.5 Engine cycles differ in practice.

What this means is that the cycle of operation is studied rather than any of the engines which use the cycle, but it goes further than that. It becomes a theoretical or paper study, retaining only the features of each stage of the cycle – whether that part of the cycle is a compression or an expansion, for example, with even that being idealized (Figure 13.6).

Next, the fuel (the energy source) is replaced by an ideal heat transfer. No fuel is introduced, no spark plugs are used, no carburettor is attached. It is simply the engine reduced to its basic operations for calculation purposes. The engine does not even have to exist – the cycle is drawn on a p–V diagram or something similar. As no fuel is introduced, air is used as the pretend working fluid. It is trapped permanently by the piston in the cylinder of the engine, for the engine valves have also been eliminated.

● **PQ 13.9** How can the work offtake from an engine be simulated for this air cycle? Recall that engines change heat energy into work energy and the second law says that we cannot get complete conversion.

Real engines work by changing fuel energy into work, through the intermediate step of the fuel energy being converted into working fluid enthalpy. A simple equation can therefore be written:

enthalpy transferred to working fluid
= work done by engine + enthalpy losses,

or,

ingoing enthalpy = work + reject enthalpy.

Thus as one part of this idealized cycle is an enthalpy addition, simulating the fuel energy supply, then an enthalpy removal in another part of the cycle would simulate the reject enthalpy of the real engine. The difference between these two enthalpy terms **must** be the useful work done.

So, with nothing to introduce practical distractions, this air standard cycle technique has several advantages:

● the limits to real engine efficiency can be found because the cycle will show the best theoretically possible performance;
● any basic cycle can be compared to any other very clearly because it is what the cycle is doing with air as the working fluid that matters;

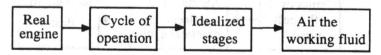

Figure 13.6 The air standard cycle is the real engine, idealized.

- any fundamental changes – not detail but fundamental – can be studied before a new engine is built or any major modification introduced;
- but note clearly that this is a way of looking at basics, not a way of dealing with detail design or of choosing components.

- **PQ 13.10** Very briefly, what is meant by engine efficiency?

   To investigate real engine efficiencies it is necessary to ask first 'what is the limit to efficiency and what would be the features of an engine with the highest possible efficiency?' Any real or ideal engine cycle is going to be made up of successive stages of the whole process. For the ideal cycle, it follows that each of the stages must also be an ideal one – an isothermal or adiabatic curve for instance – and for **maximum efficiency** the curves should enclose the biggest loop possible. The enclosed loop indicates the work done – therefore, for a given enthalpy supply, the bigger the better. Only in later considerations of real cycles will it be necessary to look at imperfections.

   Many years before the first practical internal combustion engine was built, the researcher Carnot was looking at this question of the best possible cycle of operations for converting heat into work. Engines or conversion machines of various sorts were around long before the petrol engine, incidentally. He was able to show that a cycle made up of two adiabatic and two isothermal curves was the most efficient possible. This, naturally, acquired the name *Carnot cycle*. Note that this is talking about the thermodynamic limits to operation and not some idealistic statement saying that 100% is the perfect limit (Figure 13.7).

   Our next step is to translate this Carnot conclusion into something that compares to our real engines and then to link that with the air standard cycle. A real automotive engine, whatever its practical features, has four steps in its operation. To illustrate, for a four-stroke spark ignition engine these are:

- induction of a working fluid (petrol and air mixture) into an engine cylinder – the piston moving down the cylinder does this;
- compression of the working fluid – the piston moving back up the cylinder;
- ignition (spark plug) so that the working fluid expands and does

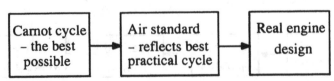

Figure 13.7 Carnot, air standard and real cycles are linked.

work – the piston being pushed down again;
- ejection of the used working fluid (the exhaust gases) – the piston moving up ejects these gases.

In a real engine, the stages are not quite so clear-cut but you see how they give rise to the four-stroke name.

To be applicable to automotive engines then, these stages have to be identified in the Carnot cycle, but take care here. It is easy to say that the real engine has four stages of its operation and the Carnot best possible cycle has four stages (the two isothermals and two adiabatics) therefore each engine stage equals one bit of the Carnot cycle.

The Carnot cycle mimics the real ones overall but **not necessarily** as identifiable stage by stage. Try to look at the Carnot cycle in isolation for the present. Carnot was asking one question – 'What cycle of operation will give the best thermal efficiency?' – and his pretend engine was doing three things in trying to answer this question;

- feeding in enthalpy;
- taking out work;
- rejecting the residual enthalpy.

Our pretend Carnot engine for this cycle will have a piston moving up and down a cylinder (just like a real one) and it will extract external work using the enthalpy of a trapped working fluid (just like a real one). In the Carnot engine, there is no through flow of the working fluid, the enthalpy transfer being through the engine walls rather than being fresh fuel and air admission and used products exhaust.

Broadly, his argument was that four steps are needed in the cyclic conversion of heat to work in an engine. He started with a compressed working fluid, introduced enthalpy, took work out, rejected enthalpy and recompressed the working fluid, ready to repeat the cycle (Figure 13.8). Referring to Figure 13.9, these non-flow steps are as follows.

(a) Introduction of enthalpy to the compressed perfect working fluid. The best process for this, where maximum work is the aim, is to do it isothermally, line AB.

- PQ 13.11 Why?

So, this stage gets the maximum work out of the enthalpy but clearly the used working fluid cannot now be ejected isothermally or the process would simply be reversed and the net change would be

Enthalpy ⇨ Expansion ⇨ Reject ⇨ Compression

Figure 13.8 The four stages of the Carnot cycle.

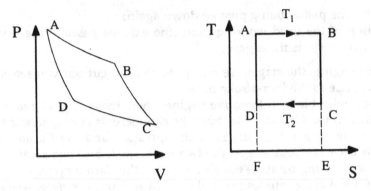

Figure 13.9 P–V and T–S diagrams for the Carnot cycle.

nil! Draw a p–V curve and you will see that the two isothermals would coincide – no loop between them, so no net work!

There has to be a stage between enthalpy in and enthalpy out.

(b)   Having introduced the enthalpy in the best possible way, then more work can be taken from the working fluid by allowing it to continue expanding. Any work output here must come from the internal energy of the working fluid, as the enthalpy input has stopped. This is best achieved by line BC.

- PQ 13.12 What sort of non-flow process is this?

This stage has taken the work generation as far as possible and the rejection process can now start.

In a real engine, stages (a) and (b) here would be seen as the compressed fuel and air mixture being ignited, burning and doing work. In a real engine also, this would be followed by the ejection of exhaust gases, the induction of more fuel and air and the compression prior to ignition and the repeat of the cycle. The next Carnot steps then have to mimic this.

(c)   Work has to be done on the used working fluid to complete the cycle. Recall that the second law says that this rejection must take place. In doing work to reject enthalpy, then the most efficient way is the reverse of stage (a) – isothermally.

(d)   Now the process continues to complete the compression of the working fluid and this is the reverse of stage (b) – reversible adiabatic compression.

While these four steps – two adiabatic and two isothermal – are easy to represent on a p–V diagram, they are even more convincing when seen on a temperature–entropy, T–S, diagram. Remember that the area of the enclosed loop on the p–V diagram shows the work done, so the area of the enclosed loop on the T–S diagram should also be as big as possible for a given energy supply.

Entropy is an indicator of the availability of energy to be applied to any purpose. Quantity is not enough. The quality of the energy has to be taken into account and this is exactly what Carnot is saying – of the energy supplied to do work, how much can be used under the best possible conditions for that purpose? That is, what is the highest possible efficiency?

- PQ 13.13 Does the T–S diagram show why the Carnot cycle gives the maximum efficiency? Is there any other realistically possible cycle that may do better?

Stay with the T–S diagram of Figure 13.9. The rectangle ABCD which the diagram generates indicates the external work developed. There are, however, two other rectangles in the diagram – ABEF and DCEF.

- PQ 13.14 What do these represent?

So comparing work rectangle ABCD with the enthalpy supply rectangle ABEF must give the **thermal efficiency** for the process – the proportion of work generated to enthalpy supplied. It can be simplified even further. As these rectangles are of the same width EF, then, by inspection, the ratio of the areas ABCD to ABEF must also equal the ratio of their heights, AD to AF.

These heights are also the maximum and minimum temperatures of the cycle of operations. A word of clarification – as we are dealing with the best possible theoretical cycle, then absolute values **must** apply. We cannot be constrained by any arbitrary temperature scales. Equally, on the p–V diagram, pressure would be absolute, not an arbitrary gauge pressure. For the next step then, these temperatures T_1 and T_2 of Figure 13.9 are absolute temperatures.

Temperature T_1 is the temperature of enthalpy introduction, the highest temperature in the cycle. Temperature T_2 is the temperature of enthalpy rejection, the lowest temperature in the cycle. The temperature difference $T_1 - T_2$ is indicative of the amount of work done. Thus the **Carnot efficiency**, this assessment of the most efficient cyclic conversion of enthalpy to work when operating between two stated (maximum and minimum) temperatures, is given by the temperature ratio

$$\text{Carnot efficiency, } \eta = \frac{T_1 - T_2}{T_1}.$$

Example 13.1 Show that these must be absolute temperatures by using some arbitrary scales. Take T_1 as 200°C and T_2 as 50°C.

First, taking the usual Celsius scale. With T_1 at 200°C and T_2 at 50°C, then the temperature difference is 150 deg C or 150 K. The difference is numerically the same whether Celsius or Kelvin is used but individual values are not the same on the two scales. The efficiency assessed incorrectly using the Celsius value would be (200 − 50)/200 or 75%.

Now convert the temperatures to Farenheit. They become 392°F and 122°F which gives an efficiency of (392 − 122)/392 or 68.9%.

These are exactly the same temperatures but on different arbitrary scales. The efficiency cannot change – that is governed by the process, so it must be the choice of temperature scale which is incorrect. The efficiency assessed correctly on the Kelvin scale is (200 − 50)/(200 + 273) which is 31.7%. The values are very different and this gives a good example of the importance of using absolute values in calculations if there is any doubt.

Scales of temperature like Celsius or Fahrenheit are arbitrary – they happen to be based on the freezing and boiling properties of water. If the most important liquid in the world was something else, then the basis of the temperature scales would be different. The absolute scale, whatever the important liquid, would still start from a real zero.

Exactly the same Carnot efficiency can be obtained from the p–V diagram. Identify p_a, V_a, p_b, V_b and so on.

The enthalpy supplied during the isothermal change at T_1 (A to B) is

$$Q_1 = p_a V_a \ln (V_b/V_a) = m R T_1 \ln (V_b/V_a).$$

and the reject enthalpy at T_2 (C to D) is

$$Q_2 = p_c V_c \ln (V_d/V_c) = m R T_2 \ln (V_d/V_c).$$

● **PQ 13.15** If the ratio of final volume to initial volume for the isothermal heat supply phase is r_v, show that the initial volume to final volume for the isothermal heat rejection stage is also r_v.

Thus,

$$Q_1 = m R T_1 \ln (V_b/V_a) = m R T_1 \ln (r_v)$$

$$Q_2 = m R T_2 \ln (V_d/V_c) = m R T_2 \ln (1/r_v) = -m R T_2 \ln (r_v).$$

So the Carnot efficiency is

$$\frac{m R T_1 \ln (r_v) - m R T_2 \ln (r_v)}{m R T_1 \ln (r_v)} = \frac{T_1 - T_2}{T_1}.$$

You should refer back to Chapters 2, 3 and 4 if any of the above gives difficulty. **Never** let misunderstandings go unresolved.

The temperatures – absolute, recall – in the efficiency assessment are those of the actual operational cycle, the maximum and minimum temperatures reached by the working fluid. They are not the temperatures of some adjacent surfaces or hardware. If this Carnot efficiency was now to be used as a point of comparison for a real engine, these temperatures are the maximum and minimum temperatures inside the real engine, where the heat-to-work conversion is taking place. They are not the temperatures of the engine casing or the cooling water or the exhaust pipe.

This simple Carnot temperature ratio is then an excellent measure of the *best* that can be done according to the laws of thermodynamics. Even allowing for important differences between ideal and real engines, it goes a long way to explaining why no real engine (using the term in its everyday sense) can be anywhere near 100% thermally efficient. Having determined the best that can be achieved theoretically, irrespective of the engine and with no practical constraints, the next step is to look at some air standard cycles. Recall that these are *idealized* versions of *practical* real engine cycles – the important features of real cycles but with a perfect working fluid and perfect operational stages. As this is a next step, the stages of these cycles will be idealized to see their best possible performance.

The **Otto cycle** (named after the investigator who devised it) is the idealized cycle of the gasoline or petrol engine, the one found in most family cars. It consists of two constant-volume and two adiabatic changes as shown in the p–V diagram of Figure 13.10. Note that the points in the Otto cycle 1–2–3–4, do not coincide with A–B–C–D in the Carnot cycle. This has been done deliberately to show that they are separate cycles and are not to be confused with each other.

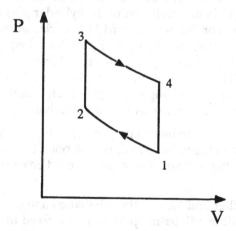

Figure 13.10 Otto cycle.

Recall that the air standard cycle is not just a single cycle which covers everything. There are several, each being an idealized representation of whatever real or possible cycle is under consideration, in exactly the same way that there is not a single ideal gas. Instead of drawing into an engine cylinder an air/fuel mixture by means of a piston and cylinder arrangement and then burning it, with valves opening and closing to admit and reject the working fluid, the cylinder is sealed and contains a fixed amount of air. The air is heated or cooled to simulate fuel energy input and hot exhaust output.

Start the Otto cycle at point 1 with the air being compressed adiabatically to point 2. This is equivalent to the air/fuel mixture in the practical engine being compressed by the piston rising up the cylinder prior to ignition. When point 2 is reached, heat is added at constant volume to raise the air pressure. This is equivalent to the fully compressed fuel/air mixture being ignited and burning instantaneously. It certainly does not burn instantaneously in the real engine but this is an idealization.

The pressure rises to point 3, the heat source is removed and the mixture expands adiabatically as the piston moves back down the cylinder. This is equivalent to the piston in the real engine being pushed back down the cylinder, thus turning a crank and doing external work – driving the car, for example. This adiabatic expansion stops at point 4 and the non-flow air trapped inside this Otto engine is cooled at constant volume back to the starting condition, point 1. In the real engine, this would be the hot exhaust gases leaving the engine and being replaced by a new cool intake of fuel/air mixture. It recognizes a limit to the usability of the working fluid.

Whichever adiabatic operation is being followed, the extreme volumes are the same. They represent the limits of the piston travel and thus the smallest and largest volumes of the cylinder plus cylinder head space of the engine. For the air standard Otto cycle, let this ratio of the larger to the smaller volume, V_4/V_3 or V_1/V_2, be called r_v. In the real engine, this would be the **compression ratio**. Heat is supplied in the constant volume change, from point 2 to point 3 and heat is rejected in the other constant volume change from point 4 to point 1. There is no heat transfer in the other strokes as they are adiabatic.

● PQ 13.16 Write down the heat supply and heat reject equations for the two constant volume stages by identifying the point temperatures as T_1, T_2, T_3 and T_4. From this write down the air standard efficiency for the Otto cycle.

Go one step further and look at the adiabatic changes. The ideal gas laws say for reversible adiabatic operation on a fixed mass of gas that $p V^\gamma$ = constant and combining this with $p V = m R T$ gives . . .

● PQ 13.17 What does it give? Write an expression which combines V and T for an ideal adiabatic change. You have met it!

From this PQ conclusion then, for the reversible adiabatic stages

$$T_1 V_1^{\gamma-1} = T_2 V_2^{\gamma-1}$$

and

$$T_3 V_3^{\gamma-1} = T_4 V_4^{\gamma-1}.$$

Thus,

$$\frac{T_3}{T_4} = \frac{V_4^{\gamma-1}}{V_3^{\gamma-1}} = r_v^{\gamma-1}$$

and

$$\frac{T_2}{T_1} = \frac{V_1^{\gamma-1}}{V_2^{\gamma-1}} = r_v^{\gamma-1}.$$

This gives

$$T_4 = \frac{T_3}{r_v^{\gamma-1}}$$

and

$$T_1 = \frac{T_2}{r_v^{\gamma-1}}$$

so that

$$T_4 - T_1 = \frac{T_3}{r_v^{\gamma-1}} - \frac{T_2}{r_v^{\gamma-1}} = \frac{T_3 - T_2}{r_v^{\gamma-1}}$$

and

$$\frac{T_4 - T_1}{T_3 - T_2} = \frac{1}{r_v^{\gamma-1}}.$$

The **air standard thermal efficiency** of the Otto cycle is thus

$$\eta = 1 - \frac{T_4 - T_1}{T_3 - T_2} = 1 - \frac{1}{r_v^{\gamma-1}}$$

Be quite clear that this is the air standard efficiency which implies that everything has been assumed perfect or idealized in some way, so at first sight this may seem of academic interest only. However, recall what has been done. The basic operational Otto cycle has been used and nothing has been changed in that. Practical difficulties or short-comings have been eliminated but again nothing fundamental has been changed. All the features of this Otto cycle would be recognizable when studying a real petrol fuelled spark ignition engine.

Thus the overall conclusions from this study of the air standard cycle must be valid and the prime one is that the efficiency of the cycle must rise as the engine compression ratio r_v rises. The higher the value of $r_v^{\gamma-1}$, the higher must be the efficiency and this is borne out in practice. As an aside, there are practical limits to the compression ratio because of combustion phenomena and their mechanical demands but the design guidance is clear – raise the compression ratio as high as possible without getting into practical difficulties. This type of simple thermodynamic study has been of great real value to the worldwide motor industry.

Example 13.2 What is the potential increase of thermal efficiency of a spark ignition (gasoline fuel, Otto cycle) engine if the compression ratio is raised from 8:1 to 10:1? Take γ for air as 1.4.

This is the common way of representing compression ratios in engines, as a simple numerical ratio. It is the ratio of the maximum (cylinder volume plus cylinder head or combustion space) volume to the minimum (cylinder head or combustion space alone) volume.

At 8:1 the air standard thermal efficiency is

$$1 - 1/(8^{\gamma-1}) = 0.5647 \text{ or } 56.47\%.$$

At 10:1 this increases to

$$1 - 1/(10^{\gamma-1}) = 0.6019 \text{ or } 60.19\%.$$

These values are very much the theoretical ones and real engines are maybe two-thirds as efficient as the air standard cycle. However, real advantages are gained by increasing compression ratios, subject to fuel properties and mechanical strength, as far as possible.

Example 13.3 A spark ignition engine has maximum and minimum cycle temperatures recorded as 1700 K and 600 K respectively. What is the Carnot efficiency and what would be the compression ratio of an air standard cycle to give the same efficiency, assuming nothing else changes?

This example is for illustration only. It does not mean that all engine temperatures are the same because they depend upon design and upon the operational conditions. The temperatures would be very different if a car engine was running permanently at maximum power on a race circuit compared to a gentle countryside cruise.

Beyond that, no change is ever in isolation. There is always a
knock-on effect, so in this case the compression ratio could not be
changed without other parameters – the working fluid temperatures,
for example – also changing.

Those temperatures give a Carnot efficiency of (1100/1700) or
64.71%. The calculation is as in Example 13.1.

Thus if all else is constant, where r_v is the air standard cycle's
compression ratio

$$1 - \frac{1}{r_v^{\gamma-1}} = 0.6471.$$

With γ for air being 1.4, this gives a compression ratio of 13.5:1.

This value is not an impossible one in real engines. However, here
it is the value for the air standard cycle and the comment was made
previously that real thermal efficiencies are maybe two-thirds or so of
the air standard value. A little arithmetic shows that a real spark
ignition engine compression ratio then would have to be rather high
– I leave it to you. The important point, however, is that both the
Carnot and the air standard efficiencies are **ideals** against which to
compare and they explain very clearly why no heat-to-work engine
can even approach 100% thermal efficiency.

To complete this, look at the Otto cycle, Figure 13.11, and compare the
real four-stroke spark ignition engine cycle to an idealized Otto one.
The same general outline holds good but there are some practical
differences.

Perhaps most noticeable is the extra bit between point 1 and point 5.
The air standard Otto cycle had two strokes of the piston – the

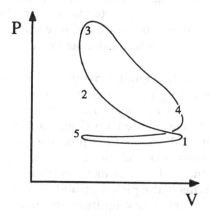

Figure 13.11 Real spark ignition cycle p–V diagram.

compressing of the mixture and the work stroke. The real four-stroke engine has two more induction and exhaust strokes where the fresh charge of fuel and air is admitted and the old, used mixture ejected. If the real engine could be perfect, these two extra strokes would be constant pressure ones, simply sucking in or pushing out the working fluid. This cannot be, of course, so there is a little bit of low pressure during induction and a higher pressure during exhaust. A good engine keeps this induction and exhaust loop as small as possible, though.

- PQ 13.18 Why?

Next, the changes from one operation to another are not sharp or instantaneous but take a finite time. Chapter 12 on reacting mixtures made reference to the time taken for fuel/air reactions. Here is a good practical example, with the piston still moving as the fuel and air burn. This is shown by the corners being rounded (no instant changes but progressive ones) and the lines for heat input and rejection not being straight and not being vertical. They are not true constant volume stages because the whole process takes time, during which the piston continues to move, so the volume changes.

The other parts of the engine operation are not truly adiabatic, for heat is being transferred to the engine casing and thence to the atmosphere at all times. Any real engine gets warm more or less all over when in use. However, the principles of the Otto cycle remain clear in the real family car engine and it is a good reference point for the comparison of designs.

The second cycle example is that of the **Diesel** (the originator's name) or **compression ignition** engine. There is some difference between the design actually proposed by Diesel and the engine which is commonly called the Diesel engine. Any confusion can be avoided by referring to the compression ignition engine if necessary. As this cycle is a second illustration, we will look at its air standard cycle almost exclusively because the real-life imperfections are broadly the same as the spark ignition engine. The real cycle has a similar extra bit, rounded corners and so on.

Mechanically and visually there are many similarities between spark ignition and compression ignition engines as used in cars and lorries and so on. They both have various numbers of cylinders and pistons. They both draw in or have injected a fuel/air mixture. They both burn that mixture and they both exhaust the products of combustion. They both have so-called two-stroke and four-stroke versions (Figure 13.12).

Concentrating on the real four-stroke compression ignition engine first, the four movements of the piston up and down a cylinder follow a very similar pattern to the spark ignition engine, but there are two important differences. On the first down stroke or induction stroke, air

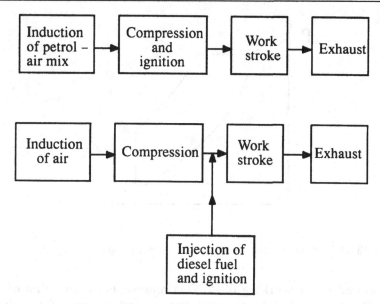

Figure 13.12 Similar but different.

alone is drawn into the cylinder, whereas the spark ignition engine is dealing with an air/fuel mixture.

The piston starts to move back up the cylinder and compresses the air, which heats up because of the work done on it during this compression. Fuel is now injected into this hot air and the mixture burns – hence the compression ignition name. The fuel properties are carefully chosen for this to happen and trying to do the same with other fuels could be somewhere between dangerous and disastrous! There is no spark used for ignition, just the effect of the hot compressed air.

Those are the differences and the rest of the cycle is as the spark ignition one. The burning mixture pushes the piston back down the cylinder and the mechanical linking of the piston to the crankshaft allows external work to be done. The final stroke of the piston back up the cylinder exhausts the burnt gases, to make way for the fresh charge. The movement of the piston is continuous and the cycle of operations is repeated.

The air standard cycle for the diesel engine is a little different from the Otto one, reflecting the operational differences. However, it is still made up of four ideal stages (Figure 13.13).

(a)   The adiabatic compression of the working fluid, reflecting the movement of the piston up the cylinder to compress the air in the real engine, point 1 to point 2.

(b)   A constant pressure stage, point 2 to point 3, where enthalpy is

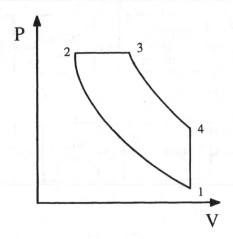

Figure 13.13 Diesel or compression ignition air standard cycle.

added to the working fluid. This represents the injection of fuel and combustion in the real engine and is part of the work-extracting stroke.

(c) An adiabatic expansion for the rest of the work extraction, as in the Otto cycle and accompanied by the real piston continuing to move down the cylinder.

(d) Finally, a constant volume heat rejection stroke, exactly as in the Otto cycle and reflecting the same real engine function.

While the compression ratio V_1/V_2 or r_v is the same as before, there is an extra ratio to identify, called the 'cut-off ratio' or something similar. On the diagram this is the ratio V_3/V_2 and it shows the portion of the piston movement back down the cylinder as the fuel is being injected and burnt. In the air standard cycle it is the portion during which enthalpy is added to the working fluid, air. This ratio is often identified by the Greek letter ρ.

The use of the air standard cycle is still that of looking at the factors which may improve the efficiency and operation in real engines. Efficiency is still the ratio of useful work done to energy supplied. The steps in determining this cycle efficiency are similar to those explored for the Otto cycle. Remember, though, that the heat addition is at constant pressure.

• **PQ 13.19** Derive an expression for the air standard efficiency of the compression ignition cycle.

Example 13.4 What is the air standard efficiency of a compression ignition engine with a compression ratio of 8:1 and a cut-off ratio of 2:1? What is the efficiency if the cut-off ratio rises to 3:1? Values vary according to design in practice but these are not unrealistic.

In the first case, using the efficiency expression derived in PQ 13.19,

$$\eta = 1 - \frac{(\rho^\gamma - 1)}{r_v^{\gamma-1}\ \gamma\ (\rho - 1)}$$

$$= 1 - \frac{(2^{1.4} - 1)}{8^{0.4} \times 1.4 \times (2 - 1)}$$

$$= 0.49 \text{ or } 49\%.$$

In the second case, the efficiency changes to

$$\eta = 1 - \frac{(3^{1.4} - 1)}{8^{0.4} \times 1.4 \times (3 - 1)}$$

$$= 0.432 \text{ or } 43.2\%.$$

These diesel engine values compare with 56.47% of the Otto cycle.

● **PQ 13.20** Diesel engines are generally quoted as having greater efficiencies than petrol engines. So what can be changed to get this greater efficiency in the air standard cycle for given cut-off ratios?

Example 13.5 For those cut-off ratios of Example 13.4, what would be the new compression ratio to give the same thermal efficiency as the Otto air standard cycle?

Using

$$\eta = 1 - \frac{(\rho^\gamma - 1)}{r_v^{\gamma-1}\ \gamma\ (\rho - 1)}$$

and putting $\eta = 0.5647$, the Otto value, with the cut-off ratio of 2 and, of course, the principal specific heat capacities ratio for air of 1.4, gives the new compression ratio r_v of 11.86:1.

Repeating this for the cut-off ratio of 3:1 gives a new compression ratio r_v of 15.57:1.

For the same thermal efficiency as a spark ignition engine, the diesel engine compression ratio has to be significantly higher, as this example shows. The fuel combustion characteristics in compression ignition engines are, however, such that high compression ratios can be used and the engine design capitalizes on this. Broadly, compression ratios are around twice the values found in spark ignition engines.

● **PQ 13.21** So what is the air standard thermal efficiency for a compression ignition engine of 20:1 compression ratio and 2.5:1 cut-off ratio?

Thus again, a basic lesson of an air standard cycle can be applied to real designs, for the efficiency advantage of real diesel engines is related to the high compression ratios which should be used. Recall though that, also again, this air standard cycle is the idealized cycle. If a practical cycle was to be superimposed on the air standard one, then a similar sort of rounding of corners and the addition of an extra piece for the induction and exhaust strokes as seen in the Otto cycle would be apparent.

● **PQ 13.22** Sketch the real cycle on top of an ideal one.

The area enclosed by the p–V graphs, ideal and real, indicate the external work which should be available. So when designs are to be improved, anything which brings the real cycle closer to the air standard has to be investigated. The air standard cycle gives a good measure against which to judge different designs. As a side exercise, it is worth drawing the Otto and Diesel cycles on a T–S diagram and then comparing them with the rectangle of the Carnot cycle. You already know the shapes of the individual stages on T–S graphs!

Just a last word about the names of the air standard cycles. The Otto cycle has, as its four stages, two adiabatic and two constant volume changes. Sometimes then, the casual name constant volume cycle is met for this one. That is fairly clear cut but, just occasionally, you may meet the casual name constant pressure cycle attached to the Diesel or compression ignition one. While there is a constant pressure change (the others of the four being a constant volume and two adiabatics, of course) this does not really describe the cycle properly.

● **PQ 13.23** The gas turbine engine air standard cycle is one which could be more accurately called a constant pressure one. Go back to Chapter 5 where the gas turbine set was met and attach the air standard ideas to it. What are the four stages of its air standard cycle?

To summarize, then, any practical engine has many influences on its efficiency, both in terms of the basic engine design and its important ancillaries, such as the fuel injection system. To effect improvements, it is vital to get to the heart of the operation and, for engine or gas cycles as they are more properly called, this means separating the principles from the details, the basic cycle from the ancillaries.

Two stages have been introduced here for that. First, the absolute thermodynamic limit to efficiency has been considered by way of the Carnot cycle. This shows the maximum attainable efficiency for given operational limits and is the proper measure of a real engine's absolute performance. Next, real cycles have been stripped of their practical aspects by means of the air standard cycle approach to show their relative thermodynamic limits and thus to identify fundamental improvements. Now, should you need to study engine or cycle practice, the true influence of practical changes can be considered properly.

PROGRESS QUESTION ANSWERS

PQ 13.1

Yes. They are laws which deal with material properties and it does not matter what the material is doing. Laws are laws.

PQ 13.2

First, the air is being compressed, so work is being done on it as it is trapped inside the compressor cylinder. Since we are dealing with a real operation, it is unlikely that the compression will be isothermal. It will follow some sort of $p V^n$ change. Some of the work done on the air will appear as increased internal energy, so its temperature will rise. Recall $T_1^n/p_1^{n-1} = T_2^n/p_2^{n-1}$.

Second, again as we are dealing with a real machine, friction will be present and friction usually shows up as heat somewhere. If surfaces inside the compressor get hot, they must influence the temperature of the air which contacts them.

PQ 13.3

A motor car engine has a gas as the dominant working fluid, even if the petrol is liquid. The reasoning is the same as for the boiler. The amount of air needed to burn the petrol is fourteen times or so that of the fuel and as the fuel burns, the products are definitely gaseous. Go back to Chapter 12 if necessary.

A refrigerator is a good example of the vapour-based cycle. The 'fridge works by the heat of the 'fridge contents being used to vaporize a liquid refrigerant. The food in the 'fridge provides the latent heat. The vapour is then compressed to a liquid by the built-in compressor to repeat the cycle and the heat from the contents and from the compressor working is dissipated by a simple heat exchanger behind the refrigerator.

PQ 13.4

Consult Figure 13.14.

PQ 13.5

AB – Constant volume, no work, no area under curve.
BC – Constant pressure, area under curve is BCDE.
CA – pV^n, area under curve is CAED.

PQ 13.6

This net expenditure of work is the difference between the two areas –
the area ABC.

PQ 13.7

Design improvements usually refer to changes to existing designs –
making an engine work better, for example. However, if the old design
is bad, then the improvement is an improvement to a bad design but it
is still a bad design. It is a relative improvement.

However, if an engine's operation is compared to the ideal (in this
case the basic engine cycle) then the real shortcomings are highlighted.

PQ 13.8

Strictly, looking at a cycle of operation means ignoring everything in
the real world! However, the cycle study has to be applied to real
engines in the end, so it is the differences between engines and the

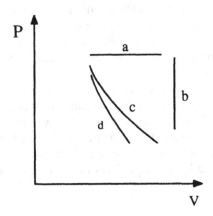

Figure 13.14 The four p–V lines: (a) constant pressure; (b) constant volume; (c)
isothermal; (d) adiabatic.

detailed practical interpretations that are to be avoided.

Your list may include, for instance, the brand of fuel, the way that the fuel/air mixture is ignited, the size of the engine valves, the mechanical energy losses (friction) in the engine.

PQ 13.9

The work that is done by this type of pretend engine must come from a reduction of enthalpy of the working fluid. How this is actually changed into work does not matter for the present.

PQ 13.10

The ratio of useful work done to initial energy input.

PQ 13.11

Go back to the NFEE. In isothermal non-flow changes, there is no increase of internal energy, therefore all the enthalpy introduced at this stage must go into external work. $\triangle Q = \triangle W$ because $\triangle U = 0$.

PQ 13.12

Reversible adiabatic – no external heat transfer, so a direct usage of internal energy to provide work, $\triangle U = \triangle W$ because $\triangle Q = 0$.

PQ 13.13

This Carnot engine is working between stated limits of pressure, volume, temperature and entropy, just as any real engine would. To compare the Carnot to any other cycle, these same limits must apply. No-one yet, incidentally, has built a real Carnot engine.

On the T–S diagram, the biggest possible work loop, fixed by the operational limits, is a rectangle. Try any other shape. Superimpose a circle or some random curves but stay within the T and S limits. By inspection, no shapes confined by these limits can give a better efficiency value.

PQ 13.14

ABEF is the enthalpy supply and DCEF is the enthalpy rejected.

PQ 13.15

The two gas law relationships that matter are the isothermal one, $p \times V$ is constant and the adiabatic one, $p \times V^\gamma$ is constant. Thus:

$$p_a V_a = p_b V_b; \quad p_c V_c = p_d V_d; \quad p_b V_b{}^\gamma = p_c V_c{}^\gamma; \quad p_a V_a{}^\gamma = p_d V_d{}^\gamma.$$

from which

$$\frac{V_b}{V_a} = \frac{p_a}{p_b} = \frac{p_d V_d{}^\gamma V_b{}^\gamma}{V_a{}^\gamma p_c V_c{}^\gamma}$$

But

$$\frac{p_d}{p_c} = \frac{V_c}{V_d}$$

thus

$$\frac{V_b}{V_a} = \frac{V_c (V_d{}^\gamma V_b{}^\gamma)}{V_d (V_a{}^\gamma V_c)^\gamma)}$$

hence

$$\frac{V_b{}^{1-\gamma}}{V_a{}^{1-\gamma}} = \frac{V_c{}^{1-\gamma}}{V_d{}^{1-\gamma}}$$

or

$$\frac{V_b}{V_a} = \frac{V_c}{V_d} = r_v \text{ , the required ratio.}$$

PQ 13.16

The air in this ideal Otto cycle is trapped in the engine and flows neither in nor out. The constant mass ideal gas equations apply and since the heat supply and the heat rejection phases are at constant volume, then these will be something like $q = mc_v \Delta T$.

For heat addition, $q_a = mc_v (T_3 - T_2)$.
For heat rejection, $q_r = mc_v (T_4 - T_1)$.

Any efficiency is the ratio of the usable something to the supplied something, so air standard efficiency is the ratio of the useful heat to the total heat supplied. The useful heat is, of course, the total heat supplied less that rejected. Thus

$$\text{efficiency} = \frac{m\, c_v (T_3 - T_2) - m\, c_v (T_4 - T_1)}{m\, c_v (T_3 - T_2)}$$

$$= 1 - \frac{(T_4 - T_1)}{(T_3 - T_2)}$$

PQ 13.17

From the characteristic equation $p = (mRT)/V$,

thus

$$\frac{mRTV^{\gamma}}{V} = \text{constant}$$

and since m and R are constant in this case, they can be absorbed in the 'constant' (little 'c' constant, remember) right-hand side of the equation as

$$\frac{TV^{\gamma}}{V} = \text{another constant}$$

or, $TV^{\gamma-1} = \text{constant}$.

PQ 13.18

Because any loop on a p–V diagram shows work done. The less work used in sucking in or pushing out the fuel, air and exhaust gases, the better.

PQ 13.19

With the usual nomenclature, the heat input is at constant pressure

$$Q_1 = m \, c_p \, (T_3 - T_2),$$

and the reject at constant volume

$$Q_2 = m \, c_v \, (T_4 - T_1).$$

$$\eta = \frac{m \, c_p \, (T_3 - T_2) - m \, c_v \, (T_4 - T_1)}{m \, c_p \, (T_3 - T_2)}$$

$$= 1 - \frac{(T_4 - T_1)}{\gamma(T_3 - T_2)},$$

and by substituting the ratios ρ, V_3/V_2 and r_v, V_1/V_2, similar to the way used for the Otto cycle, this becomes

$$\eta = 1 - \frac{(\rho^{\gamma} - 1)}{r_v^{\gamma-1} \gamma (\rho - 1)}.$$

PQ 13.20

The only variable which can be changed is the compression ratio, r_v.

PQ 13.21

$$\eta = 1 - \frac{(\rho^\gamma - 1)}{r_v^{\gamma-1} \gamma (\rho - 1)}$$

$$= 1 - \frac{(2.5^{1.4} - 1)}{20^{0.4} \times 1.4 \times (2.5 - 1)}$$

$$= 0.6255 \text{ or } 62.55\%.$$

PQ 13.22

See Figure 13.15.

PQ 13.23

The gas turbine set compresses air (ideally a reversible adiabatic process), then adds fuel in the combustion chamber and burns it (ideally a constant pressure process), then extracts work by expansion in the turbine (ideally another reversible adiabatic process) before rejecting the exhaust gases. For an air standard cycle, this last stage would be the recycling of the cooled (heat rejected) working fluid as a constant pressure process because it will be at atmospheric pressure. The air standard cycle of the gas turbine, also called the Joule cycle, is thus two constant pressure and two (reversible if we are very correct) adiabatic processes.

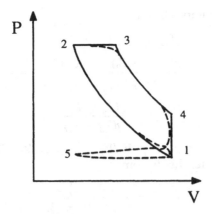

Figure 13.15 Diesel engine cycle in practice.

TUTORIAL QUESTIONS

It may be helpful to sketch cycle diagrams when attempting some of these questions. Unless otherwise indicated, use a value of 1.4 for the ratio of principal specific heat capacities. Recall that thermodynamic relationships apply both to parts of cycles and to complete cycles.

13.1 The data below were recorded during a test on a spark ignition gasoline engine. Determine (a) the air standard (Otto) cycle efficiency, (b) the test bed efficiency, (c) the energy losses due to incomplete combustion during the cycle.

Ambient temperature	20°C	Compression ratio	9:1
Fuel flow rate	10 kg/h	Net calorific value	44 MJ/kg
Average air/fuel ratio	16:1	Exhaust gas temperature	420°C
Cooling water flow rate	12 kg/min	Water temperature rise	50K
Work output rate	48 kW	Casual losses	5 kW

Mean specific heat capacities – exhaust gas 1.05, water 4.18 kJ/kgK.
[58.5%; 39.3%; 7.59 kW]

13.2 For an air standard Otto (sometimes called constant volume) cycle, the air at the beginning of the compression stroke is at 25°C, 1.1 bar and the compression ratio is 8:1. The maximum cycle pressure is 35 bar. What is the maximum cycle temperature? [1185K]

13.3 An air standard Diesel (or compression ignition) cycle has a compression ratio of 17:1 and a cut-off ratio of 2.0:1. What is the air standard efficiency? [62.3%]

13.4 For the cycle of question 13.3, what are the values of the maximum temperature and pressure in the cycle if the air is at 1.1 bar, 25°C at the start of the compression stroke? What are the values of temperature and pressure at cut-off? [58.1 bar; 925K; 58.1 bar; 1850K]

13.5 For the cycle of question 13.4, what is the pressure at the end of the adiabatic expansion? What new compression ratio would be needed to keep the same cycle efficiency if the cut-off ratio was changed from 2.0:1 to 1.9:1? What would be the new maximum cycle temperature and pressure values if other parameters were unchanged?
[2.9 bar; 16.45:1; 55.5 bar; 1736K]

13.6 In a Carnot cycle with air as the working fluid, the air is at 2 bar, 20°C at the beginning of the compression and the energy intake is 800 kJ/kg air to yield a maximum temperature of 1700K. What is the

specific work output and the maximum specific change of air entropy
during the cycle?   [662.1 kJ/kg; 0.47 kJ/kg K]

Cycles of operation – vapours 14

14.1 DIFFERENCES COMPARED TO GAS CYCLES

More accurately, this chapter should be headed 'Cycles of operation – vapours, liquids and gases' because the use of a vapour in a cycle is almost inevitably associated somewhere with the liquid and the gas from which it arises or into which it is transformed.

● PQ 14.1 Name an industrial example. You have met one.

This transformation or phase change is one of the reasons for gas cycles and vapour cycles to be separated even though the basic rules are the same. Any cycle is a series of connected and repeated events. Vapour cycles can be studied fundamentally by using the ideal forms of those events, just like the Otto cycle for the family car spark ignition engine. For these vapour cycles though, there are the **phase changes** to include and there is the special property of vapours wherein they can yield or absorb energy by adjusting their **dryness**. These changes of enthalpy without necessarily changing temperature or pressure make an important difference compared with gas cycles (Figure 14.1).

A p–V or a T–S diagram can still be used for vapour cycles but with the different effects of enthalpy at each phase, diagrams which take

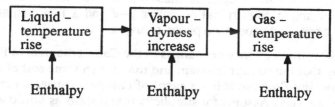

Figure 14.1 An important difference.

account of enthalpy are also useful when representing a vapour cycle.

First, one thing that does not change – the Carnot cycle is still the most efficient that can be defined between set working limits, and this applies here also. Whatever happens to the liquid, the vapour or the gas as phases change and enthalpy is absorbed or rejected, the efficiency limit is still the Carnot efficiency as defined by the upper and lower absolute temperature limits, T_1 and T_2, of the cycle:

$$\eta = \frac{T_1 - T_2}{T_1}.$$

One thing that does alter with a vapour compared to a gas though is the shape of graph produced by some changes.

● PQ 14.2 Sketch a p–V graph for a constant pressure change and an isothermal change for an ideal gas. Now put on the same graph the same changes for a vapour – just vapour, no phase changes.

The adiabatic (remember it should be described as reversible and the word is only omitted for brevity) change graph is much the same general shape for both gases and vapours. Take care though – recall that equations or relationships derived for gases earlier, such as pV^γ. in Chapter 4, were for ideal gases. They are applicable to real gases well away from their liquefaction or condensation point. In no way are vapours (say wet steam here) real gases well away from their condensation temperature!

For perfect gases and almost exactly for real diatomic gases, such as oxygen or nitrogen, the ratio of principal specific heat capacity γ is 1.4. For triatomic gases which includes gaseous water substance, H_2O, superheated steam, it is 1.3 or very close. Values are quoted for wet steam but the value must vary with the dryness fraction.

● PQ 14.3 Why?

Generally therefore, avoid trying to put ideal gas relationships to vapours. For most vapours – steam, ammonia, refrigerants – there are sufficient data available for most calculation purposes without having to go to first principles each time. Now back to the Carnot cycle.

● PQ 14.4 The Carnot cycle is made up of two isothermal and two adiabatic changes. Sketch that on both a p–V and a T–S diagram for a vapour – again, just vapour, no phase changes.

Just follow the sequence of changes in the Carnot diagram of PQ 14.4. Let us say that the vapour is steam and that it is in some sort of engine. It is a **working fluid**, so it is an **agent of change** – it uses an enthalpy supply to do work. Assume for simplicity that it starts as liquid water at its boiling point (100% wet) at point a and is to be converted to 100% dry steam. All the sensible heat has been absorbed and only the latent heat remains to be added.

At point a, latent heat h_{fg} is applied so that the steam dries and expands isothermally at temperature T_1 to point b. At this point, it is 100% dry. The dry steam at T_1 now expands adiabatically, a work stroke, to point c, falling to temperature T_2 as it does work in the engine. As this is an adiabatic change, bc, there is no heat transfer to or from the steam so any work must come from a change of the vapour's own energy.

- PQ 14.5 What will happen to the steam dryness?

Example 14.1 Steam 100% dry expands adiabatically and reversibly from 230°C to 130°C. What is the change of dryness fraction?

If this was a gas, then the pV^γ = constant equation would be usable. Wet steam is not a gas, so some other route has to be found.

- PQ 14.6 What is the other name for reversible adiabatic change?

This expansion is an **isentropic expansion,** so the entropy of the dry steam at 230°C will be the same as the entropy of the wet steam at 130°C. Reading from the thermodynamic property tables (steam tables) and using the nomenclature with which you familiar,

specific entropy of dry steam at 230°C s_{g1} = 6.213 kJ/kg K
specific entropy of liquid water at 130°C s_{f2} = 1.634 kJ/kg K
specific entropy of phase change at 130°C s_{fg2} = 5.393 kJ/kg K.

Thus for an isentropic change,

$$6.213 = 1.634 + x_1\, 5.393$$
$$\text{and } x_1 = 0.85.$$

So the steam after the isentropic expansion is 85% dry. You see – entropy is useful after all!!

This is an isentropic expansion. There is no entropy change, so it represents the perfect operation. Refer again to the steam tables and work out that the enthalpy change of the steam because of the work done is 409.1 kJ/kg. Remember now the derivation and reasoning of the Clausius inequality in Chapter 9.

Since this steam is expanding between two fixed vapour temperatures, then it is also expanding between two fixed vapour pressures – in this case of Example 14.1 between 28 and 2.7 bar. If this is the design of the engine or machine in which the steam is doing work, then those pressures are to be achieved however well or badly the device is working. It cannot, however, expand isentropically – there must always be this **inequality of entropy** in real systems, with more at the end of a process than at the beginning.

This gives rise to the term 'isentropic efficiency' – the real amount of work done or enthalpy used compared to the ideal, isentropic amount.

Example 14.2 For the same operation outlined in Example 14.1, the real process has an isentropic efficiency of 77%. What is the real enthalpy change and the real final dryness fraction?

The real change of enthalpy is 77% of the isentropic value, 0.77 × 409.1 kJ/kg or 315 kJ/kg.

The real final dryness fraction is thus 89.3%. The calculation is straightforward steam table use, so the details are not included here.

To complete the picture, this now tells us the final entropy, again straightforward steam table use with this new dryness fraction. The final entropy is 6.450 kJ/kg K, an increase of 0.238 kJ/kg K. So the Clausius inequality conclusion actually works – there is more coming out than going in. A reminder – make sure that data and calculation accuracies are compatible. Do not put too many decimal places in if the data are not that accurate.

Following the scheme of the Carnot cycle – the argument is exactly the same as for gas cycles – the next stage is heat rejection, an isothermal compression from point c to point d at T_2. The cold sink (the heat rejection) stops at point d and the cycle is completed by an adiabatic compression back to the start, point a. If you refer back to Chapter 13, you will see that this parallels the gas cycle changes, with only the diagram shape being different because of the isothermals.

Example 14.3 In the Carnot cycle used above and with the data of Example 14.1, the wet steam at the end of the heat rejection stage is compressed adiabatically back from 130°C, 2.7 bar to the starting point, saturated water at 230°C which means also 28 bar. If the compression is isentropic, what is steam dryness fraction at the start of the compression stage?

This is another application of the entropy data in the thermodynamic property tables. It is stated in the question that we are dealing with an isentropic process. The specific entropy of the saturated water at the end of the compression stage must thus equal the specific entropy of the wet steam at the start of the compression stage.

The specific entropy of saturated water at 230°C is 2.611 kJ/kg K. Other entropy values at 130°C are as in Example 14.1, thus

$$2.611 = 1.634 + x_2\, 5.393$$
$$x_2 = 0.18$$

that is, the steam at the start of isentropic compression must be no less than 18% dry to yield saturated water at 230°C. The same real-life rules apply about isentropic efficiency here as to the adiabatic expansion. The practical implication for compression is that more energy will be required to do the real task.

Since the heat addition to the vapour, ab, is latent heat h_{fg}, and the cycle efficiency is still the Carnot efficiency $(T_1 - T_2)/T_1$, then the specific external work done by the cycle, efficiency × energy input, is

$$w = h_{fg}\,\frac{(T_1 - T_2)}{T_1}.$$

● **PQ 14.7** What would the specific work output be if the steam started at dryness fraction x_1 at point a and was heated to dryness fraction x_2 at point b?

As a little diversion, if a Carnot cycle works between very close pressure limits and thus very close temperature limits, then the p–V diagram is quite narrow (Figure 14.2).

Now take a vapour which has just left the liquid stage and is about to start absorbing latent heat. Add latent heat to it right to the point of 100% dryness but no further, that is, h_{fg}. Then allow it to expand adiabatically through the minute temperature change dT and continue to follow a very thin Carnot cycle. Since T_1 and T_2 are very close, only the infinitessimally small difference dT apart, they can both be identi-

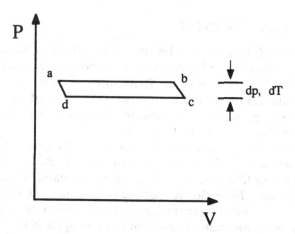

Figure 14.2 Carnot cycle between close limits.

fied simply as T in the usual way of the calculus:

$$\text{work done} = h_{fg}\frac{dT}{T}$$

Work done is also the area of the p–V diagram which is ab $\times$ dp. The points a and d are virtually coincident as are c and b, since dp is very small. As we started with the vapour having just left the liquid phase, its specific volume will be that of the liquid at its boiling point v_f. Finishing with the vapour just at its saturation point, its volume will be that of the dry vapour v_g. So the volume change ab is the vapour's volume change from beginning to end, $v_g - v_f$. Thus, the work done is also dp $(v_g - v_f)$. Hence from these two expressions of work done by the cycle,

$$h_{fg}\frac{dT}{T} = dp\,(v_g - v_f)$$

or

$$\frac{dp}{dT} = \frac{h_{fg}}{T\,(v_g - v_f)}$$

and this is called **Clapeyron's equation**. It has practical use, for instance in finding accurate values of v_g and of determining h_{fg} from pressure, volume and temperature information. Since in most practical cases v_g is very much greater than v_f, the expression can often be used in its simplified form

$$\frac{dp}{dT} = \frac{h_{fg}}{T\,v_g}.$$

14.2 THE RANKINE CYCLE

Just as engines (gas power cycles) may be studied by starting with an ideal such as the Otto or Diesel cycle, so vapour cycles may be studied using an idealized cycle called the **Rankine cycle**. Since there is so much ready information on steam, then steam will be used to illustrate the principles but, as when first looking at vapours in Chapter 7, the lessons from steam apply to all vapours.

The cycle can be applied to vapour usage of all types – engines, plant processes and so on. The principles can thus be shown by looking at one scheme – in this case a boiler providing steam, which then does work somewhere. Note clearly that following the cycle of the working fluid – steam – and its generation and use, rather than a cycle of a machine, is only one example and not the sole application.

It is worth reviewing then, briefly, steam generation in industry to set the scene for this part of the study, even though boilers and steam have been met previously. The word '**boiler**' is used for what is more accurately called a steam or hot water generator. Industrially and commercially, boilers cover a very wide range of physical sizes and water throughputs, the ratio of biggest to smallest throughput design being more than 100:1. The details of boiler design are not applicable here but the principles are.

Boiler designs vary but all are essentially water containers and fuel users. The example of a fossil-fuel fired (coal, oil, gas) steam generator will be used but the principles apply to all. Fuel is supplied to a combustion zone where it is burnt to release heat. This is available as flame and hot gases which transfer heat by radiation and convection to the water in the surrounding boiler structure. The water boils to make steam as a vapour, which may then be passed through other heat transfer zones to be dried and superheated.

The steam thus generated is at high temperature and high pressure. Everyday commercial boilers – hospitals, manufacturing – may operate up to say 20 bar and power station boilers perhaps ten times this. The steam then is carrying high quality energy (enthalpy) and is capable of doing work. It is very much a working fluid, a valuable transporter of energy. As the steam is used for its purpose, it loses energy. It cools and loses pressure also. At the end of its working stage, it is finally condensed and returned, as liquid water, to the boiler to repeat the cycle (Figure 14.3)

The boiler may have other heat transfer surfaces, where yet more energy is absorbed from the combustion products. At the beginning of the water cycle, the returning, relatively cool, condensed water may be preheated in a unit called an '**economizer**'. This is a heat exchanger which uses some of the last dregs of heat in the combustion gases because the economiser temperature is fairly low as it is dealing with the cool water.

During the working stage of the usage cycle, the steam may be interrupted in its work to have further high temperature energy added in exchangers called '**reheaters**'. The steam is quite literally reheated at

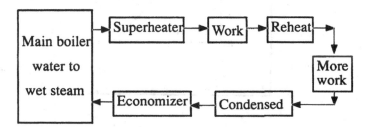

Figure 14.3 Water cycle in a boiler.

an appropriate time to do more work before continuing on the cycle. The overall aim of any of these extra heat exchange surfaces is to take out of the combustion products as much energy as possible and thus reduce wastage. Figure 14.3 shows an example of the water cycle. Not every stage is found in every cycle of every boiler and others are more complex, but this sets a scene in which to fit what follows.

The Rankine cycle is to vapour-based working fluids as the Otto and Diesel cycles are to gas working fluids. It can be compared with but is different from the Carnot cycle and it is an idealized version of a cycle that has widespread industrial and commercial use. While steam is mentioned here, the Rankine cycle applies to vapours and their associated phase changes at large. It applies as much to refrigerants at or below room temperature as it does to water boiling a long way above room temperature. The cycle can be illustrated by reference to the boiler and subsequent steam work.

Note that this is a simple illustration. You would see the same important features on all real vapour cycles but there may well be detail differences, practical changes, design improvements and so on (Figure 14.4)

Say there is a boiler generating steam from liquid water and this steam does some work in a process or engine. For a start, say this steam is not superheated, so the only phases for the present are liquid and vapour. To simplify further, say that any liquid water is always present at its boiling point relative to its pressure.

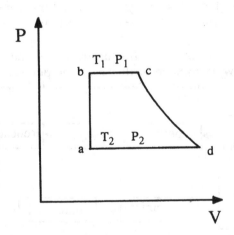

Figure 14.4 A Rankine cycle for liquid and vapour.

• PQ 14.8 Looking at Figure 14.4 and without knowing anything of the Rankine cycle, what sort of processes are ab, bc and da? What do you expect for cd?

We started to look at the Carnot cycle at the point where energy was being put into the steam, thus to begin its work cycle. Doing the same for the Rankine cycle, the reflection of practice, means starting a little earlier. Note very carefully that in both, the letter 'a' is used as the starting point but they do not coincide.

Look at Figure 14.4 and follow the water substance through its Rankine cycle. Note that the diagram is not to any scale so that the relative sizes of portions are of no consequence. It is a diagram to illustrate important features. As a brief reminder of real boiler operation, when the steam has done work in the engine or process which it feeds, it has lost temperature and pressure and has been condensed back to liquid water to start the whole cycle again.

(a) The liquid water is at point a where it has just been condensed from the previous process cycle and is returning to the boiler at temperature T_2 and pressure p_2. The water's boiling point enthalpy here is h_{f2}.

In a real boiler, the *feed pump* would push the water into the boiler, and in so doing would raise its pressure to the working pressure of the boiler. Through the feed pump and in the boiler, therefore, the water is raised to p_1 and T_1, enthalpy h_{f1}, point b on the diagram. Recall that volume changes of liquid water substance are quite small compared to vapour and gas. Very closely, then, this is a constant volume stage, ab. Enthalpy added is $(h_{f1} - h_{f2})$.

(b) The water is now at boiling point T_1 for pressure p_1. Further enthalpy addition (heat transfer from the flame and hot gases of the boiler) starts the change into and through the vapour phase to dryness fraction x_1, point c. In a real boiler system, this would mean that the steam was being taken out of the boiler to the point of work. The steam has absorbed enthalpy $x_1 h_{fg1}$ during this isothermal stage bc.

(c) Now the steam expands isentropically, along cd, which would be the best way of using it to do work since there would be no heat losses. It expands down to p_2 and dryness fraction x_2 at point d. Its enthalpy is now $(h_{f2} + x_2 h_{fg2})$.

(d) Finally it is cooled isothermally (heat rejection), along da, until it again becomes liquid water at point a. It has just condensed and has not cooled down at all in its liquid state. It is still at the boiling point for pressure p_2 but is liquid. Its specific enthalpy is h_{f2} so the heat rejected during this stage is is $x_2 h_{fg2}$.

● **PQ 14.9** What then is the efficiency of the cycle and the work done? The same ground rules as for Otto apply.

Example 14.4 As in Example 14.1, dry steam is generated at 230°C and expands isentropically during a Rankine cycle process to 130°C. At all times, any liquid water is at its boiling point for the pressure involved. What is the cycle efficiency?

The specific enthalpies required from the tables are

specific enthalpy of dry steam at 230°C $h_1 = 2803$ kJ/kg
specific enthalpy of water at 130°C $h_{f2} = 546$ kJ/kg
specific enthalpy of steam 85% dry at 130°C $h_2 = 2394$ kJ/kg

$$\eta = \frac{h_1 - h_2}{h_1 - h_{f2}} = \frac{2803 - 2394}{2803 - 546}$$

$$= 18.1\%.$$

This is only an illustrative example – it does not mean that all steam processes are less than 20% efficient!
    Thus the specific work output, efficiency × energy added per cycle, is

$$w = 0.181 \times (2803 - 546)$$
$$= 408.5 \text{ kJ/kg or kNm/kg.}$$

This compares of course to (2803 − 2394) kJ/kg, the enthalpy yield for the work output. In calculations of this type using tabular data, small discrepancies are almost inevitable because of rounding of values.
    Note, too, how close this is (same discrepancy comment applies) to the work done under the reversible adiabatic expansion of Example 14.1, 409.1 kNm/kg. While this may not be apparent from the way the Rankine cycle was drawn, if you redraw it to scale then you will see that the areas involved are dominated by the expansion curve. In real life, that is the stage where most of the steam work is done, such as in an engine or a turbine.

Example 14.5 If the condensed water was cooled below the boiling point at pressure p_2, 2.7 bar, how would this affect the efficiency and the work done? Say it was cooled to 125°C, for which $h_f = 525$ kJ/kg.

All that happens is that more heat has been rejected, so more heat has to be added to get the water back to its boiling point at 230°C. Thus, to make dry steam at 230°C, the enthalpy increase is (2803 − 525) kJ/kg and the efficiency is, where h is the enthalpy of the condensed water at 125°C

$$\eta = \frac{h_1 - h_2}{h_1 - h} = \frac{2803 - 2394}{2803 - 525}$$

$$= 17.95\%$$

• PQ 14.10 What is the difference between the work of the cycle in Example 14.4 and the work under the isentropic expansion curve?

Sometimes, the Rankine cycle is drawn with the liquid line ab coincident with the vertical axis of the p–V diagram because the liquid volume is so small compared with the vapour volume. The golden rule is, though, if in doubt, do **not** simplify anything until you are quite sure.

• PQ 14.11 Sketch the Rankine cycle of Figure 14.4 on a T–S diagram

The Rankine cycle then shows the theoretical maximum efficiency of a steam generator and usage system under normal operational conditions. It can still be compared to the Carnot efficiency but in exactly the same way as the Otto cycle, for example, may be compared. When looking at practical efficiencies, a good first step is to compare the real efficiency with the Rankine one. As with the gas cycles, practical operation means finite times for things to happen, changes which are not sharp, stages which are not ideal and so on.

Figure 14.5 has a practical cycle superimposed upon the Rankine one and you will see the rounded corners and deviation from the perfect lines. Again, this is not to scale – it is a representation to identify some points.

One particular point is seen at the bottom right of the p–V diagram, where the sharp toe of the adiabatic expansion is cut off, the triangle

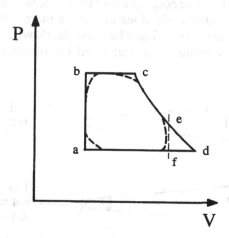

Figure 14.5 A practical cycle superimposed.

edf. In most real working fluid applications, not just vapours, there comes a point where the last bit of work available from the working fluid is not worth recovering. It may be because the increased size of equipment to do this is too costly or because extra frictional losses associated with the recovery outweigh its value – there will be a good technical or economic reason (Figure 14.6).

This superimposing compares the thermal efficiency of the practical cycle with that of the theoretical cycle and the term 'relative efficiency' is sometimes used for this ratio. The area enclosed by the real graph shows the energy available from the real cycle for doing whatever work is intended. This is the work available from the conversion of the supplied heat energy – a basic thermal efficiency. This work is then divided between various inevitable practical losses, such as friction and useful work output to give a mechanical efficiency, just as with the gas cycles.

A list of thermal efficiencies and their order may be useful (Table 14.1). The overall efficiency of the real system then is made up of four contributors:

- the fundamental thermodynamic limit as imposed by Carnot;
- the best that can be achieved within the Carnot limits by a practicable cycle, such as the Otto or Rankine cycle;
- what is achieved by the real-life cycle compared to that latter, practicable cycle (the superimposed diagrams);
- the losses imposed or work absorbers of the real machine components on top of the real-life cycle limitations.

The motor car engine is a good example. What really matters is how much power is available to drive the car along. Some power is taken in driving the engine's necessary ancillaries. Some is never available because the real engine cycle does not quite match the Otto and so on. Thermodynamics is important because that's where the technical improvements may come from, but good engineers remember that

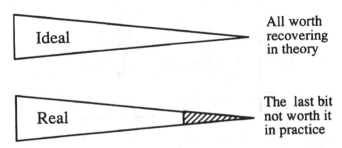

Figure 14.6 There are limits to the value of energy recovery.

Table 14.1 Some thermodynamic efficiencies

Carnot	The best possible, $(T_1 - T_2)/T_1$
Idealized cycle	The best possible for a given practical cycle, such as Otto or Rankine
Relative	The efficiency of the real cycle compared to the idealized, a comparison of enclosed areas
Mechanical	The one that matters in practice, to include practical losses – how much work for how much supplied energy

economics tends to overshadow this at times.

14.3 THE RANKINE CYCLE WITH SUPERHEAT

● PQ 14.12 Looking at the Carnot efficiency, what are the two fundamental ways open to improve a cycle's efficiency?

Since in a boiler system the condensing temperature is either likely to be fixed by the design or can only be reduced by a fairly small amount – freezing point is a limit! – then any efficiency gains can only come from raising the upper cycle temperature. For steam, this means **superheating**. Boilers can be designed for higher operational pressures, that is, boiling at a higher temperature and producing vapour at a higher temperature. Broadly though, superheating is cheaper and there are less problems in using a gas (superheated steam) than using a vapour (wet steam) at very high temperatures and pressures to gain extra work.

Example 14.6 A steam cycle operates with maximum and minimum temperatures of 200°C and 40°C. What is the Carnot efficiency? If the upper temperature is raised by 50K and later by another 50K, what are the new efficiencies?

The Carnot efficiency is the ratio $(T_1 - T_2)/T_1$ so the efficiencies for the three options are

$$\frac{200 - 40}{200 + 273} = 33.83\%, \quad \frac{250 - 40}{250 + 273} = 40.15\%, \quad \frac{300 - 40}{300 + 273} = 45.38\%.$$

Remember that these are Carnot efficiencies, not real cycle values. Even so, there is a clear and worthwhile improvement in efficiency by raising the upper temperature. That gain is a fundamental one because it has been shown on the Carnot cycle, which is independent of practice. It must be reflected in real cycle efficiencies.

Technically, the upper temperature could be raised continually but there are material limits – the materials available to do the job have to withstand the increased demands – and there has to be an economic return also.

The Rankine cycle operation to include superheat can be shown on a T–S diagram (Figure 14.7). The basic diagram is the same as before but in this new case, the steam is completely dried to point c, which is on the saturation line. Then the superheating process, adding extra heat to the dry steam which is now a true gas, carries it on to point c'. The adiabatic expansion follows the new isentropic line from c' down to d', which takes the steam back into the vapour phase. The rest of the cycle is as before.

● PQ 14.13 Looking at Figure 4.7, which points on the diagram show the extra work extracted by superheating the steam? What – words not numbers – is the extra energy demand compared to a dry vapour at point c?

Example 14.7 The steam cycle of Example 14.4 showed a Rankine efficiency of 18.1% and it used dry steam at the start of the expansion process. What would be the new efficiency if the steam is now superheated to 300°C but still expanding to 130°C?

By interpolation from the tables, the specific enthalpy of steam at 28 bar, 300°C is 3001 kJ/kg and the specific entropy 6.5864 kJ/kgK. Using the technique of Example 14.3, the steam dryness fraction at 130°C after the isentropic expansion is 0.918. The steam enthalpy at this point therefore is 2542 kJ/kg, again to data table accuracy.

The general expression for Rankine efficiency must still be the

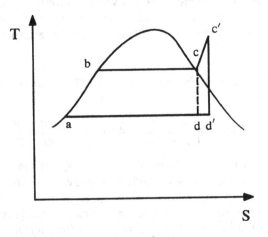

Figure 14.7 The Rankine cycle with superheat.

same, so that only the enthalpy of the superheated steam h_1 and of the different dryness fraction after expansion h_2, need to be accommodated:

$$\eta = \frac{h_1 - h_2}{h_1 - h_{f2}} = \frac{3001 - 2542}{3001 - 546}$$
$$= 18.7\%.$$

The efficiency is increased from 18.1% to 18.7% by the use of superheat and greater degrees of superheat would give greater increases of efficiency. Superheat then is a valid way of improving cycle efficiency.

As in Example 14.4, these are numbers for an illustrative example. They in no way suggest that all superheated steam machinery is less than 20% efficient.

● **PQ 14.14** What is the new work output and what is the efficiency of conversion of the superheat addition alone?

Superheating is worthwhile but that does not mean that the other way to increase efficiency – reducing the reject temperature – has to be ignored. Boiler plant working with, say, a turbine will be designed to get as much heat (enthalpy) out of the steam as possible for the work part of the operation. One way of doing this is to reduce the steam pressure as far as possible during the working stage because this is accompanied by a low temperature and thus a low ultimate heat rejection (heat exhaust from the cycle).

To achieve this, low pressure condensers are quite common in boiler plant. They are the devices which accomplish the da or d'a stage (Figures 14.4 and 14.7). For instance, a condenser operating this stage at 0.1 bar gives a condensation temperature T_2 of 45.8°C. A liquid's boiling point is inseparable from the ambient pressure, you will recall.

Example 14.8 A steam process cycle operates with a condenser at 0.1 bar and boiler pressure of 20 bar. The steam generated is superheated to 300°C, it expands isentropically and the return steam is condensed to but not below the return boiling point. Compare the Rankine and Carnot efficiencies.

The *Carnot efficiency* comes from the maximum (superheated to 300°C here) and minimum cycle (the 0.1 bar saturation temperature of 45.8°C here) temperatures

$$\frac{300 - 45.8}{300 + 273} = 44.36\%.$$

The *Rankine efficiency* is the ratio of heat usage to heat supply during the cycle

$$\eta = \frac{h_1 - h_2}{h_1 - h_{f2}}$$

From the isentropic expansion (refer back to Example 14.1 if necessary), the dryness fraction at the condenser pressure of 0.1 bar is 79.2%, thus, the specific enthalpy h_2 using steam table values is 2085.6 kJ/kg. All calculations are used in previous examples – nothing new.

The specific enthalpy of the return liquid water at condenser pressure, (0.1 bar, 45.8°C), h_{f2} is 192 kJ/kg and the specific enthalpy of the steam at 20 bar, 300°C is 3025 kJ/kg.

Thus, the thermal efficiency, the work yield or usable enthalpy compared to the enthalpy supplied is

$$\eta = \frac{3025 - 2143.6}{3025 - 192} = 31.1\%.$$

This Rankine efficiency is a fairly realistic reflection of the levels of efficiency which real steam cycles give when superheat and low condenser pressures are used. In major steam usage, such as the electricity supply industry, maximum temperatures and pressures have been taken to the usable limits of constructional materials and condenser pressures are as low as practicable. Even there though, power plant overall efficiencies are commonly in the thirty-something per cent range.

Example 14.9 For the Rankine cycle of Example 14.8, with superheat to 300°C and a condenser pressure of 0.1 bar, estimate the cycle efficiency and work yield if the expansion stage had an isentropic efficiency of 95%.

What would be the steam dryness fraction and specific entropy at the end of that stage?

Recall that isentropic efficiency, in terms of yielding usable energy from an expansion process, is the ratio of actual yield to isentropic yield. In Example 14.8, the isentropic yield (because the expansion stage was an isentropic process in that example) was given by the difference between the stage starting and finishing enthalpies, (3025 − 2143.6) kJ/kg or 881.4 kJ/kg of steam used. With an isentropic efficiency of 95%, the expansion stage can only make 95% of this isentropic amount available, that is 0.95 × (3025 − 2143.6) kJ/kg or 837.3 kJ/kg steam used. The cycle efficiency thus reduces to 29.6%.

If the steam on expanding yields 837.3 kJ/kg, then its enthalpy at the end of that stage is (3025 − 837.3) kJ/kg, 2187.7 kJ/kg. The end of the expansion is still at 0.1 bar because the condenser pressure has not changed – that is a practical operation, not an isentropic ideal. Thus for wet steam at 0.1 bar, a specific enthalpy of 2187.7 kJ/kg indicates (by the usual calculation route) a dryness fraction of 0.83 and hence a specific entropy of 6.906 kJ/kg K.

This specific entropy value is greater than the value at the start of the expansion stage (6.768 kJ/kg K) which is exactly what you would expect from the Clausius inequality and the Second Law – entropy increases as energy degrades. **Any** real process has to be accompanied by a system entropy increase!

One final point, just a brief mention for completeness, is about **reheat** in operational steam cycles. Examples 14.7 and 14.8 have shown how worthwhile it is to superheat the steam. Reheat takes matters one step further in that the superheated steam is used until it is at or near the saturated vapour level – it has used its superheat to do useful work.

Instead of now carrying on down the isentropic expansion line into the wet vapour phase, (Figures 14.7 and 14.8) the steam is returned to the boiler when it reaches point r and reheated back to point p in the superheat region. Not all boilers are equipped to do this and it is usually limited to big ones, as in electricity generation. This reheated steam now yields more work, following the second isentropic line pq before completing its cycle back to the starting point. There may be more than one reheat stage.

The efficiency advantages are much as those for superheat and the

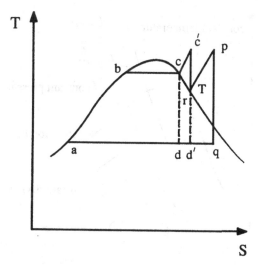

Figure 14.8 Reheat added to the cycle.

calculations simply include the the extra enthalpy provided by the reheating. Similarly, isentropic efficiencies can be included in any of the calculations – the principles are just the same for expansion whether reheat is used or not. The basic Rankine efficiency expression still holds good as well.

● PQ 14.15 Looking at the reheat part of that cycle, is it all at the same pressure? In words not numbers, state clearly what is happening to the temperatures and pressures at each stage of the cycle.

The very last point for these steam cycles – but the point applies to other vapours, since steam is only an example – is one of practical help. Throughout the vapour properties Chapter 7 and this chapter, use has been made of the thermodynamic tables. They contain much data and they thus simplify calculations in general. The practical information goes one step further in that the data have also been transferred to diagrams, so that information can be read in that way too. While they are not included here, quite deliberately, they may be useful to you later in your thermodynamics studies.

You are familiar with the temperature – entropy diagram, as in Figure 14.7. These are available with the data overprinted. A similar diagram which links enthalpy and entropy, often called the Mollier chart, is also overprinted with relevant data. Figure 14.9 gives an indication of the Mollier chart layout. Having said all that, much of the information is being transferred to computer programs, either dedicated or as part of others, to make data seeking even easier.

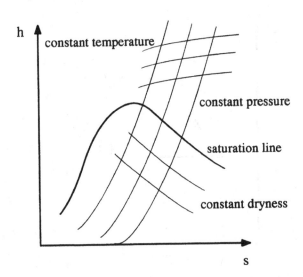

Figure 14.9 Schematic Mollier chart; h = enthalpy; s = entropy.

14.4 SOMETHING ABOUT REFRIGERATORS AND HEAT PUMPS

Steam cycles have been used to illustrate in this chapter but the rules apply to all vapour cycles – it is only the numbers that change. There is one other everyday use of vapour cycles worth mentioning – the **refrigerator** and its close relative the **heat pump**. They are technically the same with only design details and practical applications differing. These few words are very much a summary and the topic is not explored to any depth – just pointing out that the rules apply!

The widest used refrigeration or heat pump cycle is the **vapour compression** one. In this, the refrigerant working fluid (ammonia is one example, various organic fluids also) is vaporized progressively as it takes in heat from whatever is to be cooled. Normally this is done through a simple heat exchanger located inside the refrigerator cabinet, with the refrigerant being at a lower temperature than the 'fridge contents. The refrigerant is of course chosen so that its vaporization characteristics or pressure–temperature properties are correct for the task (refer to Figure 14.10).

The refrigerant, which may have entered the cabinet heat exchanger or 'evaporator' as a very wet vapour and left as quite a dry vapour, is

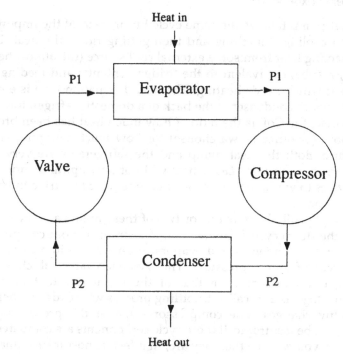

Figure 14.10 The vapour compression refrigerator or heat pump.

next compressed, which means that its temperature and pressure rise. If the gathered heat is to be dissipated to the general atmosphere, it must be available above the general atmosphere temperature – Second Law at work! If the general atmosphere temperature was below the refrigerator contents temperature, you would not need a 'fridge, would you? The *compressing pump* also circulates the refrigerant. The enthalpy gain from the refrigerator contents and the pumping/compressing action is dissipated to the external surroundings via an external heat exchanger or 'condenser' which, if you put your hand near the back of a domestic 'fridge, feels quite warm. Take great care though – do not touch the working parts!

The cooled refrigerant finally passes through an *expansion valve* from which it emerges as a wet vapour in the proper state to repeat the cycle. This valve is often adjustable, thus controlling the condition of the refrigerant entering the cabinet heat exchanger at the beginning of the cycle, hence the rate of heat removal and temperature of the contents.

Since there is no intentional work-to-heat change, as in the steam cycle, the conventional definition of efficiency does not apply.

● PQ 14.16 So how could different machines be compared in terms of their performance? Just simple words – what is the refrigerator doing and what is being expended?

The heat pump follows the same cycle but instead of the importance lying with cooling something and then getting rid of the heat, it lies with extracting heat from some external cool source (the atmosphere or a stream, maybe, equivalent to the 'fridge contents) and feeding it to somewhere which needs heating. The heat dissipation here is equivalent to the warm condenser at the back of a domestic refrigerator.

In this case, the CoP is the ratio of how much heat has been brought into a room (or office or warehouse) for how much pumping energy expenditure. Both the heat pump and the refrigerator are very good applications of the *Second Law* – heat will not flow uphill unaided. For these devices to work, the aid is the energy required to drive the pump or compressor.

An example will illustrate one or two of these points and a comparison with the steam cycle is interesting. In a simple vapour compression refrigerator, the condenser and evaporator operate at close to constant (but different, of course) pressures. The expansion valve, which reduces the refrigerant pressure from that of the condenser to that of the evaporator stage, is a so-called throttling process where ideally there are no enthalpy changes. The compressor stage of the process would, ideally again, be isentropic. If these cycle components are compared to a steam cycle, you will see that they are, in effect, almost mirror images.

The boiler feed pump raises the returning fluid to the higher pressure

– the refrigerator expansion valve reduces the pressure. In theory and closely in practice, there are no (or only insignificant) enthalpy changes solely due to pressure change.

The boiler operation adds energy to the steam at high pressure – the refrigerator rejects energy at high pressure. For both cycles this is effectively a constant pressure stage.

The boiler cycle rejects energy at low pressure – the refrigerator cycle takes in energy at low pressure. Again for both cycles these are effectively constant pressure stages.

The boiler cycle yields work by the steam expanding from high to low pressure – the refrigerator cycle takes in work energy by the fluid being compressed. Both stages are, ideally, isentropic.

Example 14.10 During a refrigerator test, ammonia entered the evaporator at 1.9 bar, 15% dry and left it 85% dry to enter the compressor. If the compressor operates with 90% isentropic efficiency to take the fluid to 14.7 bar, how much energy is dissipated by the condenser?

The condenser and evaporator stages can be taken as constant pressure ones and the throttling valve expansion as constant enthalpy. For this example, discount casual external losses.

The common ammonia data tables are not as detailed as, say, the common steam tables. For instance, h_{fg} and s_{fg} may be omitted (so they have to be calculated) and intermediate data steps may be large. There are more complete tables available but, for this example, less detailed ones have been used with linear interpolation. As, by now, you are familiar with the use of such tables (same techniques, different numbers), only the outcomes are reported here, no details.

For the evaporation stage

enthalpy rise, 15% to 85% dry at 1.9 bar = 931.1 kJ/kg
outgoing enthalpy = 1220.5 kJ/kg.

Consider now the compression stage. In expansion as with steam, the isentropic efficiency of 90% would mean that only 90% of the isentropic expansion enthalpy would actually be available for another use.

● PQ 14.17 So how will the isentropic efficiency be reflected in the compression stage? What will it mean more of or less of? Where does the energy used in the compressor end up in this cycle? And what is its ultimate destination?

The relevant numbers which your calculations should produce are:

specific entropy of ammonia going into compressor = 4.835 kJ/kg K
thus dryness at end of isentropic compression = 0.98

thus enthalpy at end of isentropic compression = 1453 kJ/kg
thus isentropic increase of enthalpy = 232.5 kJ/kg
thus actual increase (232.5 / 0.9) = 258.3 kJ/kg.

This is still at 14.7 bar even with non-isentropic operation because, like the previous steam example in Example 14.9, the pressure is a practical demand, not an ideal or isentropic one.

Ammonia enthalpy leaving compressor, enthalpy going in plus compressor enthalpy,

(1220.5 + 258.3) = 1478.8 kJ/kg.

The ammonia leaving the compressor therefore has about 2K superheat.

Energy dissipated by condenser is the sum of that gained in the evaporator and the compressor,

931.1 + 258.3 = 1189.4 kJ/kg.

The condenser, operating at the pressure of the compressor outlet, 14.7 bar, dissipates this enthalpy to return the fluid to the enthalpy level entering the evaporator, and the expansion valve or throttle returns the fluid to the proper pressure and dryness condition.

Enthalpy at condenser exit,

1478.8 − 1189.4 = 289.4 kJ/kg.

Now, h_f for ammonia at 14.7 bar is 362.1 kJ/kg, thus the working fluid at the end of compression, about to enter the expansion valve is liquid at about 23°C. As a check, the enthalpy of ammonia at 1.9 bar, 15% dry is 289.4 kJ/kg. Any discrepancies are data table roundings.

So, the rules are just the same – it is only the temperatures, pressures and thermodynamic properties which are different. If any of the calculations just done for ammonia give you problems, then check back through the steam ones for the step-by-step treatment. It's all very logical provided you do not skip bits!

PROGRESS QUESTION ANSWERS

PQ 14.1

Steam generation, as in boilers. Liquid water substance is converted to the vapour – wet steam – and then to the gas – superheated steam. In most boiler operations, the steam is condensed and the liquid reused repeatedly.

PQ 14.2

See Figure 14.11.

PQ 14.3

If the steam was 100% dry, then it would be at the beginning of the superheat range with a value of around 1.3, as stated. If the wet steam was at the other end of its range, then its properties would be those of liquid water. For liquids, there is to most practical intents only one specific heat capacity, so the ratio is 1. Wet steam between these limits must therefore have a value which varies progressively between 1 and 1.3 or thereabouts.

PQ 14.4

See Figure 14.12.

PQ 14.5

The steam energy is the only source of energy as it does work, so it will get wetter. The actual change depends upon the temperatures involved.

PQ 14.6

Isentropic

PQ 14.7

For simplicity, this vapour Carnot cycle was introduced by having saturated water at point a being heated to saturated vapour at point b. This does not have to be the case, though. It could, for instance, be wet

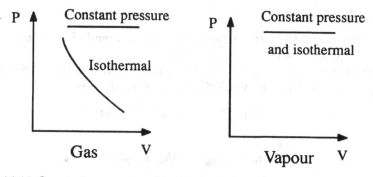

Figure 14.11 Constant pressure and isothermal changes.

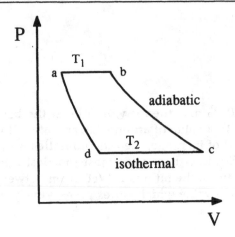

Figure 14.12 Carnot cycle for a vapour.

steam at point a and drier steam at point b.

Work output is (heat supply) × (efficiency) irrespective of starting and finishing conditions. The initial specific enthalpy is $x_1 h_{fg}$ and the final enthalpy is $x_2 h_{fg}$ for the stage ab, thus the work

$$w = (x_2 - x_1) h_{fg} \frac{(T_1 - T_2)}{T_1}.$$

PQ 14.8

Common sense rules. As these are on a p–V diagram, then ab is constant volume, bc is constant pressure, da is constant pressure. By looking at other power cycles, cd is likely to be a reversible adiabatic (isentropic) change.

PQ 14.9

The efficiency is the ratio of the energy put to intended use compared to the energy supplied, as with other cycles

The total heat input is $(h_{f1} - h_{f2}) + x_1 h_{fg1}$ and the total heat rejection is $x_2 h_{fg2}$. Thus the Rankine work done, the difference, is

$$[(h_{f1} - h_{f2}) + x_1 h_{fg1}] - x_2 h_{fg2}$$

or

$$(h_{f1} + x_1 h_{fg1}) - (h_{f2} + x_2 h_{fg2}),$$

commonly,

$$h_1 - h_2.$$

That is, the enthalpy of the vapour at the end of heat addition less the the enthalpy of the vapour at the end of the expansion stroke.

The specific enthalpy input per cycle is thus $h_1 - h_{f2}$, since the returning liquid water had some residual enthalpy. The Rankine thermal efficiency is therefore

$$\eta = \frac{h_1 - h_2}{h_1 - h_{f2}}.$$

PQ 14.10

The work of the cycle is indicated by the area enclosed by the cycle on the p–V diagram, abcd. The work of the isentropic expansion in isolation – that is, not part of any cycle – is indicated by the area bounded by the curve and the horizontal axis, with verticals dropped from points c and d.

PQ 14.11

The isothermal and the adiabatic changes are straight lines but the water pressurizing and heating is a curve. Since change of entropy with temperature is significant for a liquid, this cannot be ignored on a T–S diagram (Figure 14.13).

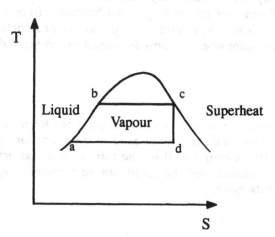

Figure 14.13 The Rankine cycle on a T–S diagram.

PQ 14.12

The Carnot efficiency is the ratio $(T_1 - T_2)/T_1$ so any change must come from a change of T_1 or T_2. That is, either the upper cycle temperature must be increased or the lower cycle temperature must be decreased. Remember that these temperatures are the temperatures of the working fluid, not of some surface or anything else in the system.

PQ 14.13

The extra work yielded is shown by the area bounded by cc'd'd. The extra energy demand to get this work is the superheat enthalpy.

PQ 14.14

The new specific work output is $0.187 \times 2445 = 459.1$ kNm compared to 408.5 kNm for the dry steam cycle, an extra 50.6 kNm. This extra is won by the expenditure of the enthalpy of superheat, which is

superheat enthalpy – saturation enthalpy $= 3001 - 2803 = 198$ kJ.

The efficiency of conversion of this part is therefore 25.56% – a better value but, unfortunately, the less efficient parts of the cycle are always there. It is still only part of the overall cycle.

PQ 14.15

As the steam does work, it must lose enthalpy for that is the measure of the energy which the steam is carrying as a working fluid. As it does work, it is likely to lose both temperature and pressure and so both are replenished during the reheat stages. Additionally, in real boiler plant, as steam flows down pipes and through nozzles, so friction demands some of the pressure energy – flow is caused by pressure difference, of course.

PQ 14.16

What really matters is how much heat is extracted from the refrigerator contents and how much energy is used in the pump or compressor to achieve this. The measure of this, the ratio of heat extracted to pump energy used, is often called the coefficient of performance, CoP, of the refrigerator or its cycle.

PQ 14.17

It means that more energy is required to do the job, not less. The amount will thus be

$$\text{actual compressor demand} = \frac{\text{isentropic amount}}{\text{isentropic efficiency.}}$$

For the cycle, this enthalpy must end up in the working fluid but in real life its ultimate destination is to be lost as heat transfer from the refrigerator to the surroundings.

TUTORIAL QUESTIONS

As with Chapter 7, you will need access to property tables for some of these questions. You may also find it helpful to sketch cycle diagrams for some of the questions. Any numbers here are for the immediate purposes only because real operation is often quite detailed. The answers are to data table accuracy, so yours may vary a little.

14.1 Steam initially at 2 bar, 150°C expands with an isentropic efficiency of 75% down to 1 bar. What is its new specific enthalpy and specific entropy? [2677 kJ/kg; 7.362 kJ/kg K]

14.2 On a vapour-based Carnot cycle for water substance, the temperature limits are 110°C and 250°C, with the entropy limits being 3.45 kJ/kg K and 5.416 kJ/kg K. What are the pressure and volume limits? For simplicity, ignore v_f in the calculations.
[39.8 bar, 0.010 m^3/kg, 0.040 m^3/kg;
1.433 bar, 0.422 m^3/kg, 0.831 m^3/kg]

14.3 For the cycle of question 14.2, assess the thermal efficiency by (a) determining the specific enthalpies at the four points of the diagram, hence finding the enthalpy supply and the enthalpy reject and (b) by the Carnot efficiency relationship using the quoted temperatures. What is the specific work output of the cycle?
[Enthalpy in 1029.3 kJ/kg; reject 753.3 kJ/kg, work 276 kJ/kg;
Efficiency values 26.8% by either method]

14.4 In a Rankine cycle operation, the lower operating and condenser pressure is 3 bar. What are the specific work outputs, for an isentropic efficiency of expansion of 80%, if dry steam is generated in the cycle at (a) 20 bar and (b) 30 bar? [272 kNm/kg; 326 kNm/kg]

14.5 For the cycle of question 14.4 operating at 30 bar, the steam is generated at 325°C but other parameters remain unchanged. What is the new specific work output?   [379 kNm/kg]

14.6 Ammonia circulates through a commercial refrigerator cycle at a rate of 3.6 kg/h. In the evaporator section, the ammonia enters at −4°C, 10% dry and leaves at 95% dry. The pump or compressor uses 200 W and all casual energy losses are negligible. What is the ratio of heat extracted to compressor work input?

If the compression is isentropic, what is the refrigerant condition at the pump exit?   [5.43:1; Approx 16.5 bar, 76°C]

Appendix A The Kinetic Theory

Imagine a hollow sphere, radius r, containing n molecules each of mass m. Remember the simplifications of the kinetic theory and think of a single molecule moving inside the sphere at velocity c, as in Figure A.1. It hits the sphere wall at A and bounces across to B and then continues bouncing around the sphere. Since the molecule is perfectly elastic, then its velocity is always c. Assume all measurements are in self-consistent units.

It will also, as it is perfectly elastic, always bounce around at angle to θ to the radius, so imagine it travelling a distance equivalent to AB between bounces,

$$\text{Distance AB} = 2r\cos\theta.$$

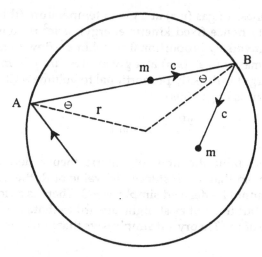

Figure A.1 The kinetic theory.

So, time between bounces $= \dfrac{2r \cos\theta}{c}$.

At each impact point, the change of momentum by this molecule is

Change of momentum $= 2mc \cos\theta$
Force $=$ rate of change of momentum.

Change of momentum per unit time, the rate of change of momentum is

$$\text{Force} = \frac{2mc \cos\theta}{2r \cos\theta/c} = \frac{mc^2}{r}.$$

Pressure $=$ force per unit area, and sphere surface area $= 4\pi r^2$, so that

$$\text{Pressure} = \frac{mc^2}{r4\pi r^2} = \frac{mc^2}{4\pi r^3}.$$

This is for one molecule, so for all n molecules, where p is the pressure,

$$p = \frac{nmc^2}{4\pi r^3}.$$

$$\text{Volume of sphere, } V = \frac{4\pi r^3}{3}$$

therefore

$$p = \frac{nmc^2}{3V}.$$

Temperature is a measure of the molecular kinetic energy, $nmc^2/2$, thus:

(a) for a given mass of gas (nm) at a given temperature (this mass at a fixed velocity hence fixed kinetic energy), nmc^2 is constant, thus pressure is inversely proportional to volume – Boyle's law;
(b) for a given mass of gas (nm) at a given pressure p, temperature T (kinetic energy) is directly proportional to volume – Charles' law;
(c) for a given mass of gas (nm),

$$\frac{pV}{T} = \text{constant.}$$

Note that this is about the limit of kinetic theory derivation. The simplifications mean that, for instance, the value of R, the gas constant for a given gas, cannot be derived simply as $n/3$. There is a clear relation between n and R but it is not as straightforward as that. This in no way reduces the value of the Theory – it simply says that you must respect its limitations.

Appendix B Some calculus derivations

As a general comment, the level of calculus used in this appendix is adequate for anything in the main body of the text and for most introductions to thermodynamics.

B1 NON-FLOW EXTERNAL WORK

Think of a sealed but flexible container of gas, such as a party balloon that can be bent into different shapes. The size and shape of the balloon are a completely free choice, so that the results of this analysis apply to any size and shape. Look at Figure B.1 but ignore the broken line for a moment. The gas pressure is p, its volume is V and the surface area of the balloon is A, all in self-consistent units.

Suppose, in the usual calculus fashion, that the gas expands by an infinitessimally small amount so that the balloon skin moves out a uniform but infinitesimally small distance dx. The broken line in Figure B.1 shows this minute expansion. Being an infinitesimally small change, it can be taken that the area A is substantially unchanged. Thus the change of volume dV consequent upon this expansion is $A \times$ dx.

By definition, work is done when a force moves through a distance. In this case, the work done is that of pushing back the surrounding atmosphere. Also, pressure is force per unit area or force is pressure $\times$ area. Thus in this gas container, the force at its surface or skin is pressure $\times$ area, $p \times A$. When the minute expansion occurs, this force $p \times A$ moves through the expansion distance dx and so a small amount of work dW is done

$$dW = p \times A \times dx$$

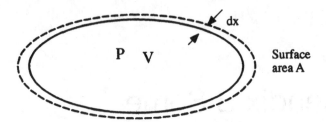

Figure B.1 Non-flow work.

However, A × dx is the minute change of volume, dV thus

$$dW = pdV$$

or non-flow work W is the integral of this;

$$W = \int pdV.$$

Since this has been derived by reference to a container of totally arbitrary shape and size, then there is no constraint to its application.

B.2 WORK AND THE GENERAL CHANGE pV^n = CONSTANT

For a general non-flow change which may be represented as pV^n = constant, say c for convenience, then $p = c\,V^n$. Recall that this is a 'small c' constant and that it has no special significance beyond its immediate use.

In B.1 it was shown that non-flow external work is given by

$$dW = pdV \text{ or } W = \int pdV.$$

So, for the external work done on or by the gas for a general change, say with a change of volume from V_1 to V_2, we have

$$W = \int \frac{cdV}{V^n}$$

$$= \frac{c}{1-n}\,[V^{1-n}]_1^2$$

$$= \frac{c}{1-n}\,(V_2^{1-n} - V_1^{1-n}).$$

Now

$$c = pV^n = p_1V_1^n = p_2V_2^n.$$

Thus,

$$W = \frac{1}{1-n}(cV_2^{1-n} - cV_1^{1-n})$$

$$= \frac{1}{1-n}(p_2V_2^nV_2^{1-n} - p_1V_1^nV_1^{1-n})$$

$$= \frac{1}{1-n}(p_2V_2 - p_1V_1).$$

Conventionally, this is usually written to give an $(n-1)$ term rather than $(1-n)$ because, commonly, n is larger than unity in real-life processes. Thus

$$W = \frac{p_1V_1 - p_2V_2}{n-1}$$

and since $pV = mRT$ this can also be written

$$W = \frac{mR(T_1 - T_2)}{n-1}.$$

If this was the special case of a reversible adiabatic change, for which the expression $pV^\gamma = $ constant applies, then the external work expression would be

$$W = \frac{p_1V_1 - p_2V_2}{\gamma - 1} = \frac{mR(T_1 - T_2)}{\gamma - 1}.$$

B.3 WORK DONE IN A NON-FLOW ISOTHERMAL CHANGE

An isothermal (constant temperature) change follows Boyle's law which says that for a given mass of gas at a given temperature, pressure is inversely proportional to volume or the product pV is constant. Thus, under isothermal conditions for a fixed mass of gas

$$pV = p_1V_1 = p_2V_2 = \ldots$$

so that

$$p = \frac{p_1V_1}{V} = \frac{p_2V_2}{V} = \ldots$$

At all times for non-flow external work, $W = \int p\,dV$

and substituting $W = p_1V_1\int\frac{dV}{V}$

$$= p_1V_1\ln(V_2/V_1) = p_2V_2\ln(V_2/V_1).$$

B.4 WORK AND FLUID FLOW

The term external work has been used carefully in the text, to mean the useful work which has been extracted from a system. There were expressions in both the non-flow energy equation, NFEE and the steady flow energy equation, SFEE, to deal with this external work. In systems where a fluid flows, however, there is another work to consider – the actual work required to make a fluid flow through the pipework or the machinery or whatever. This is not external work – it is not work which will be seen as a useful output.

When a fluid (liquid, slurry, vapour, gas) flows along a pipe, then there is a frictional resistance to the fluid flowing. This is exactly the same sort of frictional resistance that acts if you try to push a box across a floor. This frictional resistance needs a force to overcome it, so work (force × distance) is done as the box is pushed. Similarly, work is done if a fluid is pushed down a pipe, as frictional resistance is overcome.

The usual way of overcoming this resistance is to give the fluid a high pressure, like water pressure in a hose-pipe. If a fluid enters a pipe at pressure p, then the work done in pushing the fluid along the pipe results in the fluid pressure falling. Stay with the hose-pipe.

If I have three hose-pipes, say 20 m, 10 m and 5 m long and each is connected to the tap, then the water coming out of the 5 m hose will squirt further than the water from the 10 m hose which in turn will squirt further than the water from the 20 m hose. The reason is that more of the tap water pressure is used up getting the water to the end of the longer hoses, so less of that pressure is available as kinetic energy (squirt distance) than for the shorter hoses.

Look at Figure B.2 and say that the pressure drop (the amount of pressure lost from the original value p) is dp to move the fluid through a distance d along the pipe of cross sectional area A. The volume V of fluid moved along for this loss of pressure is $A \times d$. Pressure is force per unit area, so force is pressure × area. If an element of pressure dp is lost in moving the element of volume V then the force expenditure is (element

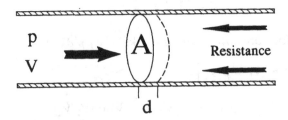

Figure B.2 Frictional resistance to flow.

of pressure) × area, $dp \times A$. Work done is force × distance, so the element of work done dW is

$$dW = (dp \times A) \times d.$$

As $A \times d$ is the volume moved for this expenditure of work, then $dW = V\,dp$. Take care to understand the difference between external work, the work available from a process or change to do something useful, and the other work – the work needed to make the process or change take place. This latter work usually ends up as a casual heat loss from the system.

LOG MEAN TEMPERATURE DIFFERENCE, LMTD

Think of a parallel flow heat exchanger, total heat transfer area A and mean heat transfer coefficient h. The hot fluid, flow rate m_1, specific heat capacity c_1, enters at temperature T_{h1} and leaves at temperature T_{h2}. The cold fluid conditions are similarly m_2, c_2, T_{c1} and T_{c2}.

At some intermediate point between inlet and outlet, an element of heat transfer dq takes place across an element of area dA. Let the temperatures of the hot and cold fluids at that intermediate point be T_h and T_c. As a result of this element of heat transfer, the hot fluid temperature falls by $-dT_h$ and the cold fluid temperature rises by $+dT_c$. Thus

$$dq = m_1 c_1(-dT_h) = m_2 c_2(+dT_c) = h\,dA(T_h - T_c),$$

from which

$$dT_h = \frac{-dq}{m_1 c_1} \quad \text{and} \quad dT_c = \frac{+dq}{m_2 c_2}.$$

In this element of heat transfer, the temperature difference between the fluids has changed according to dT_h and dT_c (which are not the same) by an amount $dT_h - dT_c$. Call this $d(\triangle T)$ and call the temperature difference at this arbitrary intermediate point, $T_h - T_c$, $\triangle T$. Then

$$dT_h - dT_c = d(\triangle T) = \frac{-dq}{m_1 c_1} - \frac{dq}{m_2 c_2} = -dq\left(\frac{1}{m_1 c_1} + \frac{1}{m_2 c_2}\right),$$

but

$$dq = hdA(T_h - T_c) = hdA\triangle T$$

thus

$$d(\triangle T) = -hdA\triangle T\left(\frac{1}{m_1 c_1} + \frac{1}{m_2 c_2}\right).$$

Overall, for total heat transfer rate $\dot{q}$,

$$\dot{q} = m_1 c_1(T_{h1} - T_{h2}) = m_2 c_2(T_{c2} - T_{c1})$$

thus

$$m_1 c_1 = \frac{\dot{q}}{T_{h1} - T_{h2}}$$

and

$$m_2 c_2 = \frac{\dot{q}}{T_{c2} - T_{c1}}$$

so

$$d(\Delta T) = -h dA \Delta T \left(\frac{T_{h1} - T_{h2}}{\dot{q}} + \frac{T_{c2} - T_{c1}}{\dot{q}} \right)$$

$$= -h dA \Delta T \frac{(T_{h1} - T_{c1}) - (T_{h2} - T_{c2})}{\dot{q}}.$$

If the temperature difference at entrance is T_1 and at exit T_2, then

$$d(\Delta T) = -h dA \Delta T \frac{(T_1 - T_2)}{\dot{q}} = h dA \Delta T \frac{(T_2 - T_1)}{\dot{q}}$$

Now, cross-multiplying and integrating for ΔT between T_1 and T_2, also for dA for the whole area A, gives

$$\ln (T_2/T_1) = hA \frac{(T_2 - T_1)}{\dot{q}}.$$

or

$$\dot{q} = \frac{hA (T_2 - T_1)}{\ln (T_2/T_1)},$$

and the term

$$\frac{(T_2 - T_1)}{\ln (T_2/T_1)}$$

is the log mean temperature difference, LMTD. While this has been derived here for a parallel flow heat exchanger, the use of a counterflow exchanger for the derivation would give exactly the same answer.

Appendix C The last round-up

C.1 SOME FORMAL STATEMENTS, DEFINITIONS AND LAWS

Each of the following, whether it is a statement, a definition or a law, is explored in the text and the list is not exclusive. These are reminders, not replacements and it is assumed that some simple definitions will be pursued through elementary texts. Quite deliberately, there has been no attempt at rigid separation.

C.1.1 LAWS

Boyle's law For a given mass of a given gas held at constant temperature, volume is inversely proportional to pressure.

Charles' law For a given mass of a given gas held at constant pressure, volume is directly proportional to temperature. These were observation laws, seen to be accurate within the measurement ability of the day. It is now better to say 'For a given mass of ideal gas . . . ' although the laws are applicable to real gases over small changes of condition. This means applicable for many industrial purposes, within the limits of routine measurement accuracy. The laws are progressively less accurate for real gases as they approach liquefaction or if they are used at very high pressures. These comments also apply to other long-standing, observation-based thermodynamic laws.

Dalton's law Dalton's original statement implied that any gas in a mixture acted as though that mass of the individual gas were in isolation. The law is often applied by name to partial pressures of gas mixtures, saying that any gas in a mixture within a container contributes to the total pressure an amount equivalent to that which it would exert if it alone, in its mixture quantity, occupied the container.

Gibbs–Dalton law This extended Dalton's ideas to a range of gas properties, both intensive and extensive, covering for instance internal energies and entropies. Note that the thermodynamic properties can be treated with reasonable surety this way but that other properties are not necessarily additive – thermal conductivity, for instance.

Avogadro's hypothesis At a given temperature and pressure, the number of moles of gas in a given volume is independent of the gas. That is, for those stated conditions, there are as many molecules of hydrogen in 1 m^3 of hydrogen as there is molecules of nitrogen in 1 m^3 nitrogen and so on. The comments for Boyle's and Charles' laws apply broadly here.

Zeroth Law of Thermodynamics If two bodies are in thermal equilibrium with a third body, then they are in thermal equilibrium with each other. This Law is the basis of temperature measurement.

First Law of Thermodynamics If a closed system changes from one state to another, the net sum of heat and work transfer will be the same, whatever route that change takes. While the important consequence of the Law as stated is that heat and work are interchangeable, it could be extended nowadays to say that all energy forms are interchangeable. This assumes the availability of equipment so to do, and take note that it does not say that they will be completely interchangeable.

Second Law of Thermodynamics There are various formulations because of the similar or parallel work of several investigators. Perhaps the two which say (a) that heat cannot flow of its own accord against a temperature gradient and (b) that no work cycle can operate with only one temperature level are of most immediate practical importance.

Non-flow energy equation, NFEE Where a fixed, contained mass of working fluid (notably gases) is used as an agent of energy change or transfer, there must always be a balance between the heat transfer, external work done and internal energy change. Commonly but not exclusively associated with the First Law in forms such as $dq = dw + du$. As a simple extension, any energy forms appropriate to fixed, contained masses of working substances may be included.

Steady flow energy equation, SFEE Where there is a steady flow of working fluid through a system, all the energy forms should be recognized and quantified, such that a statement of the type 'energy flowing into the system = energy flowing out of the system' may be applied. The SFEE is a common general application of the law of conservation of energy and is often shown as dq = dw + de or similar, where de encompasses all the other energy form changes. Note the similarity of equation form to NFEE, where there was only one other energy form change, du, that of internal energy.

Continuity equation For a steady flow rate of working fluid through any system, the mass flow rate across any plane dealing with the whole of the working fluid must equal that across any similar plane.

C.1.2 PROPERTIES

Property Any identifiable feature of a substance or system. Normally the word is attached to measurable quantities such as density, but this is not a qualification. Colour or smell are properties which are not necessarily quantified.

Density Mass per unit volume, kg/m^3, the importance here being that the density of a gas varies with the applied temperature and pressure and thus density cannot be considered in isolation.

Enthalpy A measure of total energy capacity, the capacity for doing work or transferring heat in any combination. It does not include the kinetic or gravitational potential energy of the material in question. Mathematically for it is the product mc_pT, or mcT where there is effectively only one significant specific heat capacity, as liquids and solids in practice. For convenience and by convention, usually referred to a datum of 0°C. Units J (joules) or, more likely in practical circumstances, kJ.

Entropy An abstract property but still a thermodynamic property, related to the quality of energy and thus its usefulness. The practical value of the concept of entropy is as an assessment, analytical or calculation tool. Basic units are J/kgK and quantities are normally referred to an arbitrary datum such as 0°C but, as with other properties, kJ/kgK are common and entropy changes are likely to be more important than actual amounts. There is no direct measure of the property, no entropy meter.

Internal Energy Essentially the kinetic energy of a material's molecules. As temperature is a measure of this kinetic energy, then temperature and internal energy are linked directly. For gases, internal energy is the product mc_vT and the comments on datum and units are as for enthalpy. As with energy transfer, it is the change of enthalpy and of internal energy which figure more prominently than the absolute amounts, commonly.

Pressure Force per unit area. Units are N/m^2 or its equivalent, Pa. The bar (100 000 Pa) is an everyday unit of convenience.

Principal specific heat capacities As the application of heat to a gas can result in a range of combinations of temperature change and external work transfer, a gas can have numerous specific heat capacities. Two which are readily identifiable are under conditions of constant pressure and constant volume. These are often called the principal specific heat capacities and are identified as c_p and c_v respectively. Units still J/kgK or kJ/kgK. These properties provide the gas constant R for any specified ideal gas in the form $R = c_p - c_v$, so the units of R are those of specific heat capacity.

Specific heat capacity The amount of heat energy required to raise unit mass of a specified material through unit temperature. Commonly, in SI units, 1 kg through 1 K and in casual conversation the word 'capacity' is often omitted. The symbol c is usual and the units are J/kgK or kJ/kgK. In some special and therefore clearly defined cases, a volume specific heat capacity is recorded.

Temperature A measure of molecular kinetic energy, its measurement being the best application of the Zeroth Law. Units degrees Kelvin but Celsius (and Fahrenheit in some cases) commonplace.

Viscosity A fluid's property of resistance to flow. Assessed as the ratio of shear stress applied to shear rate, it is also called dynamic viscosity and absolute viscosity with units kg/ms or Ns/m^2. The ratio of viscosity divided by density is useful in some transport process calculations and this ratio is called kinematic viscosity with units of m^2/s.

C.1.3 GENERAL DEFINITIONS

Adiabatic Without external heat transfer. The word is sometimes used alone when 'reversible adiabatic' is intended, so care has to be taken in its use and interpretation.

Characteristic gas equation Also the term 'ideal gas equation of state' or similar. For any ideal gas, the gas's mass, pressure, volume and temperature are related in the form $pV = mRT$ where R is the gas constant for the specified gas. It is applicable to real gases within carefully considered limits.

Dryness Per cent or fraction. A measure of the degree of conversion of a liquid to its gas as it passes through the vapour phase. A vapour which has just become gas by having no remaining liquid content but being still at the relevant boiling point is said to be a dry vapour or saturated vapour or even dry saturated vapour.

Energy A capacity for doing work, transferring heat or some similar transport process. In theory (and to a great extent in practice), all forms of energy are interchangeable.

Energy Transfer While amounts of energy (heat, work) are important, it is usually the energy transfer (heat transfer, work transfer) that is of practical value. As with Nm and J, the watt (Nm/s, J/s) is small and kW is more likely to be met.

Gas constant The constant R (a valuable Constant, not a casual one) in the Characteristic Gas Equation and having an individual value for an individual ideal gas, being the difference between the principal specific heat capacities. As with the equation, the value of R applies to real gases for many practical purposes.

Heat An energy form easier to recognize than to define and easier to recognize because of the effects of heat transfer. Most formal definitions are either obscure or inadequate. Perhaps the neatest is that heat is an interaction between systems consequent upon their temperature difference. As with work, the basic unit of the joule is small, hence kJ is usually met.

Heat of change of phase An alternative name to latent heat, the terms have enjoyed academic favour at various times.

Ideal gas An idealized form of real gases. Thus there is not one ideal gas but as many as there are real gases. Ideal gases follow the simple observational laws of Boyle and Charles and are acceptable representations of real gas behaviour for many everyday industrial purposes.

Isentropic At the same entropy, hence 'isentropic efficiency', meaning a measure of operational efficiency compared to the parallel operation performed isentropically. Often associated with expansion and compression stages in various processes. Note also 'reversible adiabatic'.

Isobaric At the same pressure.

Isothermal At the same temperature.

Latent heat Usually with added words, as in 'latent heat of evaporation'. Heat energy which, when added to a substance, changes the phase of the substance, such as in the conversion of liquid to gas. Temperature does not change during this operation, assuming that nothing else is altered.

Mole, mol, kmol Various names are attached, almost casually. For SI, the name kilogram-mol avoids confusion, kgmol, kilomole or kmol being usable. It is the relative mass of an atom or a molecule, whichever is being considered, when compared to a reference mass. The original reference was the mass of an atom of the lightest element, hydrogen, and this was given a relative mass of 1 unit. The point of reference has changed over the years as measurements have improved and the current one is the carbon-12 atom. A definition is that 1 kgmol of any material or substance contains as many basic (smallest recognizable) portions of that substance as there are atoms in 12 kg of carbon-12. Its practical interpretation is that the mass of a kgmol of anything is its molar (for compounds) or atomic (for elements) mass in kilograms. Sometimes meet the units kg/mol.

Open and closed systems If everything of importance to the operation of a system happens within the system boundary, then the system can be called closed. This is quite artificial since nothing is totally isolated in the real world but it is a recognition of the important items and the negligible ones. Where inputs and outputs arrive from and depart to locations outside the system boundary, then such a system is said to be open. The true difference is one of application and common sense.

Perfect and semi-perfect gases Variations of the ideal gas concept. A perfect gas follows the simple laws and has fixed property values. A semi-perfect gas follows the simple laws but has properties which vary with temperature, notably specific heat capacities. Temperature is the biggest single influence on many properties, so the semi-perfect gas keeps the simplicity of ideal gases but includes an important refinement.

Phase A condition of homogeneity of a substance, shown by a continuity of chemical and physical properties. Many properties vary, especially with temperature, so progressive minor changes are acceptable within the formal definition. By convention, since vapour is a recognizable condition, it is common to speak of a vapour phase even though true homogeneity is not seen.

Process Anything which causes a change of state, whether a chemical or physical change. The everyday use of the word is in its attachment to some industrial or commercial operation such as manufacture.

Reversible adiabatic A process without external heat transfer, performed perfectly so that it could be reversed back to its starting condition without overall energy or property loss. In practice impossible, but a good comparison measure. The word isentropic is an alternative for this term.

Saturated vapour See *dryness*.

Saturated liquid The converse of saturated vapour, being liquid at its boiling point but not yet having started the conversion to gas via vapour.

Sensible heat Heat energy which, when added to a substance, raises its temperature.

State The condition of a system or substance as described or defined completely by its properties, chemical and physical, to give an unique identity. Strictly, all properties should contribute but, as many properties are interdependent, this is not necessary for the identification of a state. For instance, many gaseous state considerations can be made by recording the gas pressure, volume and temperature.

System Any coherent, interdependent grouping of artefacts, substances, working fluids or process conditions. Examples may include the human body, a clock or a petroleum refinery.

System boundary The location limits of a system. These are not necessarily physical limits, such as a wall, for they may be quite imaginary and generated for convenience. Any arbitrary boundaries may be introduced to a system to study local inputs, outputs and consequences.

System inputs and outputs Factors of importance to the system, crossing the system boundary. May be tangible, such as a flowing fluid, or intangible such as heat transfer.

Superheat Sensible heat when added to a gas. The term is used when vapours are or have been involved and is the heat added to the dry vapour to raise its temperature beyond the relevant boiling point.

Units and dimensions Elements of measurement. Dimensions are the fundamental features, such as time, length or mass. Units are their useful interpretations such as seconds, metres and kilograms.

Universal gas constant The product of (molar mass) × (gas constant) for any specified ideal gas gives the universal gas constant, often designated R_m but there are variations. Its value is commonly recorded as 8314 J/molK but there are various greater accuracies in various reference sources. A knowledge of the molar mass for any ideal gas gives its gas constant readily as $R_m = R \times MM$ for any gas.

Vapour The transitional stage between the liquid and the gaseous phases. Not a true phase, since its properties alter rapidly with temperature and since it is a progressively changing mixture of liquid and gaseous material. However, as vapours are recognizable and their properties are measurable, the term 'vapour phase' is accepted commonly.

Work Work is done when a force moves through a distance. Work transfer is the term used in conjunction with work being done. Note external work as a useful product of changes in system states. The unit the Newton-metre is quite small for practical use, hence kNm is common.

Working fluid An embracing term to cover without individual identity any fluid, whether individual, compound or mixture, which may be used in a process. If thermodynamic relationships are derived by reference to some undefined working fluid, then the relationships are general ones.

C.1.4 CYCLES

Cycle A sequence of connected changes which are ultimately repeated. The seasons are cyclic, the working of an engine is cyclic. A cycle is a process, as states of the involved substance change; each consequential step of the cycle is also a process in its own right.

Carnot cycle A theoretical engine cycle which predicts the maximum thermodynamic performance and efficiency for given cycle parameters.

The Carnot efficiency is the theoretical maximum efficiency for any given cyclic operating conditions.

Air standard cycle A theoretical cycle which reflects the main thermodynamic features of a real work cycle using a gas as the working fluid. These cycles thus set realistic goals for spark ignition and compression ignition engines, for example. There are several such air standard cycles to suit the several real engine cycles, these often being identified by an investigator's name, such as Otto, rather then application, such as spark ignition engines.

Rankine cycle A theoretical cycle for operations involving vapours, parallel in aims and applicability to the various air standard cycles for gaseous working fluids.

C.2 SOME COMMON EQUATIONS

This again is a reminder list. It is neither exhaustive nor exclusive and in no way supplants the proper understanding which should be developed by careful consideration of the chapter material. You must refer to the relevant chapters for any operational conditions attaching to the use of these equations.

The nomenclature is not repeated as it is as used in the relevant chapters and only one form of each equation is given.

Characteristic gas equation
or ideal gas equation of state $pV = mRT$

For a fixed mass of ideal gas $\dfrac{p_1V_1}{T_1} = \dfrac{p_2V_2}{T_2} = \ldots = \dfrac{p_nV_n}{T_n}$

Gas constant $R = c_p - c_v$

Universal gas constant $R_m = (MM) \times R$

Adiabatic index $\gamma = c_p/c_v$

Change of internal energy $\Delta U = mc_v\,\Delta T$

Change of enthalpy $\Delta H = mc_p\Delta T$

Specific enthalpy relationship $h = u + pV$

Non-flow energy equation $\Delta Q = \Delta W + \Delta U$

Non-flow work according to pV^n $\quad W = \dfrac{p_1V_1 - p_2V_2}{n-1} = \dfrac{mR(T_1 - T_2)}{n-1}$

Isothermal non-flow work $W = pV\ln(V_2/V_1)$

Steady flow energy equation $\Delta Q = \Delta W + \Delta E$

Continuity equation	$c_1\rho_1A_1 = c_2\rho_2A_2 = \ldots = c_n\rho_nA_n$
Vapour enthalpy	$h = h_f + xh_{fg}$
Vapour entropy	$s = s_f + xs_{fg}$
Vapour volume	$v = xv_g + (1 - x)v_f$
often approximates to	$v = xv_g$
Entropy change, general	$dS = dQ/T$

Entropy change, constant volume
$$\Delta S = mc_v\ln\,(T_2/T_1)$$

Entropy change, constant pressure
$$\Delta S = mc_p\ln\,(T_2/T_1)$$

Entropy change, constant temperature
$$\Delta S = mR\ln\,(V_2/V_1) = mR\ln\,(p_1/p_2)$$

Entropy change according to pV^n
$$\Delta S = \frac{m(nc_v - c_p)\,\ln\,(T_2/T_1)}{n - 1}$$

van der Waals equation
$$RT = (p + \frac{a}{v^2})\,(v - b)$$

Conduction through plane wall
$$\dot{q} = \frac{A(T_1 - T_n)}{x_1/k_1 + \ldots + x_n/k_n}$$

Conduction through pipe walls
$$\dot{q} = \frac{k2\pi L\,(T_1 - T_n)}{[\ln(r_2/r_1)]/k_1 + \ldots + \text{etc.}}$$

Convection
$$\dot{q} = hA(T_2 - T_1)$$

Overall convection coefficient
$$\frac{1}{h} = \frac{1}{h_1} + \frac{1}{h_2}$$

Combined coefficient
$$\frac{1}{h} = \frac{1}{h_1} + \frac{1}{k_1/x_1} + \frac{1}{k_2/x_2} + \frac{1}{h}$$

Log mean temperature difference
$$\Delta T = \frac{T_a - T_b}{\ln(T_a/T_b)}$$

Radiant heat transfer, based on
$$\dot{q} = \sigma\epsilon AT^4$$

Carnot efficiency
$$\eta = \frac{T_1 - T_2}{T_1}$$

Otto air standard efficiency
$$\eta = 1 - \frac{1}{r_v^{\gamma - 1}}$$

Diesel air standard efficiency
$$\eta = 1 - \frac{(\rho^\gamma - 1)}{r_v^{\gamma - 1}\gamma(\rho - 1)}$$

Rankine efficiency, variations on
$$\eta = \frac{h_1 - h_2}{h_1 - h_{f2}}$$

C.3 SOME CONVERSION FACTORS FROM SI TO IMPERIAL

1 kg	= 2.205 lb	1 kJ		= 0.948 Btu
1 m	= 3.281 ft	1 Nm		= 0.737 ft lbf
$1 m^2$	= $10.76 ft^2$	1 W		= 3.413 Btu/h
$1 m^3$	= $35.31 ft^3$	1 kW		= 0.948 Btu/s
1 N	= 0.225 lbf	1 kW		= 1.34 hp
1 Pa	= $0.127 lb/ft^2$	1 kJ/kg K	= 0.239 Btu/lb degF
1 bar	= $14.5 lb/in^2$	$1 W/m^2$	= $0.317 Btu/ft^2$ h
		$1 W/m^2 K$	= $0.176 Btu/ft^2$ h degF

C.4 SOME REFERENCE TEXTS

This is intentionally a short list. There are many more texts worthy of use by the thermodynamics student and thermodynamics specialist. These few, however, show a reasonable spread and will give either relevant data or further material treatment or possibly both.

Thermodynamic and Transport Properties of Fluids Rogers and Mayhew, 4th edn 1991, Basil Blackwell, Oxford.

Engineering Tables and Data Howatson, Lund and Todd, 2nd edn, Chapman & Hall, London.

Dictionary of Energy Slesser (Gen. Editor), 1982, Macmillan, London.

Fundamentals of Engineering Thermodynamics Moran and Shapiro, 1988, John Wiley, New York.

The Efficient Use of Steam Goodall (Gen. Editor), 1980, IPC, Guildford.

Engineering Thermodynamics Boxer, 1978, Macmillan, London.

Applied Thermodynamics and Energy Conversion Sharpe, 1987, Longman, Harlow.

Heat Transfer Holman, 4th edn, 1976, McGraw Hill, London.

Index

Page numbers appearing in *italic* refer to tables. Note that various laws, properties and definitions are listed in Appendix C, with relevant equations.

Absolute zero 15
Absorptivity 253, 256–7
Accuracy 24
Activation energy 274
Adiabatic 83, 115, 198, 326
  see also Isentropic; Reversible
    adiabatic
Adiabatic temperature rise 288
Air/fuel ratio 272, 277, 287
Air standard cycles 299
Ambient temperature 284
Ammonia 183–4, 326, 345–6
Analysis of gas mixtures 45
Arithmetic average temperature
  difference 249–50
Average values 126
Avogadro's hypothesis 37

Bar 32
Black body 253, 256
Boiler 152, 331, 333
Boiling 145, 148, 149
  point 145, 148
  range 148
Boyle's law 25–30, 85
Buckingham's theorem 261
Bulk fluid temperature 232

Calorific value 277
Carnot
  cycle 302–7, 328–30
  efficiency 193, 305–7, 326, 339
Change of state 144
Charles' law 25–30

Characteristic dimension 260
Characteristic ideal gas equation 35,
  135
  and gas mixtures 44
Clapeyron's equation 330
Clausius inequality 208–11, 328
Coefficient of performance 344
Combustion 107, 115, 269–88
Combustion chamber 105, 116
Compressing pump 344
Compression ignition 300
  see also Diesel cycle
Compression ratio 308, 310
Compressor 105, 115
Condensation 155
Condensers 339, 343–6
Conduction 218, 220–31, 242–50, 326
Configuration factor 258
Constants 27
Constant pressure cycle 316
Constant volume cycle 316
Continuity equation 95–7
Continuous combustion 269, 276
  see also Combustion
Convection 218–19, 231–6, 242–50
Convective heat transfer coefficient
  232, 259–62
Counterflow heat exchanger 245
  see also Heat exchangers
Critical point 168, 183
Cross-flow heat exchanger 245
  see also Heat exchangers
Cut-off ratio 314
Cyclic combustion 269
  see also Combustion